Process Chemistry
of Lubricant
Base Stocks

CHEMICAL INDUSTRIES

A Series of Reference Books and Textbooks

Founding Editor

HEINZ HEINEMANN
Berkeley, California

Series Editor

JAMES G. SPEIGHT
Laramie, Wyoming

Process Chemistry of Lubricant Base Stocks

Thomas R. Lynch
Mississauga, Ontario, Canada

CRC Press
Taylor & Francis Group
Boca Raton London New York

CRC Press is an imprint of the
Taylor & Francis Group, an informa business

CRC Press
Taylor & Francis Group
6000 Broken Sound Parkway NW, Suite 300
Boca Raton, FL 33487-2742

First issued in paperback 2020

© 2008 by Taylor & Francis Group, LLC
CRC Press is an imprint of Taylor & Francis Group, an Informa business

No claim to original U.S. Government works

ISBN-13: 978-0-367-57767-4 (pbk)
ISBN-13: 978-0-8493-3849-6 (hbk)

Library of Congress Cataloging-in-Publication Data

Lynch, Thomas R.
  Process chemistry of lubricant base stocks / Thomas R. Lynch.
    p. cm. -- (Chemical industries series)
  Includes bibliographical references and index.
  ISBN 978-0-8493-3849-6 (alk. paper)
    1. Petroleum products. 2. Petroleum--Refining. 3. Lubricating oils. I. Title. II. Series.

  TP690.L96 2007
  665.5'385--dc22   2007020175

**Visit the Taylor & Francis Web site at
http://www.taylorandfrancis.com**

**and the CRC Press Web site at
http://www.crcpress.com**

Table of Contents

Preface

The purpose of this book is to provide the reader with an introduction to the chemistry of lubricant base stock manufacturing processes which use petroleum as feedstock and to the development work that has gone into this area over the past century and a half. I believe there is a need for such a work and it should appeal to those involved in either process or product development. The reader will gain insight into the chemical techniques employed and an introduction to many of the most significant papers in this area.

The unifying thread here is the chemistry of the process steps and therefore the structure, reactivity, and physical properties of the compounds existing naturally in petroleum and their subsequent transformation. The connections between structure, physical properties, and reactivity have been unraveled over time through rigorous investigations from both industry and academia. The revolutionary changes which the industry has seen over the past 25 years have truly been remarkable and are a tribute to the many people involved in the petroleum, lubricants, and automotive industries. In this book I have not sought to be comprehensive, rather to introduce the main chemical concepts and provide the reader with the most important sources for the background of the chemistry involved.

Early chapters provide a background to some of the physical properties that base stocks are expected to meet, the chemical and physical means by which they are distinguished, and the relationships between structure and physical properties. The viscosity index property is a key measure of viscosity response to temperature and deserves the attention of the full chapter (Chapter 3) that it receives. Methodology to determine both petroleum and base stock composition would require several books to outline. I have chosen to restrict this subject in Chapter 4 to a number of older methods which are still applicable but I have also included some discussion of NMR methods which increasingly will play a vital role. Since oxidation during use is probably the biggest hurdle that lubricants face, Chapter 5 provides a summary of the most significant work on the oxidation of base stocks and those oxidation studies on formulated products that reflect information on base stock composition and the process.

At this stage, having outlined the trends in desirable chemical structures and properties of base stocks, subsequent chapters deal with the commercial processes that have emerged, still paying close attention to the changes at the molecular level. The separation processes of solvent extraction and solvent dewaxing are outlined in Chapter 6 together with some description of the results from a very fine study by Imperial Oil people on the chemistry of hydrofinishing, a new technology at the time which rapidly displaced clay treating. Chapter 7 provides an account of the development of hydrocracking as a lubes process, which has come to dominate base stock manufacturing in North America, now widespread throughout the world, and made possible Group II and III base stocks. In Chapter 8, I have attempted to provide a detailed account of the chemical changes due to hydroprocessing, the equilibria, rates, products and impact on physical properties. Chapters 9 and 10 focus on the important art of dewaxing by processes other than solvent dewaxing; by wax removal through formation of urea clathrates, by cracking via "cat dewaxing" or through the remarkable development of wax hydroisomerization by   Chevron's Isodewaxing™ process or that of ExxonMobil's MSDW™ process.

The penultimate chapter is on the production of White Oils, where the processes have close links to those of base stocks, and the last chapter, departing from petroleum-sourced base stocks, is focused on the processes involved in the production of highly paraffinic (and very high quality) base stocks from natural gas. This is the potential elephant in the base stock world because of anticipated quality and volumes.

My thanks go to my former colleagues at Petro-Canada from whom I learned so much, colleagues, particularly Mike Rusynyk, who assisted in this book's preparation by reading and commenting on parts of this work, to publishers, companies and authors who gave permission to reproduce figures and tables, and to my editor at CRC Press, Jill Jurgensen, who patiently dealt with all my questions.

My final thanks go to my wife who has waited patiently for this to come to an end.

Author

Tom Lynch is an independent consultant in the lubricants industry. He has 25 years of experience with hydroprocessed lubes working for Petro-Canada in its Research and Development Department on Process Development and subsequently at the company's Lubricants Refinery. He obtained his B.Sc. degree from University College, Dublin, Ireland, and his Ph. D. from the University of Toronto, both in chemistry. He is the author of papers on the chemistry of sulfur compounds, molecular rearrangements, and hydroprocessing.

1 Introduction

1.1 BASE STOCKS: GENERAL

Lubricants have been used by mankind from the very early days of civilization to assist in reducing the energy needed to slide one object against another. The first lubricants were animal fats, and much later whale oil was used. It was not until crude oil was discovered in commercial quantities in Oil Springs, Ontario, Canada, in 1858 and in Titusville, Pennsylvania, in the United States in 1859 that the concept of petroleum-based lubricants could be seriously considered on a large scale. The first petroleum refinery to produce base stocks (the petroleum distillates fractions used in lubricants) in the Western Hemisphere was built by Samuel Weir in Pittsburgh in the 1850s. One of the earliest lubricant producers (to reduce "waste" production) was the Standard Works in Cleveland, Ohio, owned in part by John D. Rockefeller, whose company subsequently became Standard Oil.

Other petroleum companies subsequently followed suit and the industry developed in size and scope over time as industrialization took hold and the demand for lubricants grew. Access to lubricants is essential to any modern society. Not only do lubricants reduce friction and wear by interposition of a thin liquid film between moving surfaces, they also remove heat, keep equipment clean, and prevent corrosion. Applications include gasoline and diesel engine oils, machinery lubrication, and turbine, refrigeration, and transformer oils and greases. In 2005 the world's production of base stocks from petroleum totaled some 920,000 barrels per day[1] (bpd), with 25% of that (231,000 bpd) being in North America. Currently ExxonMobil, at 140,000 bpd, is the world's largest producer of base stocks, followed by Royal Dutch/Shell Group (78,000 bpd). The world's largest (40,300 bpd) lube plant is Motiva Enterprise's Port Arthur plant;[2] Motiva is a 50/50 joint venture between Shell Oil and Saudi Refining. The annual world production volume is about equivalent to that of two to three large refineries, but lube production is dispersed across the world and annual production volume per plant is quite small (e.g., in North America, the average size is 10,000 bpd and in Europe it is 6600 bpd). Lube plants are usually part of fuel refineries.

The subject of this book is the chemistry of petroleum base stocks and of their manufacturing processes from crude oil fractions. Petroleum base stocks are hydrocarbon-based liquids, which are the major component (80% to 98% by volume) of finished lubricants, the remaining 2% to 20% being additives to improve performance. Therefore this book does not deal with the manufacture of nonpetroleum base stocks such as synthetics (from olefins such as 1-decene), ester-based ones, and others.

Base stocks usually have boiling ranges between 600°F and 1100°F at atmospheric pressure (some are lighter) and lube feedstocks therefore come from the high-boiling region—the vacuum gas oil fraction and residue—of crude oil. Base stock boiling ranges may extend over several hundred degrees Fahrenheit. For the purpose of engine oil quality assurance, the American Petroleum Institute (API) has defined a base stock "as a lubricant component that is produced by a single manufacturer to the same specifications (independent of feed source or manufacturer's location); that meets the same manufacturer's specification; and that is identified by a unique formula, product identification number or both...."[3] A base oil is defined as "the base stock or blend of base stocks used in an API-licensed oil," while a base stock slate is "a product line of base stocks that have different viscosities but are in the same base stock grouping and from the same manufacturer." Alternatively the "slate" is the group of base stocks from a lube process that differ in viscosities, and there may be five or six from any given plant. Although they are referenced for other applications, API base stock applications apply mainly to components used in engine oils.

Base stocks are classified into two broad types—naphthenic and paraffinic—depending on the crude types they are derived from. Naphthenic crudes are characterized by the absence of wax or have very low levels of wax so they are largely cycloparaffinic and aromatic in composition; therefore naphthenic lube fractions are generally liquid at low temperatures without any dewaxing. On the other hand, paraffinic crudes contain wax, consisting largely of n- and iso-paraffins which have high melting points. Waxy paraffinic distillates have melting or pour points too high for winter use, therefore the paraffins have to be removed by dewaxing. After dewaxing, the paraffinic base stocks may still solidify, but at higher temperatures than do naphthenic ones because their molecular structures have a more paraffinic "character." Paraffinic base stocks are preferred for most lubricant applications and constitute about 85% of the world supply.

1.2 BASE STOCKS FROM CRUDES

Within a naphthenic or paraffinic type, base stocks are distinguished by their viscosities and are produced to certain viscosity specifications. Since viscosity is approximately related to molecular weight, the first step in manufacturing is to separate out the lube precursor molecules that have the correct molecular weight range. This is done by distillation. Figure 1.1 provides a schematic of the hardware of a crude fractionation system in a refinery used to obtain feedstocks for a lube plant. Lower-boiling fuel products of such low viscosities and volatilities that they have no application in lubricants—naphtha, kerosene, jet, and diesel fuels—are distilled off in the atmospheric tower. The higher molecular weight components which do not vaporize at atmospheric pressure are then fractionated by distillation at reduced pressures of from 10 mmHg to 50 mmHg (i.e., vacuum fractionation). Thus the "bottoms" from the atmospheric tower are fed to the vacuum tower, where intermediate product streams with generic names such as light vacuum gas oil (LVGO) and heavy vacuum gas oil (HVGO) are produced. These may be either narrow cuts of specific viscosities destined for a solvent refining step or broader cuts destined for hydrocracking to lubes and fuels.

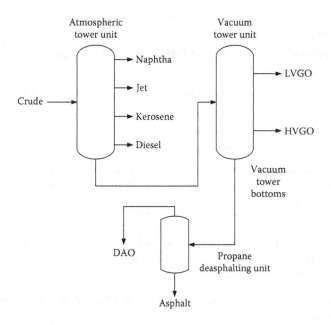

FIGURE 1.1 Schematic of a refinery crude fractionation train and deasphalting unit.

The vacuum tower bottoms may contain valuable high-viscosity lube precursors (boiling point greater than 950°F) and these are separated from asphaltic components (these are black, highly aromatic components that are difficult to refine) in a deasphalting unit. Deasphalting units separate asphalt from refinable components by solubility, and this is usually solubility in propane for lube purposes. This waxy lube feedstock is called deasphalted oil (DAO). Further refining of the DAO—dewaxing and solvent refining or hydrotreatment—produces bright stock, which is a heavy (very viscous) base stock that is a "residue" (i.e., it is not a distillate overhead). The DAO can also be part of the feed to a lube hydrocracker to produce heavier base stocks. Representative boiling and carbon number ranges for feedstocks are given in Table 1.1—they will vary somewhat from refinery to refinery and depend on the needs of the specific lube processes employed and those of fuel production.

The waxy distillates and DAO require three further processing steps to obtain acceptable base stock:

- Oxidation resistance and performance must be improved by removal of aromatics, particularly polyaromatics, nitrogen, and some of the sulfur-containing compounds.
- The viscosity-temperature relationship of the base stock (improve the viscosity index [VI]) has to be enhanced—by aromatics removal—to meet industry requirements for paraffinic stocks.

TABLE 1.1

Representative Boiling and Carbon Number Ranges for Lube Feedstocks

Fraction	Approximate Boiling Point Range (°F)	Carbon Number Range[a]
LVGO	600–900	18–34
HVGO	800–1100	28–53
DAO	950+	38+

[a] Carbon number ranges are referred to by the boiling points of the nearest n-paraffins; for example, the carbon number range of a 650–850°F fraction is C_{20}–C_{30} (651–843°F).

- The temperature at which the base stock "freezes" due to crystallization of wax must be lowered by wax removal so that equipment can operate at winter temperatures.

There are two strategic processing routes by which these objectives can be accomplished:

Processing steps which act by chemical separation: The undesirable chemical compounds (e.g., polyaromatics) are removed using solvent-based separation methods (solvent refining). The by-products (extracts) represent a yield loss in producing the base stock. The base stock properties are determined by molecules originally in the crude, since molecules in the final base stock are unchanged from those in the feed;

or

Processing steps which act by chemical conversion: Components with chemical structures unsuitable for lubes are wholly or partially converted to acceptable base stock components. These processes all involve catalysts acting in the presence of hydrogen, thus they are known collectively as catalytic hydroprocessing. Examples are the hydrogenation and ring opening of polyaromatics to polycyclic naphthenes with the same or fewer rings and the isomerization of wax components to more highly branched isomers with lower freezing points. Furthermore, the chemical properties of existing "good" components may be simultaneously altered such that even better performance can be achieved. Conversion processes are generally considered to offer lower operating costs, superior yields and higher base stock quality. In conversion processes, the eventual base stock properties reflect to some degree the molecules originally in the feed, but the extent of chemical alteration is such that products from different feedstocks can be very similar.

Separation processes are often depicted as "conventional" technologies and these solvent refining processes currently account for about 75% of the world's paraffinic base stock production. Conversion processes account for the remaining 25% and use catalytic hydroprocessing technology developed since World War II. This route has become particularly significant in North America, where more than 50% of base stock production uses this route. Some companies have chosen to combine separation and conversion, since the latter has been developed in steps and opportunities for synergism and the reuse of existing hardware have been recognized.

Figure 1.2 demonstrates how separation and conversion processes achieve the same end by different means. In the conventional solvent refining sequence, a polar solvent selectively extracts aromatics, particularly those with several aromatic rings and polar functional groups, resulting in an aromatic extract (the reject stream) and an upgraded waxy "raffinate" whose viscosity is less than that of the feed due to the removal of these polyaromatics. The major purposes of the extraction step are to reduce the temperature dependence of the viscosity (i.e., increase the VI) of the raffinate and improve the oxidation stability of the base stock. Since the raffinate still contains wax, which will cause it to "freeze" in winter, the next step—dewaxing—removes the wax. Again, a solvent-based method is used; in this case, crystallization of wax. This reduces the temperature at which the oil becomes solid—essentially the pour point. If desired, the wax can subsequently be de-oiled to make hard wax for direct commercial sale. The base stock now has almost all the desirable properties, however, in a last step it is usually subjected to clay treatment, which improves color and performance by

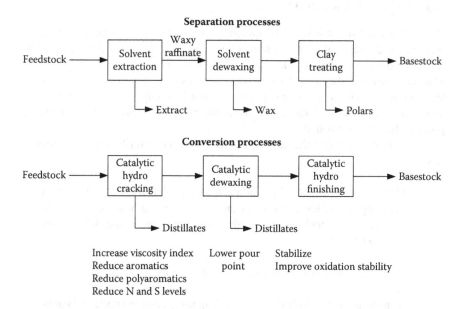

FIGURE 1.2 Comparison of process schematics for separation and conversion process routes for lubes.

taking out a few percent largely composed of polyaromatics and nitrogen, sulfur, and any oxygen compounds. This clay treating step has now been largely replaced by a catalytic hydrofinishing step.

In the conversion process, catalytic hydrogenation in the first stage lube hydrocracking unit saturates part of the feedstock aromatics by hydrogenating them to cycloparaffins and also promotes significant molecular reorganization by carbon-carbon bond breaking to improve the rheological (flow) properties of the base stock (again improving the VI). Usually in this stage, feed sulfur and nitrogen are both essentially eliminated. Some of the carbon-carbon bond breaking produces overheads in the form of low-sulfur gasoline and distillates. The fractionated waxy lube streams, usually those boiling above about 700°F, are then dewaxed, either by solvent dewaxing or, more frequently, by catalytic hydroprocessing (in which either wax is cracked to gasoline or isomerized to low melting isoparaffins in high yields and which has a positive effect on VI). The final step in conversion processes is usually catalytic hydrogenation to saturate most of the remaining aromatics to make base stocks stable for storage and to improve their performance. Base stocks produced by this route are frequently water white, whereas solvent extracted stocks retain some color. The advantages of the conversion route are many: less dependence on supplies of expensive high-quality "lube" crudes, which the solvent refining process requires and which are increasingly in short supply, higher base stock yields, and lubricants that better (and in some cases exclusively) meet today's automotive lubricant requirements.

1.3 BASE STOCK PROPERTIES

Base stocks are manufactured to specifications that place limitations on their physical and chemical properties, and these in turn establish parameters for refinery operations. Base stocks from different refineries will generally not be identical, although they may have some properties (e.g., viscosity at a particular temperature) in common. At this point it is worth briefly reviewing what measurements are involved in these specifications, what they mean, and where in the process they are controlled.

Starting with density, the most important ones that describe physical properties are

- Density and gravity, °API: Knowledge of the density is essential when handling quantities of the stock and the values can also be seen to fit with the base stock types. An alternative measure is the API gravity scale where

$$\text{API gravity} = 141.5/\text{specific gravity} - 131.5.$$

- Density increases with viscosity, boiling range, and aromatic and naphthenic content, and decreases as isoparaffin levels increase and as VI increases.

- Viscosity measured at 40°C and 100°C: Base stocks are primarily manufactured and sold according to their viscosities at either 40°C or 100°C, using kinematic viscosities (see later). Viscosity "grades" are now defined by kinematic viscosity in centistokes (cSt) at 40°C; formerly they were established on the Saybolt universal seconds (SUS) scale at 100°F. Higher viscosity base stocks are produced from heavier feedstocks (e.g., a 100 cSt at 40°C oil is produced from a HVGO and cannot be made from a LVGO since the molecular precursors are not present). As viscosity increases, so does the distillation midpoint.
- Viscosity index (VI): VI is a measure of the extent of viscosity change with temperature; the higher the VI, the less the change, and generally speaking, higher VIs are preferred. VI is usually calculated from measurements at 40°C and 100°C. The minimum VI for a paraffinic base stock is 80, but in practice the norm is 95, established by automotive market needs. Naphthenic base stocks may have VIs around zero. The conventional solvent extraction/solvent dewaxing route produces base stocks with VIs of about 95. Lower raffinate yields (higher extract yields) in solvent refining mean higher VIs, but it is difficult economically to go much above 105. In contrast, conversion processes enable a wide VI range of 95 to 140 to be attained, with the final product VI depending on feedstock VI, first stage reactor severity, and the dewaxing process. Dewaxing by hydroisomerization gives the same or higher VI relative to solvent dewaxing. To obtain a VI greater than 140, the feedstock generally must be either petroleum wax or Fischer-Tropsch wax.
- Pour point: The pour point measures the temperature at which a base stock no longer flows, and for paraffinic base stocks, pour points are usually between −12°C and −15°C, and are determined by operation of the dewaxing unit. For specialty purposes, pour points can be much lower. The pour points of naphthenic base stocks, which can have very low wax content, may be much lower (−30°C to −50°C). For very viscous base stocks such as Bright stocks, pour points may actually reflect a viscosity limit. Pour points are measured traditionally by ASTM D97,[4] but three new automated equivalent test methods are the "tilt" method (ASTM D5950), the pulse method (ASTM D5949), and the rotational method (ASTM D5985).
- Cloud point: The cloud point is the temperature at which wax crystals first form as a cloud of microcrystals. It is therefore higher than the pour point, at which crystals are so numerous that flow is prevented. The longstanding ASTM method is D2500, with three new automated methods being ASTM D5771, D5772, and D5773. Many base stock inspection sheets no longer provide cloud points. Cloud points can be 3°C to 15°C above the corresponding pour points.
- Color: Solvent extracted/solvent dewaxed stocks will retain some color as measured by ASTM D1500. Hydrocracked stocks, when hydrofinished at high pressures, are usually water white and their color is best measured on the Saybolt color scale (ASTM D156).

TABLE 1.2
SimDist of a 500N Base Stock

Percent Off	Temperature (°C)
1	354
5	422
10	443
30	480
50	505
70	534
90	572
95	589
99	630

- Distillation: At one time this would have been carried out using an actual physical distillation using either ASTM D86, a method performed at atmospheric pressure and applicable to very light lubes, or by vacuum distillation according to ASTM D1160 for heavier ones. Neither of these methods is employed much for base stocks nowadays because of their time and manpower requirements. Distillation today is usually performed by gas chromatography and the method is commonly called either simulated distillation (SimDist) or gas chromatographic distillation (GCD) using ASTM D2887. This method is capable of excellent accuracy, repeatability, and fast turnaround times and is normally automated. It is applicable to samples with final boiling points of less than 1000°F (538°C). For very heavy samples, ASTM WK2841 can analyze samples with boiling points in the range of 345°F to 1292°F (174°C to 700°C) (C10–C90). Results are usually reported as a table (e.g., Table 1.2) or graphically (as in Figure 1.3) in either degrees Fahrenheit or Celsius.
- Appearance: Base stocks should be "clear and bright" with no sediment or haze.
- Flash point: The flash point measures the temperature at which there is sufficient vapor above a liquid sample to ignite and is a significant feature in product applications where it is used as a common safety specification. Flash points are a reflection of the boiling point of the material at the front end of the base stock's distillation curve. Flash points generally increase with viscosity grade. High flash points for a given viscosity are desirable. Good fractionation and increased base stock VIs favor higher flash points. The Cleveland Open Cup method (ASTM D92) is the most often cited for North American base stocks, while the Pensky-Martens test (ASTM D93) is sometimes used.
- Volatility: This has emerged as a significant factor in automotive lubricant products from environmental and operational standpoints and again pertains predominantly to the distillation front end. Low volatility

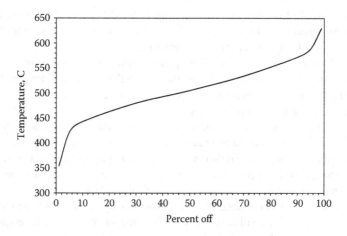

FIGURE 1.3 SimDist (or GCD) representation of a base stock's distillation profile.

(minimal losses at high temperatures) reduces emissions, is beneficial for emissions catalysts, reduces oil consumption, and helps prevent engine oil viscosity changes. Volatility is obviously affected by viscosity grade, but for a constant viscosity is established in part by sharper fractionation and in part by VI. It is measured either by the Noack method (ASTM D5800), using a thermogravimetric (TGA) method (namely ASTM D6375), or by gas chromatography (ASTM D6417 for engine oils).

- Aniline point: This is a measure of the ability of the base stock to act as a solvent and is determined from the temperature at which equal volumes of aniline and the base stock are soluble (ASTM D611). High aniline points (approximately 100°C or greater) imply a paraffinic base stock, while low aniline points (less than 100°C) imply a naphthenic or aromatic stock.
- Viscosity gravity constant: This is an indicator of base stock composition and solvency that is calculated from the density and viscosity according to ASTM D2501; it usually has a value between 0.8 and 1.0. High values indicate higher solvency and therefore greater naphthenic or aromatic content. This is usually of interest for naphthenic stocks. See Chapter 4 for further details.
- Refractive index: The refractive index is used to characterize base stocks, with aromatic ones having higher values than paraffinic ones. The value increases with molecular weight.
- Refractivity intercept: This is calculated (ASTM D2159) from the density (d) and refractive index (n) (both at 20°C) using the sodium D line (ASTM D1218), where

$$\text{Refractivity intercept} = n - (d/2),$$

and is a means of characterizing the composition of the sample. Values range from 1.03 to 1.047.

- Brookfield low temperature viscosity (ASTM D2983): This is the low temperature shear rate apparent viscosity measurement between −5°C and −40°C and is reported in centipoise (cP).
- Cold cranking simulator: The apparent low temperature viscosity of engine oils (ASTM D5293) correlates with the ease of low temperature engine cranking, measured in centipoise rather than centistokes, and the temperature is always given (e.g., CCS at −25°C = 1405 cP).

Chemical and compositional property measurements include the following:

- Sulfur: Sulfur is present in all lube plant feedstocks fractionated from crude oil and its content may be up to several percentage points. Solvent refining removes some but not all, therefore such stocks with no further treatment can contain up to several mass percent of sulfur. Hydrofinishing of solvent refined stocks can reduce this level substantially. Base stocks from conversion processes will have sulfur levels in the low parts per million (ppm) range since sulfur is relatively easily removed in severe hydroprocessing.
- Nitrogen: Like sulfur, nitrogen is present in all lube feedstocks, generally in the 500 to 2000 ppm range. These levels are reduced by solvent extraction and nearly eliminated by hydrocracking.
- Aromatics: These are predominantly monoaromatics in both feedstocks and products, but substantial levels of di- and triaromatics can be present in feedstocks. Aromatics, particularly polyaromatics, worsen base stock oxidation stability and can be virtually eliminated by conversion processes. Solvent extracted stocks still contain substantial levels of aromatics. These can be measured by several methods, including supercritical fluid chromatography (ASTM D5186), high-pressure liquid chromatography (HPLC) (ASTM D6379), chromatography over alumina/silica gel (ASTM D2549), preparative chromatography (ASTM D2007), and ultraviolet (UV) spectroscopy.
- $\%C_A$, $\%C_N$, $\%C_P$: These are the average carbon-type distributions, namely aromatic, naphthenic, and paraffinic. Aromatic carbons (C_A) are those in aromatic rings. To give a simple example, toluene has six aromatic carbons and one paraffinic carbon, and for this the $\%C_A$ is 86 and the $\%C_P$ is 14. Tetralin has six aromatic carbons and four naphthenic ones. For hydrocracked base stocks, we would expect the $\%C_A$ to be near zero. These parameters can be determined via the n-d-M method (ASTM D3238) and from viscosity-gravity constant and refractivity intercepts via ASTM D2140. $\%C_A$ can also be determined by nuclear magnetic resonance (NMR) spectroscopy (ASTM D5292). The result will depend on the method employed. See Chapter 4 for more discussion of these compositional methods.

Table 1.3 provides inspection results for a range of base stocks of different origins. Since the values are representative of those types, some commentary is worthwhile. First, the general format of these tables is to list the inspections (tests) performed in the left-hand column, with the column to the right of that identifying

TABLE 1.3
Inspection Results for Hydrotreated Naphthenic, Solvent Refined, and Hydrocracked Base Stocks

Test	Method	Source: Type: A Naphthenic	B Naphthenic	C Solvent Refined	D Hydrocracked	E Hydrocracked
API gravity, °API	ASTM D1298	29.8	24.0	29.0	32.0	38.1
Density at 15°C, kg/L	ASTM D4052	0.877	0.910	0.8816	0.865	0.8343
Viscosity, cSt at 40°C	ASTM D445	7.5	29.8	30.0	42.0	39.5
Viscosity, cSt at 100°C	ASTM D445	2.07	4.55	5.09	6.3	6.7
VI	ASTM D2270	56	35	95	95	125
Pour point, °C	ASTM D97	−54	−39	−15	−15	−18
Sulfur, mass %	ASTM D5185	0.03	0.09	0.45	15 ppm	6 ppm
Nitrogen, ppm	ASTM D5762			50	<1	<1
Aniline point, °C	ASTM D611		75			
Color	ASTM D1500	L 0.5	L 1.5	L 0.5	L 0.5	L 0.5
Appearance	ASTM D4176	C&B				
Flash point, °C	ASTM D92	151	157	—	—	—
Volatility, Noack	ASTM D5800	—	—	20	13	5
Aromatics, %	ASTM D2007	—	—	20.5	<1	<1
Hydrocarbon type analysis	ASTM D2140					
C_A, %		4				
Hydrocarbon type analysis	ASTM D3238					
C_A, %		6	—	—	—	—
C_N, %		52	—	—	—	—
C_P, %		42	—	—	—	—

the test methods used. Identification of the source of the methods used contributes significantly towards "certifying" the numbers contained in the table. This same format is usually used for feedstocks and any intermediate products.

Within the table, the two naphthenic oils on the left-hand side have high densities relative to all the others, regardless of viscosity, because they are largely composed of cycloparaffins and aromatics. These naphthenic oils have the lowest pour points of the lot, reflecting the absence of paraffinic structures, the feature they are best known for. In contrast, base stock E has the lowest density of this group because it is highly paraffinic (high VI) and paraffins and paraffin-like molecules are low density components. Base stocks D and E have extremely low sulfur levels because they have been severely hydrotreated. In contrast, the solvent extracted C has a sulfur content of 0.25%, which alone labels it as a solvent extracted oil, but more severe hydrofinishing could have reduced the levels much further. The very high VI of E shows that this cannot be a solvent extracted stock since that process route cannot economically achieve such high VIs. Finally, for D, produced by lube hydrocracking followed by dewaxing using isomerization, its very low sulfur, nitrogen, and aromatics contents are outcomes of these routes.

1.4 FEEDSTOCKS AND BASE STOCKS: GENERAL COMPOSITIONAL ASPECTS

Petroleum distillates and residues contain a complexity of hydrocarbons, some of which have already been mentioned, together with sulfur and nitrogen compounds. These were originally complex enough, being of plant origin, but after spending several million years buried at high temperatures, identification of individual structures is only possible for those with the lowest molecular weight (i.e., in the case of some naphtha components, where the number of isomers is limited). Table 1.4 illustrates the complexity using the simplest of hydrocarbon classes, the n-paraffins.[5]

As a consequence, we have to rely on the identification of groups or "lumps" of compounds that fall into similar chemical classifications. This approach has been quite successful as a means of either separating or quantifying them by instrumental methods.

Beginning with the simplest, these chemical groups are (Figure 1.4)

- n-Paraffins: These are C_{18} and greater members of the n-paraffin homologous series, which are present in significant quantities in feeds and waxy intermediate streams with a boiling range of 600°F to 850°F, depending on the wax content of the feed. As the boiling point increases beyond 850°F, they become much less common. n-Paraffins are easily identified and quantified by gas chromatography because they give sharp peaks and can be concentrated in the slack wax fraction from solvent dewaxing. They are significant because they have high melting points and therefore increase the pour point of base stocks. Base stocks with low n-paraffin contents have low pour points (e.g., naphthenics).

TABLE 1.4
Carbon Number of n-Paraffins and the Number of Isomers

Carbon Number	Isomers
5	3
8	18
10	75
12	355
15	4347
20	3.66×10^5
25	3.67×10^7
30	4.11×10^9
35	4.93×10^{11}
40	6.24×10^{13}

Source: K.H. Altgelt and M.M Boduzynski, Composition and Analysis of Heavy Petroleum Fractions (New York: Marcel Dekker 1993) With permission.

- Isoparaffins: These have n-paraffin backbones with alkyl branches; on an isoparaffin chain there may well be several branches of methyl groups or higher. Those isoparaffins most similar in structure to n-paraffins (e.g., single branches near a chain end) have higher pour points and will be removed by solvent dewaxing. Identification of individual members may be quite difficult. Isoparaffins as a group are commonly said to have high VIs and low pour points, and confer good oxidation resistance. They are therefore a sought-after component in base stocks. Polyalphaolefins (PAOs) are synthetic isoparaffinic base stocks that are of high commercial value because of their low pour points and excellent performance; they are not discussed in this book.
- Cycloparaffins (naphthenes): Cycloparaffins contain one or more cyclohexane or cyclopentane rings, or a combination thereof. If several rings are present, these are usually in the condensed form, presumably because of their natural origin. Mass spectroscopy of the saturates fraction can identify the number of rings and the percentage of the molecules having each number of rings. Alkyl substituents on the rings are branched and unbranched alkyl groups. Monocycloparaffins with 1,4 substituents are widely regarded as favorable structures, whereas polynaphthenes (3 + rings) are considered unfavorable for both VI and oxidation resistance.
- Aromatics: Basic structures have one to six or more benzene rings with some of the carbon–hydrogen bonds replaced by carbon–carbon bonds of alkyl substituents. Generally frequency declines with an increasing number of rings. Alkyl-substituted benzenes with 1,4 alkyl groups have

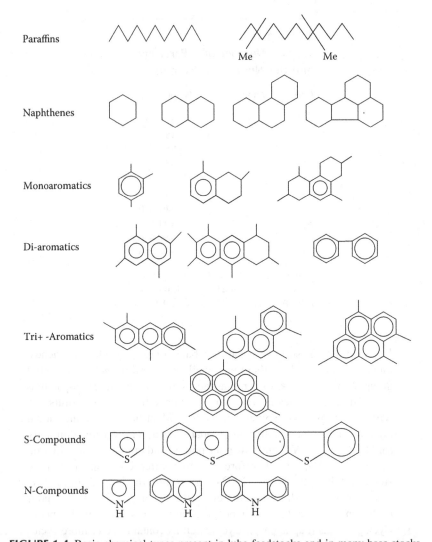

FIGURE 1.4 Basic chemical types present in lube feedstocks and in many base stocks.

high VIs and good oxidation resistance, whereas fused polyaromatic structures are undesirable.

- Sulfur-containing compounds: These may be thiols, sulfides, thiophenes, benzo- and dibenzo-thiophenes, and more complex structures. Solvent extraction reduces measured sulfur levels and therefore the content of sulfur compounds in solvent refined lubes; in solvent refined lubes, oxidation studies show that there appears to be an optimum level for sulfur compounds. Lube hydrocracking generally will reduce sulfur to about 10 ppm or less in the base stocks. The 4,6-di-alkyl substituted dibenzothiophenes are the most resistant type of sulfur compounds to hydrotreating (due to steric hindrance), while thiols and sulfides are the most easily hydrogenated. Sulfur compounds constitute a poison to hydroisomerization dewaxing catalysts and to nickel and noble metal catalysts and must be reduced to low levels in the feeds to those catalyst types.

- Nitrogen-containing compounds: Nitrogen largely appears in pyrroles, benzo- and dibenzo-carbazoles, pyridines, and quinolines. Nitrogen compounds are best minimized in finished lubes since they contribute to color formation. Lube hydrocracking reduces nitrogen levels to a few parts per million.
- Oxygen-containing compounds: Compounds containing chemically bound oxygen (e.g., furans, carboxylic acids, etc.) in lube feedstocks are seldom an issue and as a rule are overlooked.

1.5 API BASE STOCK CLASSIFICATIONS

A framework in which base stocks are differentiated from one another for the purpose of base stock interchanges came with the development of base stock categories by the API in 1993 (Table 1.5).[6] There are now five categories, three of which apply to paraffinic stocks and one to naphthenics. Paraffinic base stocks fall into categories I–III, and there are just three criteria involved—sulfur content, the percentage of saturates, and the VI—in determining the category of a particular stock. It should be noted that groups II+ and III+ are not official categories. These terms are employed in marketing, but are frequently referred to, hence their inclusion in Table 1.5.

The criteria for group I stocks places no limitations on sulfur content or percent aromatics, and in practice these are essentially all solvent refined. To reduce the aromatics content to less than 10% generally requires catalytic hydroprocessing equipment. The VI range here is broad (80 to 120) and in practice

TABLE 1.5
API Paraffinic Base Stock Categories

Group	Sulfur, Mass %		Saturates, Mass %	VI
I	>0.03	and/or	<90	≥80 to <120
II	≤0.03	and	≥90	≥80 to <120
II+	≤0.03	and	≥90	≥110 to <119
III	≤0.03	and	≥90	≥120
III+	≤0.03	and	>90	≥130<150
IV	All polyalphaolefins [PAOs]			
V	All stocks not included in groups I–IV (e.g., esters, pale oils)			

Source: The American Petroleum Institute, Publication 1509. With permission.

most solvent refined base stocks produced have VIs of 95 to 105. Generally a VI of 95 is about the market minimum. High VIs are expensive to obtain by solvent extraction since yields decline rapidly as VI rises.

Base stocks falling in the group II category will, in the vast majority of cases, be hydrocracked stocks, since the low sulfur and high saturates limits (low aromatics of less than 10%) are otherwise difficult to attain. The majority of group II stocks produced have VIs of 95 to 105. Group II+ is a commonly used industry subset (not a formal part of the API classification) defined by a VI in the range of 110 to 120 and created because of the current demand for the low volatility that accompanies these VIs.

Group III base stocks are differentiated by their very high VIs, which defines them as being products from either fuel hydrocracking units (which operate at high severities and low lube yields) or by hydrocracking or isomerizing wax. Group III+, like group II+, is an informal subset. Group III+ base stocks are sourced mainly from gas-to-liquids (GTL) plants.

Group IV base stocks are polyalphaolefins and originate from long-chain terminal olefins (e.g., 1-decene). Group V includes any base stocks not defined by groups I–IV, such as naphthenic base oils, esters, polyglycols, and polyinternalolefins.

1.6 VISCOSITY GRADES FOR INDUSTRIAL LUBRICANTS

Each refinery usually produces a limited number (5 to 12) of base stocks, since each base stock requires that there be economically sufficient precursors in the front-end feed and, as well, in the feed to the final fractionation after all processing is complete. In addition, the refinery tries to adjust production rates for individual base stocks to meet their demand. Today, base stocks are classified by their viscosity in square millimeters per second (mm^2/s) (or centistokes) measured at 40°C and usually employ the International Organization for Standardization (ISO) system for industrial lubricants (ASTM D2422), with a range of 2 to 3200 cSt. Twenty viscosity grades and their ranges are provided in Table 1.6.[7]

Previously it had been customary to assign viscosity grades according to SUS viscosities at 100°F. The SUS grades corresponding to the ISO grades are given in the right-hand column of Table 1.6. SUS viscosities are rarely directly measured any longer—where there is a need for those numbers, they are calculated from kinematic viscosities at 40°C and 100°C. However, SUS names and grades have had remarkable staying power and are still employed in the naming systems employed by various companies (e.g., an ISO 22 is frequently [usually in North America] called a 100N and an ISO 100 is called a 500N). This is done to avoid customer confusion in changing names. To add to the confusion, some companies, particularly with regard to base stocks for automotive applications, designate their products by viscosity at 100°C and therefore these are called 4 cSt (ISO 22) or 10 cSt (ISO 100) products. For an informative article on viscosity grades see "ISO Viscosity Grades" by M. Johnson.[8]

TABLE 1.6
ISO Viscosity System for Industrial Fluid Lubricants

Grade Identification	Midpoint, cSt at 40°C	Viscosity Limits, cSt at 40°C Minimum–Maximum	American SUS Grades (Range)
ISO VG 2	2.2	1.98–2.4	32 (32.0–34.0)
ISO VG 3	3.2	2.88–3.44	36 (35.5–37.5)
ISO VG 5	4.6	4.14–5.06	40 (39.5–42.5)
ISO VG 7	6.8	6.12–7.48	50 (46.0–50.5)
ISO VG 10	10	9.00–11.0	60 (55.5–62.5)
ISO VG 15	15	13.5–16.5	75 (71.5–83.5)
ISO VG 22	22	19.8–24.2	105 (97.0–116)
ISO VG 32	32	28.8–35.2	150 (136–165)
ISO VG 46	46	41.4–50.6	215 (193–235)
ISO VG 68	68	61.2–74.8	315 (284–347)
ISO VG 100	100	90.0–110	465 (417–510)
ISO VG 150	150	135–165	700 (625–764)
ISO VG 220	220	198–242	1000 (917–1121)
ISO VG 320	320	288–352	1500 (1334–1631)
ISO VG 460	460	414–506	2150 (1918–2344)
ISO VG 680	680	612–748	3150 (2835–3465)
ISO VG 1000	1000	900–1100	4650 (4169–5095)
ISO VG 1500	1500	1350–1650	7000 (4169–5095)
ISO VG 2200	2200	1980–2420	10,200 (9180–11,221)
ISO VG 3200	3200	2880–3520	14,840 (13,355–16,324)

Source: ASTM, ASTM D2422. With permission.

1.7 SOCIETY FOR AUTOMOTIVE ENGINEERS VISCOSITY CLASSIFICATION FOR ENGINE OILS

The Society for Automotive Engineers (SAE) has developed a viscosity classification system for finished (i.e., not just the base stock) engine lubricants that defines viscosity ranges as well as low temperature properties. The 2004 SAE J300 grades are shown in Table 1.7.[9]

Single-grade winter oils, 0W through 25W, specify minimum hot viscosities (at 100°C) and maximum viscosities at low temperatures to ensure easy starting. SAE grades 20 through 60 are not intended for winter use. Multigrade oils define low and high temperature properties to provide additional engine protection (e.g., a 5W30 oil meets the low temperature requirements of a 5W oil and its viscosity at high temperature falls within the viscosity range for a 30 grade oil; that is,

TABLE 1.7
SAE Viscosity Grades for Engine Oils, J300 (2004)

SAE Viscosity Grade	Viscosity (cP) at Temperature (°C), Maximum		Viscosity, mm²/sec (cSt) at 100°C	
	Cranking	Pumpability	Minimum	Maximum
0W	6200 at –35°C	60,000 at –40°C	3.8	—
5W	6600 at –30°C	60,000 at –35°C	3.8	—
10W	7000 at –25°C	60,000 at –30°C	4.1	—
15W	7000 at –20°C	60,000 at –25°C	5.6	—
20W	9500 at –15°C	60,000 at –20°C	5.6	—
25W	13,000 at –10°C	60,0000 at –15°C	9.3	—
20	—	—	5.6	<9.3
30	—	—	9.3	<12.5
40	—	—	12.5	<16.3
50	—	—	16.3	<21.9
60	—	—	21.9	<26.1

Source: Society of Automotive Engineers/SAE J300. With permission.

overall the viscosity of the oil changes less with temperature than that of a single-grade winter oil).

1.8 API ENGINE OIL CLASSIFICATIONS

As both gasoline and diesel engine technology advances, new demands are placed on lubricant performance and it becomes important to clearly distinguish the appropriate applications of engine lubricants from different manufacturers. Thus, in the United States, the API together with the SAE and ASTM identify lubricant standards at intervals as engine technology needs require. This classifies engine oils according to their performance and related to their intended type of service. In Japan, the lubricant specification organization is the Japanese Automotive Standards Organization (JASO), while in Europe it is the Association des Constructeurs Europeéns d'Automobiles (ACEA). A further significant body is the International Lubricant Specification and Approval Committee (ILSAC), which is a body composed of U.S. and Japanese engine manufacturers (General Motors, Ford, DaimlerChrysler, Toyota, and Honda) whose specifications apply mainly to North America and Japan.

Both gasoline and diesel engine classifications are letter grades, preceded by "S" (service) for gasoline engine lubricants and by "C" (commercial) for diesel engine lubricants. Gasoline engine oil classifications and the time periods they have been in force are shown in Figure 1.5. ILSAC gasoline engine certifications are indicated by GF-X, where X is a number. So far these have corresponded to API certifications, namely GF-1 = SH, GF-2 = SJ, GF-3 = SL, and GF-4 = SM.

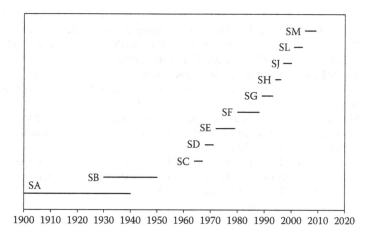

FIGURE 1.5 API gasoline engine oil classifications since 1900.

The significance of these classifications to base stock manufacturers is that they influence the selection of base stock types (API groups I–V) and the properties that are employed to blend engine oils. Since about 60% of worldwide lubricant demand[10] is for engine oils, the classifications tend to drive new manufacturing technology for all paraffinic base stocks. And of course the reverse occurs, as base stocks are produced by new technologies, engine lubricant developers use these to develop more advanced engine designs, which then provides drivers for new manufacturing technology. Figure 1.5 shows that in the 1900 to 1960 period, engine oil development was very slow, but since then the pace has picked up noticeably such that engine, environmental, and marketplace developments have required new categories every three to four years.

REFERENCES

1. Lubes 'N' Greases, *2005 Guide to Global Base Oil Refining* (Falls Church, VA: LNG Publishing).
2. T. Sullivan, "The Oncoming Train," *Lubes 'N' Greases*, January:14 (2006).
3. API 1509, *Engine Oil Licensing and Certification System*, 15th ed., April 2002, Appendix E. API Base Oil Interchangeability Guidelines for Passenger Car Motor Oils and Diesel Engine Oils, 2004, Section E.1.2, Definitions (Washington, DC: American Petroleum Institute).
4. Most measurements on petroleum products are by industry-accepted methods that are established and certified by independent organizations. One of those is the American Society for Testing and Materials (ASTM), whose petroleum methods are identified as ASTM DXYZ. Other such organizations include the Institute of Petroleum (IP, now the Energy Institute) in the United Kingdom, and the Japanese Standards Association (JSA).
5. K. H. Altgelt and M. M. Boduszynski, *Composition and Analysis of Heavy Petroleum Fractions* (New York: Marcel Dekker, 1993).

6. API Publication 1509, *Engine Oil Licensing and Certification System*, 15th ed., April 2002, Appendix E. API Base Oil Interchangeability Guidelines for Passenger Car Motor Oils and Diesel Engine Oils, 2004, Section E.1.3, Base Stock Categories (Washington, DC: American Petroleum Institute).

7. ASTM D2422, "Standard Classification of Industrial Fluid Lubricants by Viscosity System," *Annual Book of ASTM Standards*, vol. 05.01 (West Conshohocken, PA: American Society for Testing and Materials, 2005).

8. M. Johnson, "ISO Viscosity Grades," *Machinery Lubrication*, July 2001.

9. *J300 Engine Oil Viscosity Classification*, revised May 2004 (Warrendale, PA: Society of Automotive Engineers).

10. G. Agashe and M. Phadke, "Global Lubricant Base Stocks Industry: Kline's 20/20 Vision," presentation to the Independent Commodity Information Service/London Oil Report (ICIS/LOR), September 20, 2005.

2 Viscosity, Pour Points, Boiling Points, and Chemical Structure

2.1 VISCOSITY

2.1.1 INTRODUCTION

The viscosity of a base stock is a quantitative measure of its resistance to flow. It is the key property of base stocks since it is a major factor in determining their application; for example, low viscosity stocks can be used for automotive transmission oils, while higher viscosity stocks are employed in diesel engine oils. Base stocks are usually named according to their viscosity. Viscosity measurements on base stocks assume that the liquids are Newtonian (i.e., that shear stress and shear rate are linearly related).

The mathematical relationships involved can be developed from Figure 2.1,[1] which represents two identical plates (of liquid) of area A, one stationary, the other being moved by a force F at velocity V, and separated by a fluid film of thickness h. Viscosity, μ, is then defined as the force per unit area required to move the plate at unit velocity when the plates are separated by a unit distance:[2]

$$\mu = (F/A)/(V/h).$$

This equation provides the *dynamic* viscosity of a liquid, which is used in engineering work. Base stock viscosities are almost invariably measured in *kinematic* viscosity, which corresponds to the time for flow under the influence of gravity.

$$\text{Kinematic viscosity} = \text{dynamic viscosity/density,}$$

or

$$\nu = \mu/\rho.$$

2.1.2 VISCOSITY UNITS

Viscosity measurements are expressed in either systematic or empirical units; both are used, but the current trend is very much toward using systematic units.

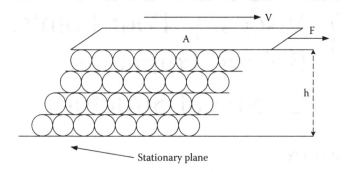

FIGURE 2.1 Definition of viscosity.
Source: D. Godfrey and R. L. Peeler, "Explanation of SI Viscosity Units," *Lubrication Engineering* 38:613–614 (1982). With permission.

However, empirical units have left a lasting impression on terminology, and while still employed from time to time in providing names for base stocks (e.g., 100N, 250N), their actual viscosity measurements and their ranges on specification sheets are measured in systematic units and converted to empirical ones.

2.1.2.1 Systematic Units

Increasingly used are those corresponding to the SI system (International System of Units), where the newton (N, $kg \cdot m/s^2$) is the unit of force, and the pascal (Pa, N/m^2) is the unit of pressure. Therefore,

$$\text{Dynamic viscosity, } \mu = Pa/(m/s \cdot m) = Pa \cdot s \text{ (units are pascal seconds)} = N/m^2 \cdot s$$
$$= kg \cdot m/s^2/m^2 \cdot s = kg/m/s$$

and

$$\text{Kinematic viscosity, } \nu = \mu/\rho = kg \cdot m/s/kg/m^3 = m^2/s, \text{ but is usually expressed}$$
$$\text{in } mm^2/s,$$

where $1 \ m^2/s = 10^6 \ mm^2/s$ to bring units and experimental measurements to the same magnitudes.

Therefore, to summarize, in the SI system, dynamic viscosity is expressed as pascal seconds and kinematic viscosity is expressed as square meters per second. In practice, dynamic viscosities are reported as millipascal seconds (mPa·s). Base stock viscosities are usually reported as kinematic viscosities.

In the centimeter-gram-second (CGS) system, where the smaller units of the centimeter and gram are employed, the unit of dynamic viscosity is the poise:

$$\text{Dynamic viscosity, } \mu = (F/A)/(V/h) = ((\text{dyne/cm}^2)/(\text{cm/s/cm}))$$
$$= \text{g·cm/s}^2/\text{cm}^2/\text{cm/s/cm} = \text{g/cm·s} = 1 \text{ poise (P)}.$$

For practical use, the poise is too large a unit; the unit normally seen is the centipoise (cP).

For kinematic viscosity in the CGS system,

$$\text{Kinematic viscosity, } \nu = \mu/\rho = \text{g/cm·s/g/cm}^3 = \text{cm}^2/\text{s} = 1 \text{ stokes (St)}.$$

Again, in practice, the stokes is too large a unit and it is as centistokes (cSt) that kinematic viscosities are normally reported.

The relation between the SI and CGS units of viscosity is 1 mPa·s = 1 cP and 1 mm^2/s = 1 cSt. North American companies report base stock kinematic viscosities on the centistokes scale at 40°C and 100°C. Occasionally they are given as square millimeters per second (mm^2/s); the ASTM method (D445) reports units this way. At 40°C, base stock kinematic viscosities range from about 4 to 500 mm^2/s^{-1}.

2.1.2.2 Empirical Units

2.1.2.2.1 Saybolt Universal Viscosity

This is measured as the time (in seconds) required for a sample to flow through the orifice of a Saybolt universal viscometer, according to the conditions specified in ASTM D88. Viscosities measured are expressed as, for example, Saybolt universal seconds (SUS), Saybolt seconds universal (SSU), or Saybolt universal viscosity (SUV), and are usually measured at both 100°F and 210°F. This is an outdated method, but as mentioned earlier, its memory lingers in the industry terminology.

Viscosity grades were originally based on SUS viscosities at 100°F and are still very much in use today (e.g., 40N, 100N, 500N, etc.)[3] but are being replaced by International Standards Organization (ISO) grades. Since it is now rare to measure viscosities directly by the SUS procedure, kinematic viscosities can be converted to Saybolt via the equation[4]

$$SUS_t = (1.0 + 0.000061(t - 100))[(4.6324\nu + (1.0 + 0.03264\nu)/\{(3930.2 + 262.7\nu + 23.97\nu^2 + 1.646\nu^3) \times 10^{-5}\}],$$

where t is the temperature in degrees Fahrenheit and ν is the viscosity in centistokes. Web sites of some companies include viscosity conversion programs.[5]

2.1.2.2.2 Saybolt Furol Viscosity

This is very similar to the method above, but is applied to very high viscosity samples such as asphalts. Results are reported as Saybolt seconds furol (SSF).

2.1.2.2.3 Redwood Viscosity

This empirical method is used in Europe but has little use in North America. Like the Saybolt method, this method measures the time for a fixed volume to flow out of a standard Redwood viscometer by the IP70 procedure and the result is reported in Redwood seconds.

2.1.3 TEMPERATURES USED FOR MEASUREMENT

The convention now is to measure base stock viscosities at 40°C and 100°C. Waxy intermediates usually have their viscosities measured at 65°C instead of at 40°C, a temperature at which many of these will have solid wax present. These measurements are made according to ASTM D445[6] or its equivalent, and when performed correctly give results that are sufficiently accurate to determine the viscosity index (VI)[7] of the sample. Historically, particularly when viscosities were measured in SSU units, the temperatures employed were 100°F and 210°F. It should be mentioned that it is not uncommon to find that viscosities measured for other purposes (e.g., for vacuum gas oil fractions in crude assays) are not sufficiently accurate to provide good waxy VIs (which are useful for assessing crudes for lube use).

2.1.4 HYDROCARBON VISCOSITIES AND COMPOSITION

Hydrocarbon viscosities increase with increasing molecular weight and with structural complexity. The n-paraffins exhibit a steady increase with molecular weight[8–10] (Table 2.1). From the comparisons in Table 2.2 of hydrocarbons with the same carbon numbers, it can be seen that cycloparaffins containing two or more fused rings have higher viscosities than n-paraffins for the same carbon number. Highly naphthenic base stocks will therefore have higher viscosities at the same carbon number than paraffinic ones. Relative to n-paraffins, branched paraffins[8,11] have very similar viscosities, as can be seen in Table 2.3.

Introduction of a single aromatic ring does not change viscosities significantly (Table 2.4), but polyaromatic rings do. Their conversion to monoaromatics or perhydro- structures by hydrotreatment or removal by extraction from a lube feedstock should be accompanied by a decrease in viscosity. When structures with a single aromatic ring are involved, reduction of the benzene ring to a cyclohexyl structure will cause little change in viscosity.

From these examples, it can be seen that almost invariably among isomers the n-alkane has the lowest viscosity at both 40°C and 100°C. The table below (Table 2.5) demonstrates that when aromatic compounds which have a total of three or more aromatic rings are hydrogenated, the perhydro- products are of lower viscosity than the aromatic types—this is particularly true when polyaromatic hydrocarbons are hydrotreated without cracking or isomerization. In contrast,

TABLE 2.1
Viscosities of Some n-Paraffins

Name	Formula	Carbon Number	Viscosity, cSt at 40°C	Viscosity, cSt at 100°C	Reference
n-Decane	$C_{10}H_{22}$	10	0.98	0.56	API Project 42
n-Dodecane	$C_{12}H_{26}$	12	1.45	0.74	API Project 42
n-Tridecane	$C_{13}H_{28}$	13	1.75	0.85	API Project 42
n-Tetradecane	$C_{14}H_{30}$	14	2.09	0.97	API Project 42
n-Pentadecane	$C_{15}H_{32}$	15	2.48	1.10	API Project 42
n-Hexadecane	$C_{16}H_{34}$	16	2.93	1.25	API Project 42
n-Heptadecane	$C_{17}H_{36}$	17	3.42	1.40	API Project 42
n-Octadecane	$C_{18}H_{38}$	18	3.97	1.56	API Project 42
n-Nonadecane	$C_{19}H_{40}$	19	4.56	1.73	API Project 42
n-Eicosane	$C_{20}H_{42}$	20	5.25	1.90	API Project 42
n-Uneicosane	$C_{21}H_{44}$	21	6.06	2.07	API Project 42
n-Tricosane	$C_{23}H_{48}$	23	7.66	2.54	J. Denis
n-Tetracosane	$C_{24}H_{50}$	24	8.68	2.74	J. Denis
n-Hexacosane	$C_{26}H_{54}$	26	10.73	3.24	J. Denis
n-Octacosane	$C_{28}H_{58}$	28	13.14	3.75	J. Denis
n-Dotriacontane	$C_{32}H_{66}$	32	18.95	4.92	J. Denis
n-Pentatriacontane	$C_{35}H_{72}$	35	22.60	5.88	J. Denis
n-Hexatriacontane	$C_{36}H_{74}$	36	26.60	6.27	J. Denis
n-Triatetracontane	$C_{43}H_{88}$	43	48.87	9.15	G. W. Nederbragt and J. W. M. Boelhouwer
n-Tetratetracontane	$C_{44}H_{90}$	44	51.30	9.44	J. Denis
n-Tetranonacontane	$C_{94}H_{190}$	94	183.30	39.32	API Project 42

Sources: "Properties of Hydrocarbons of High Molecular Weight," Research Project 42, 1940–1966, American Petroleum Institute, New York; J. Denis, "The Relationship Between Structure and Rheological Properties of Hydrocarbons and Oxygenated Compounds Uses as Base Stocks," *Journal of Synthetic Lubricants*, vol. 1(1–3):201–238 (1984); J. W. Nederbragt and J. W. M. Boelhouwer, "Viscosity Data and Relations of Normal and Iso-Paraffins," *Physica* X111(6–7):305–318 (1947); R. T. Sanderson, "Viscosity-Temperature Characteristics of Hydrocarbons," *Industrial & Engineering Chemistry* 41(2):368–374 (1949). With permission.

TABLE 2.2
Viscosities (cSt at 40°C and 100°C) of Model Saturated Hydrocarbons[8]

Carbon Number		Name	Number of Fused Rings	Viscosity (cSt at 40°C)	Viscosity (cSt at 100°C)
10	n-$C_{10}H_{22}$	n-Decane	0	0.9759	0.5586
10		cis-Decalin	2	2.509	1.083
10		trans-Decalin	2	1.780	0.850
13	n-$C_{13}H_{28}$	n-Tridecane	0	1.748	0.849
13		Perhydrofluorene	3	4.288	1.638
14	n-$C_{14}H_{30}$	n-Tetradecane	0	2.089	0.9707
14		Perhydrophenanthrene	3	4.81	1.75
16	n-$C_{16}H_{34}$	n-Hexadecane	0	2.935	1.245
16		Perhydropyrene	4	11.52	2.90
16		Perhydrofluoranthene	4	8.97	2.51
18	n-$C_{18}H_{38}$	n-Octadecane	0	3.97	1.56
18		Perhydrochrysene	4	23.85	3.98
21	n-$C_{21}H_{44}$	n-Uneicosane	0	6.059	2.071
21		Perhydrodibenzo(a,i)fluorene	5	290.6	10.66

TABLE 2.2 (CONTINUED)
Viscosities (cSt at 40°C and 100°C) of Model Saturated Hydrocarbons[8]

Carbon Number		Name	Number of Fused Rings	Viscosity (cSt at 40°C)	Viscosity (cSt at 100°C)
26	n-$C_{26}H_{54}$	n-Hexacosane	0	10.73	3.24
26	C_8	2-Octylperhydrotriphenylene	4	103.34	8.67
26	C_{10}	2-Decylperhydroindeno[2, 1-a]indene	4	47.60	6.02

Source: "Properties of Hydrocarbons of High Molecular Weight," Research Project 42, 1940–1966, American Petroleum Institute, New York. With permission.

TABLE 2.3
Viscosities at 40°C and 100°C of n- and Branched Paraffins

Name	Formula	Carbon Number	Viscosity (cSt at 40°C)	Viscosity (cSt at 100°C)
2-Methyldecane	$C_{11}H_{24}$	11	1.145	0.6144
Hexadecane	$C_{16}H_{34}$	16	2.935	1.245
2-Methylpentadecane	$C_{16}H_{34}$	16	2.916	1.236
Octadecane	$C_{18}H_{38}$	18	3.973	1.558
2-Methylheptadecane	$C_{18}H_{38}$	18	4.02	1.562
4,9-Di-n-Propyldodecane	$C_{18}H_{38}$	18	3.378	1.234
Uneicosane	$C_{21}H_{44}$	21	6.059	2.071
3-Methyleicosane	$C_{21}H_{44}$	21	6.093	2.114
10-Methyleicosane	$C_{21}H_{44}$	21	5.81	1.974
8-n-Hexylpentadecane	$C_{21}H_{44}$	21	5.455	1.761
Tetracosane	$C_{24}H_{50}$	24	8.622	2.744
2-Methyltricosane	$C_{24}H_{50}$	24	8.945	2.806
n-Hexacosane	$C_{26}H_{54}$	26	10.73	3.24
11-n-Butyldocosane	$C_{26}H_{54}$	26	9.687	2.733

(continued)

TABLE 2.3 (CONTINUED)
Viscosities at 40°C and 100°C of n- and Branched Paraffins

Name	Formula	Carbon Number	Viscosity (cSt at 40°C)	Viscosity (cSt at 100°C)
9-n-Butyldocosane	$C_{26}H_{54}$	26	9.898	2.768
7-n-Butyldocosane	$C_{26}H_{54}$	26	10.351	2.871
5-n-Butyldocosane	$C_{26}H_{54}$	26	10.701	2.97
6,11-Di-n-amylhexadecane	$C_{26}H_{54}$	26	10.914	2.679
3-Ethyl-5(2-ethyl-Butyl)octadecane	$C_{26}H_{54}$	26	10.988	2.832
11-n-Amylheneicosane	$C_{26}H_{54}$	26	9.411	2.678
11(3-Pentyl)heneicosane	$C_{26}H_{54}$	26	9.687	2.695
5,14-Di-n-Butyloctadecane	$C_{26}H_{54}$	26	11.27	2.78
7-n-Hexyleicosane	$C_{26}H_{54}$	26	9.981	2.795
11-neopentylheneicosane	$C_{26}H_{54}$	26	10.875	2.832
3-Ethyltetracosane	$C_{26}H_{54}$	26	10.831	3.225

Source: "Properties of Hydrocarbons of High Molecular Weight," Research Project 42, 1940–1966, American Petroleum Institute, New York. With permission.

TABLE 2.4
Comparison of n-Alkane Viscosities with Those of Aromatics of the Same Carbon Number

Carbon Number	Name	Number of Aromatic Rings	Viscosity (cSt at 40°C)	Viscosity (cSt at 100°C)
14	n-Tetradecane	0	2.089	0.9707
14	1-Phenyloctane	1	2.057	0.9464
14	2-n-Butyl Naphthalene	2	2.936	1.137
16	n-Hexadecane	0	2.934	1.2452
16	1-Phenyl Decane	1	2.8817	1.2202
18	n-Octadecane	0	3.97	1.56
18	9-n-Butyl Anthracene	3	62.54	4.478
19	n-Nonadecane	0	4.4563	1.726
19	7-Phenyltridecane	1	5.535	1.649
26	n-Hexacosane	0	10.73	3.24
26	2-Phenyleicosane	1	12.026	3.330
26	2-n-Octylchrysene	4	360.454	14.43
26	3-n-Decylpyrene	4	57.74	7.091

Source: "Properties of Hydrocarbons of High Molecular Weight," Research Project 42, 1940–1966, American Petroleum Institute, New York. With permission.

TABLE 2.5
Effect of Hydrotreatment on Hydrocarbon Viscosities

Carbon Number	Formula	Name	Number of Aromatic Rings	Viscosity (cSt at 40°C)	Viscosity (cSt at 100°C)
13	$C_{13}H_{12}$	Diphenylmethane	2	2.13	0.96
13	$C_{13}H_{24}$	Dicyclohexylmethane	0	3.86	1.42
14	$C_{14}H_{16}$	2-Butylnaphthalene	2	2.94	1.14
14	$C_{14}H_{20}$	2-Butyltetralin	1	3.16	1.23
14	$C_{14}H_{26}$	2-Butyldecalin	0	3.74	1.38
14	$C_{14}H_{18}$	1,2,3,4,5,6,7,8-Octahydrophenanthrene	1	7.57	2.01
14	$C_{14}H_{24}$	Perhydrophenanthrene	0	4.81	1.748
26	$C_{26}H_{34}$	2-Decyl-4b,5,9b,10-tetrahydroindeno[2.1.a]indene	2	69.96	7.88
26	$C_{26}H_{46}$	2-Decylperhydroindeno[2.1.a]indene	0	48.11	6.96
26	$C_{26}H_{32}$	9-n-Octyl[1,2,3,4-tetrahydro]naphthacene	3	668.91	18.05
26	$C_{26}H_{40}$	1,2,3,4,5,6,7,8,9,10,17,18-dodecahydro-9-n-octylnapthacene	1	297.07	12.29
26	$C_{26}H_{46}$	9-n-Octylperhydronaphthacene	0	261.52	11.86
26	$C_{26}H_{30}$	3-n-Decylpyrene	4	57.74	7.09
26	$C_{26}H_{46}$	3-n-Decylperhydropyrene	0	40.61	6.10
26	$C_{26}H_{34}$	9-n-Dodecylanthracene	3	67.19	6.91
26	$C_{26}H_{48}$	9-n-Dodecylperhydroanthracene	0	41.43	5.88

Source: "Properties of Hydrocarbons of High Molecular Weight," Research Project 42, 1940–1966, American Petroleum Institute, New York. With permission

when a single benzene ring is involved, viscosities show slight increases on hydrotreatment (Table 2.6).[9,11]

2.2 POUR POINTS AND CHEMICAL STRUCTURE

2.2.1 INTRODUCTION

For proper lubrication of mechanical equipment, petroleum-based lubricants are designed to operate in the liquid phase. Accordingly, for a base stock it is important to know the temperature at which the transition between the liquid and

TABLE 2.6
Effect of Hydrotreatment on Viscosities of Phenyl Eicosanes

Position of Substituent	Phenyl Eicosanes		Cyclohexyl Eicosanes	
	Viscosity (cSt at 40°C)	Viscosity (cSt at 100°C)	Viscosity (cSt at 40°C)	Viscosity (cSt at 100°C)
1	10.82	3.28	15.40	4.06
2	12.05	3.34	16.10	4.07
3	12.63	3.33	14.86	3.78
4	13.30	3.32	15.75	3.73
5	13.80	3.30	16.13	3.69
7	13.31	3.22	15.72	3.58
9	12.90	3.14	14.90	3.45

Source: J. Denis, "The Relationship Between Structure and Rheological Properties of Hydrocarbons and Oxygenated Compounds Used as Base Stocks," *Journal of Synthetic Lubricants*, vol. 1(1–3):201–238 (1984). With permission.

solid phases occurs and manufacture it to be as low as necessary. This temperature is known as the pour point and is defined as "the lowest temperature at which movement of the specimen is observed."[12] It is measured classically by ASTM D97. Associated with the pour point is the cloud point, the temperature at which the sample first becomes cloudy due to the initial stages of wax crystallization. This is measured by ASTM D2500[13] and is naturally higher than the pour point. Both have traditionally been measured by lowering the temperature of the sample at defined rates, then measuring the temperature at which wax crystals first appear (cloud point) and at which the sample no longer moves when the tube containing the sample is tilted (pour point). Low pour points (less than 0°C) and cloud points are obviously desirable for cold weather lubrication and particularly for refrigeration equipment. Paraffinic base stocks usually have pour points of −18°C to −6°C, while those of naphthenic base stocks may be −30°C to −40°C. The cloud point is usually 5°C to 15°C greater than that of the pour point. Synthetic base stocks usually have pour points of about 30°C or lower. Pour point precision by D97 is ± 3°C. Instrumental methods for pour point determination are ASTM D5949, D5950, and D5985.

The measurement for a pure compound corresponding to the pour point of a base stock is the melting point, which in practice is the temperature at which the liquid and crystalline phases are in equilibrium. Pure compound behavior is used to relate melting point (ergo pour point) to structure and composition and therefore provides guidance to dewaxing research. Complex mixtures such as lubricants do not have a melting point, since they melt or solidify over a temperature range, hence the development of the pour point concept.

2.2.2 Pour Points and Composition

The highest pour point components among hydrocarbons are the n-paraffins. Figure 2.2 shows that their melting points steadily increase with carbon number[14] and in the C_{20} to C_{50} interval, these are between 40°C and 90°C (in Figure 2.2, data to C_{39} are measured points, from C_{40} onward, melting points were calculated by Fisher's method[15]). To produce base stocks from intermediate waxy lubes, n-paraffins must be removed by solvent dewaxing or converted to lower melting point components in the dewaxing step to achieve the necessary pour points. Any selectivity in either process should favor removal/conversion of the higher n-paraffins, since these have higher melting points.

Of critical importance to the chemistry of dewaxing, particularly dewaxing by hydroisomerization, is the fact that branched paraffins have lower melting points than the n-isomers, and the magnitude of the decrease increases as branching moves toward the center of the molecule. Figure 2.3 demonstrates this for 2- and 3-methyl branched alkanes[16] in the C_{15} to C_{35} range. It can be seen that the melting point of a 2-methyl paraffin is about 15°C below that for a normal isomer and the melting point of a 3-methyl is reduced by about 35°C, all of the same carbon number. Wax obtained by solvent dewaxing will contain, apart from n-paraffins, decreasing quantities of 2-, 3-, 4-, etc. branched paraffins.

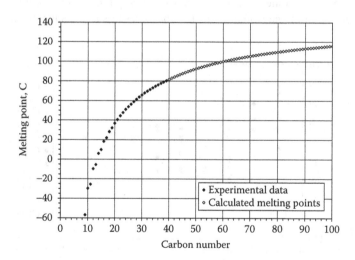

FIGURE 2.2 n-Paraffin melting points: experimental and calculated data.
Source: Advances in Chemistry Series, no. 22, *Physical Properties of Chemical Compounds*, vol. II (Washington, DC: American Chemical Society, 1959) and C. H. Fisher, "Equations Correlate n-Alkane Physical Properties with Chain Length," *Chemical Engineering*, September 20:111–113 (1982).

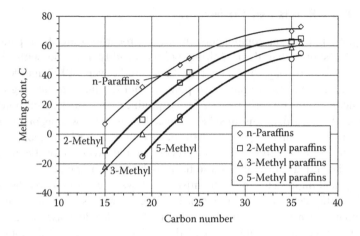

FIGURE 2.3 Melting points of normal and branched paraffins.
Source: J. J. Wise, J. R. Katzer, and N.-Y. Chen, "Catalytic Dewaxing in Petroleum Processing," paper presented at the American Chemical Society National Meeting, April 13–18, 1986. With permission.

This progression of melting points downward with movement of the methyl group toward the center of the chain can be seen more clearly in Figure 2.4 for C_{10} to C_{20} paraffin isomers.[17] For these monomethyl paraffin isomers, the melting point decrease from one isomer to the next becomes less as the substituent moves

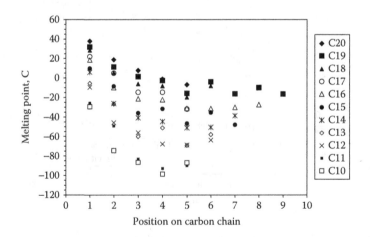

FIGURE 2.4 Methyl substituted paraffins: position of methyl group versus melting point for C_{10} to C_{20} paraffins.
Source: Data from K. J. Burch and E. G. Whitehead, "Melting Point Models of Alkanes," *Journal of Chemical and Engineering Data* 49(4):858–863 (2004).

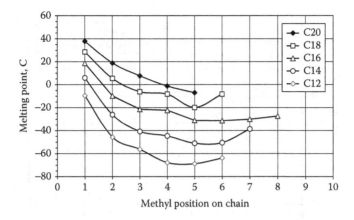

FIGURE 2.5 Dependence of melting points of methyl-branched even number paraffins on the methyl position.
Source: Data from K. J. Burch and E. G. Whitehead, "Melting Point Models of Alkanes," *Journal of Chemical and Engineering Data* 49(4):858–863 (2004). With permission.

from the 2-position to about the 4- or 5-position, where essentially no further decreases occur.

Interestingly, these authors point out that when the data are broken out by even and odd carbon numbers (Figures 2.5 and 2.6), chains with even carbon numbers show a smooth decrease as the methyl group moves inward, while those with odd carbon numbers exhibit an apparent oscillation.

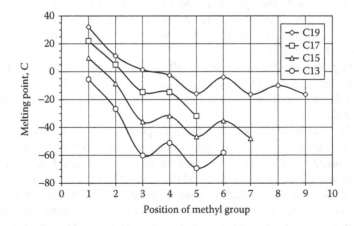

FIGURE 2.6 Dependence of melting points of methyl-branched odd number paraffins on the methyl position.
Source: Data from K. J. Burch and E. G. Whitehead, "Melting Point Models of Alkanes," *Journal of Chemical and Engineering Data* 49(4):858–863 (2004). With permission.

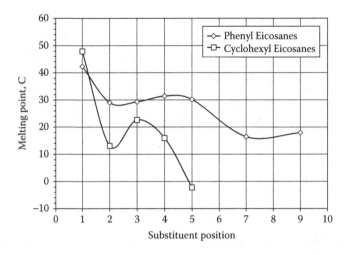

FIGURE 2.7 Melting points of phenyl- and cyclohexyl-substituted eicosanes: effect of substitution position and ring saturation.
Source: "Properties of Hydrocarbons of High Molecular Weight," Research Project 42, 1940–1966, American Petroleum Institute, New York. With permission.

Phenyl- and cyclohexyl-substituted eicosanes[8] can be considered as models for some of the nonparaffinic components of wax, and also for the components of the oil itself, where the low pour point isomers should be found. The phenyl and cyclohexyl substituents confer some of the same effects on melting points as do methyl groups. And carrying this one step further along the process chain, it is interesting to note that the melting points of these compounds are reduced overall by hydrotreating the phenyl ring to a cyclohexyl one (Figure 2.7).

Adding data for more complex paraffins from Project 42 of the American Petroleum Institute (API) to Figure 2.2—either multiple branched or with larger branches than methyl groups—indicates that melting points decrease even further (Figure 2.8).

Finally, one would expect that the pour point would be dependent on the amount of wax present and the carbon number distribution of the wax molecules—n-paraffins of higher carbon number, for example, would be expected to have a greater effect on pour point, and the more of them, the greater the effect as well. Also, n-paraffins would be expected to have a greater effect than isoparaffins. Krishna et al.[18] studied the effect of n-paraffins on pour point, not on waxy lubes, but on two gas oils (250°C to 375°C and 375°C to 500°C) from Bombay High crude oil. However, we would expect their results to describe in a general way the behavior of waxy base stocks. The broad cuts they distilled were each further fractionated into five 25°C narrow cuts and pour, cloud, and cold filter plugging points measured for each. Urea adduction (Chapter 9) was employed

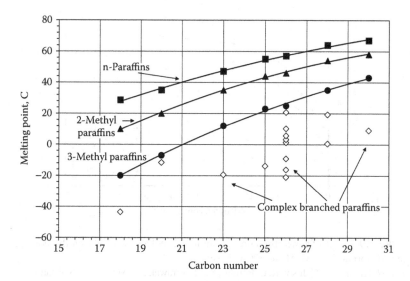

FIGURE 2.8 Effect of complex branching on the melting points of paraffin isomers. *Source:* J. J. Wise, J. R. Katzer, and N.-Y. Chen, "Catalytic Dewaxing in Petroleum Processing," paper presented at the American Chemical Society National Meeting, April 13–18, 1986. "Properties of Hydrocarbons of High Molecular Weight" Research Project 42, 1940–1966, American Petroleum Institute, New York. With permission.

to selectively obtain the n-paraffins from each fraction (these n-paraffin fractions actually contained 5% to 11% isoparaffins) and the n-paraffin distribution in each was measured by gas chromatography. They then added the narrow cut n-paraffin fractions to the broad cut gas oils whose n-paraffins had been removed ("deparaffined"), thereby obtaining curves which related pour point to the amount and molecular weights of the n-paraffins added. Figure 2.9 shows the results for the heavy gas oil case for five of the n-paraffin narrow cuts. Results with lighter gas oil were similar. In Figure 2.9 it can be seen that pour point depends on the amount of n-paraffins and their molecular weights, and the authors were able to quantify these relationships.

At low n-paraffin concentrations, the pour point increase for all the n-paraffin additions is about 3°C per percent n-paraffin addition, and beyond 10% paraffin content, this increase is reduced to about 0.25°C per percent. These curves can be linearized by recasting the equation in the form

$$\text{Pour point, °C} = A*\log(100/PC) + B,$$

where PC is the n-paraffin concentration and the intercepts on the y-axis should correspond to the melting points of the average n-paraffin for each cut and the constant B is linearly related to the chain length, n, of the average n-paraffin for each cut:

$$B = C*n + D,$$

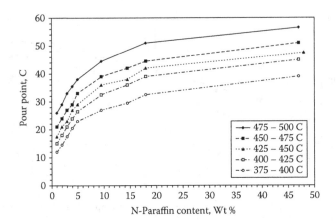

FIGURE 2.9 Effect of n-paraffin concentration and molecular weight on the pour point of the "denormalized" 375°C to 500°C base.
Source: R. Krishna, G. C. Joshi, R. C. Purohit, K. M. Agrawal, P. S. Verma, and S. Bhattacharjee, "Correlation of Pour Point of Gas Oil and Vacuum Gas Oil Fractions with Compositional Parameters," *Energy & Fuels* 3:15–20 (1989). With permission.

where C and D are constants. Therefore,

$$\text{Pour point, } °C = A*\log(100/PC) + C*n + D.$$

The results are plotted in Figure 2.10 and the authors were successful in calculating the values of these constants applicable to these samples.

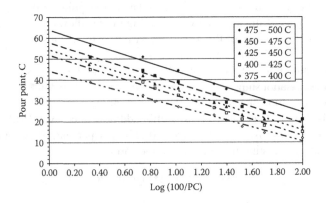

FIGURE 2.10 Linear dependence of the pour point versus the log term for the inverse of the n-paraffin concentration.
Source: R. Krishna, G. C. Joshi, R. C. Purohit, K. M. Agrawal, P. S. Verma, and S. Bhattacharjee, "Correlation of Pour Point of Gas Oil and Vacuum Gas Oil Fractions with Compositional Parameters," *Energy & Fuels* 3:15–20 (1989). With permission.

2.3 BOILING POINTS AND STRUCTURE

Hydrocarbon boiling points increase with increasing carbon number, unsaturation in the form of polycyclic aromatic rings, and polarity, the last point really referring to pyrrole-type nitrogen compounds, naturally occurring carboxylic acids, phenols, and oxidation products. Figure 2.11[19] illustrates these trends. Thus 600°F— at the bottom end of the light base stock—corresponds to the boiling point of the n-C_{18} n-paraffin, perhydropyrene (C_{16}), phenanthrene (C_{14}), and quinolones (C_9). This behavior continues through the higher boiling point ranges. Major reductions in boiling points (and viscosities) occur when polyaromatics are saturated during moderate to severe hydrotreatment. Table 2.7 illustrates the magnitude of these changes for some specific pairs of hydrocarbons.[8]

From a more microscopic viewpoint, Figures 2.12 and 2.13[20] show that branching in paraffins reduces boiling points by small but consistent amounts. For isoparaffins involving a single methyl branch, the effect is largest for a methyl group in the 2-position, and the impact decreases significantly as the methyl group moves

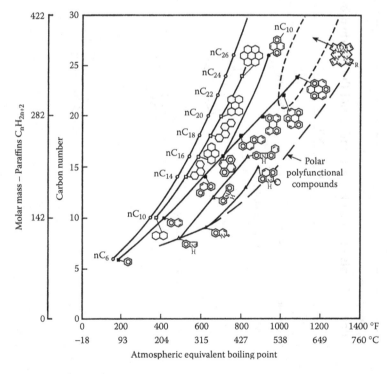

FIGURE 2.11 Illustration of the effect on boiling point of the composition and molecular weight of petroleum model compounds.

Source: K. H. Altgelt and M. M. Boduszynski, *Composition and Analysis of Heavy Petroleum Fractions* (New York: Marcel Dekker, 1993). With permission.

TABLE 2.7
Boiling Points of Some Polyaromatics and their Perhydro- Derivatives

Carbon Number	Name	Polyaromatic		Perhydro	
		Structure	Boiling Point °F	Structure	Boiling Point (°F)
25	1-n- Pentadecyl naphthalene	C$_{15}$	799	C$_{15}$	777
26	2-n- Dodecyl phenanthrene	C$_{12}$	870	C$_{12}$	799
26	9-n-Dodecyl anthracene	C$_{12}$	851	C$_{12}$	792

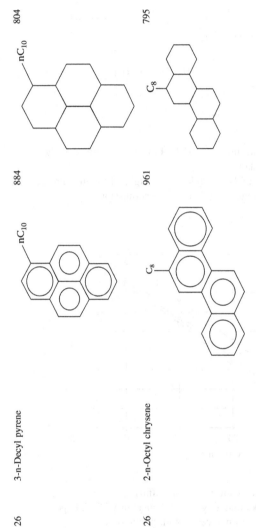

26   3-n-Decyl pyrene

26   2-n-Octyl chrysene

Source: "Properties of Hydrocarbons of High Molecular Weight" Project 42, 1940–1966, American Petroleum Institute, New York. With permission.

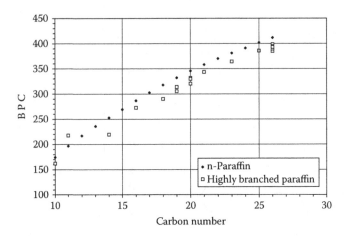

FIGURE 2.12 Boiling points of n-paraffins and highly branched paraffins of the same carbon number using API Project 42 data.
Source: "Properties of Hydrocarbons of High Molecular Weight," Research Project 42, 1940–1966, American Petroleum Institute, New York. With permission.

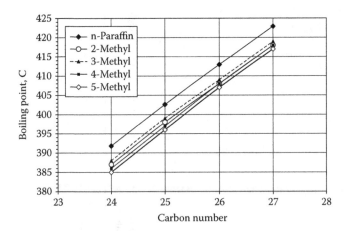

FIGURE 2.13 Boiling points of C_{24} to C_{27} n- and iso-paraffins.
Source: A. Rossi, "Wax and Low Temperature Engine Oil Pumpability," SAE Paper 852113 (Warrendale, PA: Society of Automotive Engineers). With permission.

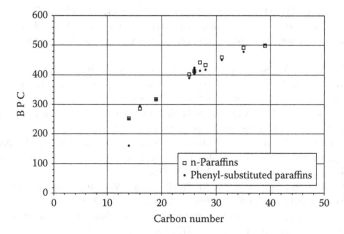

FIGURE 2.14 Boiling points of n-paraffins and phenyl-substituted paraffins versus carbon number.
Source: "Properties of Hydrocarbons of High Molecular Weight," Research Project 42, 1940–1966, American Petroleum Institute, New York. With permission.

away from the end of the molecule—there is a trend toward reduced boiling point as a methyl substituent moves away from the end of the chain (Figure 2.13). Single-ring aromatics[8] do not change boiling points significantly compared to n-paraffins with the same number of carbon atoms (Figure 2.14). Correlations have been developed to calculate n-paraffin boiling points[15,21] over the range C_{21} to C_{100} and also to successfully model those of some 74 alkanes from C_1 through C_9.[22]

REFERENCES

1. D. Godfrey and R. L. Peeler, "Explanation of SI Viscosity Units," *Lubrication Engineering* 38:613–614 (1982).
2. S. W. Rein, "Viscosity-I," *Lubrication* (New York: Texaco Inc., 1978), vol. 64, no. 1, pp. 1–12.
3. The abbreviation N is for neutral as understood in acid/base terms and is applied to paraffinic lube distillates. Solvent extracted base stocks are sometimes given the prefix SNO (solvent neutral oil) before the SUS viscosity (e.g., SNO-100). Hydrocracked base stocks are usually designated as neutrals (e.g., 40N. 650N, etc.), but the neutral is not really necessary any longer and many base stock manufacturers have dropped it. Naphthenic base stocks may have the prefix N.
4. ASTM D2161, "Standard Practice for Conversion of Kinematic Viscosity to Say-bolt Universal Viscosity or to Saybolt Furol Viscosity," *ASTM Annual Book of Standards* (West Conshohocken, PA: American Society for Testing and Materials).
5. For example, see the Cannon Instruments Web page, www.cannoninstrument. com/Classroom.htm.

6. ASTM D445, "Standard Test Method for Kinematic Viscosity of Transparent and Opaque Liquids—The Calculation of Dynamic Viscosity," *ASTM Annual Book of Standards*, vol. 05.0 (West Conshohocken, PA: American Society for Testing and Materials).

7. ASTM D2270, "Standard Practice for Calculating Viscosity Index from Kinematic Viscosity at 40 and 100°C," *ASTM Annual Book of Standards*, vol. 05.01 (West Conshohocken, PA: American Society for Testing and Materials).

8. "Properties of Hydrocarbons of High Molecular Weight," Research Project 42, 1940–1966, American Petroleum Institute, New York.

9. J. Denis, "The Relationship Between Structure and Rheological Properties of Hydrocarbons and Oxygenated Compounds Used as Base Stocks," *Journal of Synthetic Lubricants* 1(1–3):201–238 (1984).

10. J. W. Nederbragt and J. W. M. Boelhouwer, "Viscosity Data and Relations of Normal and Iso-Paraffins," *Physica* X111(6–7):305–318 (1947).

11. R. T. Sanderson, "Viscosity-Temperature Characteristics of Hydrocarbons," *Industrial & Engineering Chemistry* 41(2):368–374 (1949).

12. ASTM D97, "Standard Test Method for Pour Point of Petroleum Products," *ASTM Annual Book of Standards*, vol. 05.01 (West Conshohocken, PA: American Society for Testing and Materials).

13. ASTM D2500, "Standard Test Method for Cloud Point of Petroleum Products," *ASTM Annual Book of Standards*, vol. 05.01 (West Conshohocken, PA: American Society for Testing and Materials).

14. Advances in Chemistry Series, no. 22, *Physical Properties of Chemical Compounds*, vol. II (Washington, DC: American Chemical Society, 1959).

15. C. H. Fisher, "Equations Correlate n-Alkane Physical Properties with Chain Length," *Chemical Engineering*, September 20:111–113 (1982).

16. J. J. Wise, J. R. Katzer, and N.-Y. Chen, "Catalytic Dewaxing in Petroleum Processing," paper presented at the American Chemical Society National Meeting, April 13–18, 1986.

17. K. J. Burch and E. G. Whitehead, "Melting Point Models of Alkanes," *Journal of Chemical and Engineering Data* 49(4):858–863 (2004).

18. R. Krishna, G. C. Joshi, R. C. Purohit, K. M. Agrawal, P. S. Verma, and S. Bhattacharjee, "Correlation of Pour Point of Gas Oil and Vacuum Gas Oil Fractions with Compositional Parameters," *Energy & Fuels* 3:15–20 (1989).

19. K. H. Altgelt and M. M. Boduszynski, *Composition and Analysis of Heavy Petroleum Fractions* (New York: Marcel Dekker, 1993).

20. A. Rossi, "Wax and Low Temperature Engine Oil Pumpability," SAE Paper 852113 (Warrendale, PA: Society of Automotive Engineers).

21. A. K. Kudshaker and B. J. Zwolinski, "Vapor Pressures and Boiling Points of Normal Alkanes, C_{21} to C_{100}," *Journal of Chemical and Engineering Data* 11(2): 263–255 (1966).

22. D. E. Needham, I.-C. Wei, and P. G. Seybold, "Molecular Modeling of the Physical Properties of the Alkanes," *Journal of the American Chemical Society* 110:4186–4194 (1988).

3 Development of the Viscosity Index Concept and Relationship to Hydrocarbon Composition

3.1 VISCOSITY INDEX

3.1.1 BACKGROUND

The viscosity behavior of base stocks at temperatures between the cloud point, when crystals first begin to appear, and the much higher temperature at which volatility or decomposition commences is of central interest to lubricant formulators and users. Like that of most liquids, the kinematic viscosity of lube-range hydrocarbons decreases with increasing temperature. The *degree* of this decrease is important because in the majority of applications, particularly automotive, minimal change is most desirable. For example, the viscosities of naphthenic and paraffinic base stocks have very different behaviors with temperature change. These stocks, with the same viscosity of 4.0 cSt at 100°C, can have viscosities of 25.3 (naphthenic) and 16.8 (paraffinic) cSt at 40°C. Paraffinic base stocks have less viscosity variation than naphthenic stocks and therefore are preferred for many applications, particularly in the automotive area. This temperature susceptibility of viscosity depends on the base stock composition and is determined by the feedstock, the manufacturing process, and process operating conditions, and is a critical process target in manufacturing most lube base stocks.

3.1.2 DEVELOPMENT OF THE CONCEPT: DEAN AND DAVIS WORK

Many methods have been developed to represent changes in base stock viscosity with temperature. A useful method has to be able to express these changes for base stocks of all viscosities of interest, the meaning of the scale has to be readily apparent, the measurement must be relatively simple and cheap, and it must be widely accepted by base stock manufacturers and formulators and customers. The method that has emerged and lasted (too long for some!) is the viscosity index (VI) procedure. By this method the viscosity-temperature relationship of any base stock can be expressed by a single number, originally on a 0 to 100

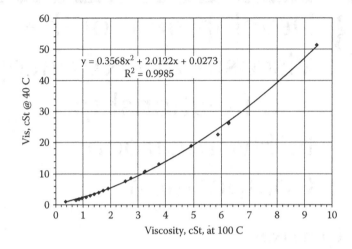

FIGURE 3.1 Viscosities (cSt) at 40°C and 100°C of n-paraffins of chain length C_{11} to C_{44}. *Sources:* J. Denis, "The Relationship Between Structure and Rheological Properties of Hydrocarbons and Oxygenated Compounds Used as Base Stocks," *Journal of Synthetic Lubricants* 1(1–3):201–238 (1984); J. Denis and G. Parc, "Rheological Limits of Mineral and Synthetic Hydrocarbon Base Stocks," *Journal of the Institute of Petroleum* 56(556):70–83 (1973); and "Properties of Hydrocarbons of High Molecular Weight," Research Project 42, 1940–1966, American Petroleum Institute, New York. With permission.

scale, but more usually now in the range of 0 to 200, but negative numbers are also known; the latter are usually seen for extract by-products from solvent refining streams.

The typical characteristics of the variation of viscosity with temperature can be seen in a plot of hydrocarbon viscosities at two temperatures. Figure 3.1 shows the relationship for n-paraffins,[1–3] and when both axes are converted to logarithms a linear relationship results. Figure 3.2 (the temperatures employed for this type of diagram are now usually 40°C and 100°C, whereas in the past 100°F and 210°F were used; correspondingly, units for viscosity were originally in Saybolt universal seconds (SUS); they are now in centistokes. Distillation fractions through base stocks show similar behavior.

In 1929 E. W. Dean and G. H. B. Davis,[4] of Standard Oil, used this representation for viscosities to develop the VI concept for lubricant feedstocks, intermediates, and base stocks. In modified form, this method for describing the viscosity-temperature relationship for base stocks and lubricants has become one of the bedrocks of the industry. The VI of a base stock now immediately brings to mind, rightly or wrongly, other features of interest to lubricants professionals, for example, oxidation stability and chemical composition (e.g., content of iso-paraffins and cycloparaffins), and more recently, volatility. It is, however, best to regard this method as a means to express the rheological properties (the science of flow) of a base stock and use other more specific tests to measure other properties.

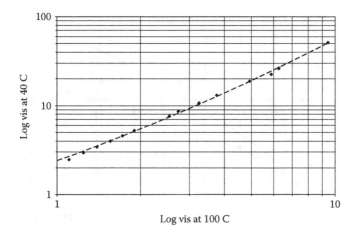

FIGURE 3.2 Log viscosities (cSt) of n-paraffins of chain length C_{11} to C_{44}.

As developed by Dean and Davis, this empirical method expresses viscosity variation with temperature numerically, initially on a simple scale of 0 to 100, based on two sets of reference distillate fractions. These oils were from two crudes whose distillates had not been refined in any manner (i.e., they had not been dewaxed or solvent refined). The viscosity changes with temperature of the "0" reference oil fractions were large, while those of the fractions from the "100" reference were small. These assignments of 0 and 100 were of course arbitrary and reflected experience at that time. It was assumed in developing the method that all distillation fractions from each of these reference crudes had the same VI (and that approximately agreed with the current knowledge) and that this was true for all other crudes and their lubricant fractions. A further assumption was that the VIs of all oils would fall between 0 and 100.

To develop this method, Dean and Davis[5] measured the viscosities (in SUS) at two temperatures (100°F and 210°F) for sets of distillation fractions from the reference crudes. The samples were, as mentioned, from two extremes of temperature dependence and were:

(a) From Pennsylvanian crude whose lubricating oils were known to change relatively little in viscosity with temperature. The fractions obtained from this crude were designated as series H(igh) and the VI assigned to all the fractions was 100. All viscosities thus fell on the 100 VI curve by definition.

(b) From a U.S. Gulf Coast crude whose lubricating oil viscosity undergoes significant changes with temperature. The fractions were identified as series L(ow) and the VI assigned was 0, and their viscosities, in turn, defined the 0 VI curve.

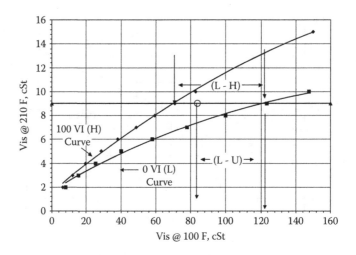

FIGURE 3.3 Viscosity index: concept development.

The VI is defined by the position of a sample's viscosity at 100°F between the 0 and 100 VI viscosities at 100°F, all measured at identical viscosities at 210°F. Figure 3.3 illustrates this, where the sample VI is expressed as its percentage position between the 0 and 100 VI curves. To determine the VI of a sample, the viscosities at 100°F and 210°F are measured and the viscosities at 100°F are determined from a figure similar to Figure 3.3 for the 0 and 100 VI reference oils. The VI of the sample is thus its distance along the viscosity at 210°F line, expressed as a percentage of the separation in viscosities (Equation 3.1):

$$VI = 100*(L - U)/(L - H), \tag{3.1}$$

where U is the viscosity at 100°F of the oil being studied, and L and H are the viscosities at 100°F for series L and H having the same viscosity at 210°F as the oil in question. In practice, a set of tables or a computer program is now employed for these calculations rather than a set of curves.

The VI concept is simple, but implementation has required significant additional work to satisfy many criticisms that have been raised. Discussions continue about its application, but a satisfactory replacement has not yet been developed.

In the illustration in Figure 3.3, where the sample whose VI is to be determined has a viscosity at 40°C of 84.0 cSt and at 100°C of 9.0 cSt (denoted by the open circle), the horizontal line through the 9.0 cSt point intersects the 100 and 0 VI curves at points H and L, respectively, on the viscosity at 40°C axis. The VI definition above therefore represents the percentage that the $(L - U)$ difference constitutes of the $(L - H)$ difference. When this concept was developed, VIs were not expected to exceed 100 or fall below 0. In practice, both have occurred and the region above 100 VI has become commercially very important, and as we will see, has forced changes in the way VI is calculated. Negative VIs do not have much practical significance.

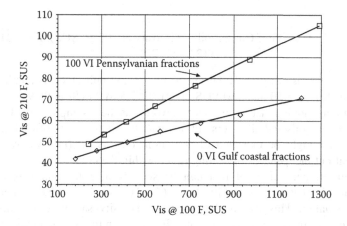

FIGURE 3.4 Original data from Dean and Davis on Pennsylvanian and Gulf Coast lubricant fractions.
Source: Data from E. W. Dean and G. H. B. Davis, "Viscosity Variations of Oils with Temperature," *Chemical and Metallurgical Engineering* 36(10):618–619 (1929).

Figure 3.4 shows the original seven data points of Dean and Davis, with viscosities measured in SUS at 100°F and 210°F. The VI was then defined by the following formula:

$$VI = 100*(\text{Viscosity}_{\text{GulfCoast}} \text{ at } 100°F - \text{Viscosity}_{\text{Sample}} \text{ at } 100°F)/(\text{Viscosity}_{\text{GulfCoast}}$$
$$\text{at } 100°F - \text{Viscosity}_{\text{Pennsylvania}} \text{ at } 100°F), \tag{3.2}$$

where the viscosities are identical at 210°F for the sample and the two reference oils. In this basic equation, the sample viscosities are actual measurements and are as accurate as was possible at the time. The other viscosities for the 0 and 100 VI "samples" are calculated values and the value of the VI method is based on how good these are. This has been an area of on-going controversy.

The Dean and Davis method required the measurement of viscosities at two temperatures followed by a simple calculation procedure. The simplicity made this method widely appealing, and it remains so today except that the calculation has become more complicated. As developed, when the viscosities at 100°F and 210°F were determined in SUS units, the viscosities at 100°F for the two reference oils with the same viscosities at 210°F as the sample were calculated using equations or tables provided by Dean and Davis and then the VI was calculated from Equation 3.2. Since inception, viscosity measurements have changed from SUS units to centistokes and temperatures from 100°F and 210°F to 40°C and 100°C.

The initial equations for the viscosities at 100°F were

Series H (100 VI, Pennsylvania oils), viscosity at 100°F (SUS), $Y = 0.0408x^2$
$$+ 12.568x - 475.4 \tag{3.3}$$

and

Series L (0 VI, Gulf Coastal oils), viscosity at 100°F (SUS), $Y = 0.216x^2$
$$+12.07x - 721.2, \hspace{2cm} (3.4)$$

where x is the viscosity at 210°F (SUS) and were applicable to samples with viscosities between 50 and 160 SUS at 210°F (7.5 to 34 cSt at 98.9°F). Again, it is worth mentioning that implicit in this work was the assumption that for these two lubricating oils and all others, all distillate fractions cut out of each had the same VI (as might be expected of a homologous series) (i.e., VI was crude dependent but independent of boiling range within the crude).

A further general point that can be seen from both figures is that as viscosities decrease, the viscosity deltas $(L - H)$ and $(L - U)$ being measured become smaller. This means that for low viscosity samples, measurement precision must increase, and even a one-digit error in the second decimal place of the viscosity can be important. Provided viscosity measurements adhere to the requirements of ASTM D445, VI results will be accurate. Waxy samples usually have their lower temperature viscosity measured at 65°C and are then recalculated to 40°C.

The invention of the VI concept led to its becoming an immediate success after its publication, since it met the criteria of simplicity, ease of determination (measurement of viscosity at two temperatures), and expression of the result as a simple number between 0 and 100 for all oils with a viscosity at 100°C of greater than 2.0 cSt. Oils with viscosities at 100°C of less than 2.0 cSt ran into measurement problems at 210°F or 100°C because of the loss by distillation of lower boiling components at this temperature during the test.

Industrially the VI method was first used to explain a variety of issues, including ease of cold weather starting, pumpability, oil consumption, and the general notion of "quality."[6] For base stocks for use in automotive applications, VI (usually greater than 95) remains one of the key specifications to be met during manufacturing. In manufacturing, base stock yield is very much related to feed and product VIs. Solvent refining plants producing paraffinic base stocks use "lube" crudes, which are those whose distillates have sufficiently high dewaxed VIs (approximately 50 or greater), while hydrocracking plants are fortunate in being able to employ feedstocks with much lower VIs.

The VI method was quickly espoused by the industry and eventually became ASTM D567.[7] It was applied largely to solvent refined base stocks whose VIs were largely in the 0 to 100 VI range. D567's existence was not trouble-free and a number of problems discussed below had to be addressed. Eventually it was succeeded in 1964 by ASTM D2270[8] after serious problems were identified for samples with VIs greater than 100.

3.1.3 VISCOSITY INDEX ISSUES: REFERENCE SAMPLES

An early difficulty faced by the use of this method was that the initial equations (Equations 3.5 and 3.6) for the reference samples covered a limited viscosity range (minima were 180 SUS at 100°F and 42 SUS at 210°F, corresponding to 39 cSt at 100°F and 5.0 cSt at 210°F). When the method was used outside of that range it gave unreliable VIs (a not uncommon event when methods are employed beyond their original scope!):

$$\text{Series H: } Y = 0.0408x^2 + 12.56x - 475.4 \tag{3.5}$$

and

$$\text{Series L: } Y = 0.216x^2 + 12.07x - 721.2, \tag{3.6}$$

where Y is the viscosity (in SUS) at 100°F and x is the viscosity (in SUS) at 210°F. For example, for a base stock with a viscosity at 100°C of 2.5 cSt (2.545 cSt at 210°F, 34.443 SUS at 210°F), a viscosity at 40°C of 9.516 cSt (10.215 cSt at 100°F, 59.585 SUS at 100°F), and a VI of 81 calculated by the current method (D2270) or by its predecessor (D567), the VI by the original equations was 198.

To resolve this, in 1940 Dean, Bauer, and Berglund[5] published low viscosity extensions for use between 2.0 and 4.2 cSt (at 98.9°C) for both the 0 and 100 VI curves together with a new set of constants for the intermediate range between 4.2 and 7.29 cSt. This meant that the VI tables constructed were broken into the following ranges (at 210°F):

- 2.00 to 4.20 cSt using the equations below.
- 4.20 to 7.29 cSt using recalculated values which gave revised values of the H, L, and D values.
- Above 7.29 cSt, no changes were made.

The equations for the low viscosity extension were, using kinematic viscosities at 100°F and 210°F,

$$\text{For 100 VI: } Y = 1.4825x + 0.91375x^2 \tag{3.7}$$

$$\text{For 0 VI: } Y = 1.655x + 1.2625x^2, \tag{3.8}$$

where Y is the kinematic viscosity at 100°F (in cSt) and x is the kinematic viscosity at 210°F (in cSt) (these authors were now measuring in kinematic viscosities because of the newfound awareness of the need for great accuracy at low viscosities, and the SUS method could not deliver this).

The three ranges meant discontinuities between the ranges, with a particular problem being the 0 VI reference samples in the regions 2.0 to 4.2 cSt and 4.2 to 7.9 cSt.. This can be seen in Figure 3.5 and with more clarity in Figure 3.6. This meant the introduction of nonlinearities in the scales in this viscosity region, particularly for the 0 VI reference oil. The effect of this can be seen in Figure 3.6, where at 210°F viscosities of less than 4 cSt, calculated 100°F viscosities are different than extrapolated from the data at greater than 4 cSt.

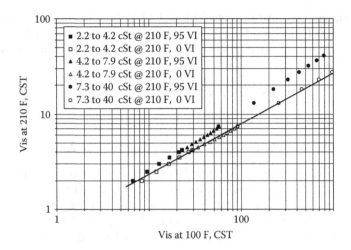

FIGURE 3.5 Data from Dean et al.[5]: viscosity at 210°F and 100°F for standard samples. *Source:* E. W. Dean, A. D. Bauer, and J. H. Berglund, "Viscosity Index of Lubricating Oils," *Industrial and Engineering Chemistry* 32(1):102–107 (1940).

3.1.4 VISCOSITY INDEX ISSUES: HIGH VI RANGE

A further issue that emerged with ASTM D567 when it was applied to samples containing polymeric VI improvers was that at VIs greater than 100, the log viscosity lines became curved, such that at high VIs there were two viscosities

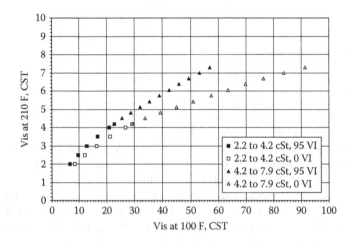

FIGURE 3.6 Results from Dean et al.[5]: viscosity at 210°F and 100°F of VI reference samples.
Source: Data from E. W. Dean, A. D. Bauer, and J. H. Berglund, "Viscosity Index of Lubricating Oils," *Industrial and Engineering Chemistry* 32(1):102–107 (1940).

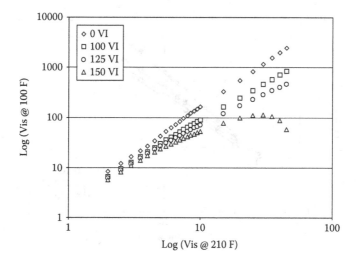

FIGURE 3.7 Curves of ASTM D567 viscosities (cSt at 100°F and 210°F) for stocks with VIs of 0, 100, 125, and 150.

at 100°F corresponding to each viscosity at 210°F (e.g., at 150 VI in Figure 3.7); that is, the log viscosity lines had clearly become definite curves at high VIs.

Following recommendations by the Third and Fourth World Petroleum Congresses[9,10] that these problems be addressed, an ASTM panel was convened to devise a solution. Discussions focused on two main alternatives, a completely new method, fundamentally based, that would have major appeal to scientific people, or, the alternative favored by commercial representatives, a modification of ASTM D567 that would correct the situation for VIs greater than 100 and retain as much as possible of the existing system. This group[11] recommended the latter approach and adopted[12] the use of the "viscosity-temperature function"[13] for VIs greater than 100, leaving the less than 100 VI region unchanged. By this method, for VI greater than 100:

$$VI = \{[(\text{antilog } N) - 1/0.00715\} + 100, \tag{3.9}$$

where

$$N = (\log H - \log U)/\log Y \text{ or } Y^N = H/U \tag{3.10}$$

and H is calculated by the same equation as for VI less than 100 and U is the kinematic viscosity at 40°C of the sample whose VI is to be determined. This methodology applies to samples with viscosities at 100°C of between 2.0 and 70.0 cSt. At greater than 70 cSt, L and H are calculated by the following equations:

$$L = 0.8353Y^2 + 14.67Y - 216 \tag{3.11}$$

$$H = 0.1684Y^2 + 11.85Y - 97. \tag{3.12}$$

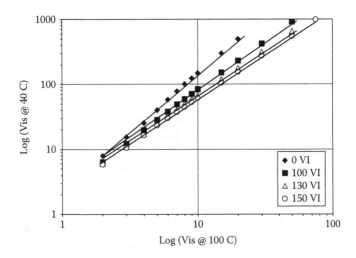

FIGURE 3.8 ASTM D2270: log(viscosity at 100°C) versus log(viscosity at 40°C) for base stocks with different VIs.

This revised form became ASTM D2270, issued first in 1964, and Figure 3.8 shows that the nonlinearity issue at high VIs was taken care of.

The deviations between VIs determined by D567 and D2270 (Table 3.1) show that at VIs greater than 100, D567 generally results in lower VIs and some of these were very large.

TABLE 3.1
Delta VIs (VI$_{ASTM\ D2270}$ VI$_{ASTM\ D567}$ at Greater Than 100 VI for 210°F Kinematic Viscosities Between 2.5 and 30 cSt)

Viscosity, cSt at 210°F	ASTM D567 VIs						
	100	102	105	110	120	130	150
2.5	0	0	−3	0	+1	+3	+11
4.0	0	−1	−2	−3	−5	−7	−8
6.0	0	0	0	0	+1	+4	+16
10.	0	0	+1	+2	+7	+14	+44
15.0	0	0	+1	+3	+9	+20	+64
20.0	0	0	+2	+4	+12	+25	+89
30.0	0	+1	+2	+6	+17	+37	+161

Source: W. A. Wright, "A Proposed Modification of the ASTM Viscosity Index," Proceedings of the American Petroleum Institute annual meeting, Section III—Refining, pp. 535–541 (1964). With permission.

The origin of the relationship used at VIs greater than 100 was developed by Wright[13] as follows:

The mean enthalpy of activation for flow over a temperature range T_1 to T_2 (100°F to 210°F in this case) is

$$\Delta H\pm^{12} = C*\log(v_1/v_2),$$

where v_1 and v_2 are the kinematic viscosities at 100°F and 210°F and C is a constant.

Also, the molecular weight (MW) of an oil is a logarithmic function of the kinematic viscosities, thus

$$A*v_2^m = v_1/v_2,$$

where m is a constant characteristic of the oil and A is a constant (approximately 2.25):

$$\log A + m*\log(v_2) = \Delta H\pm^{12}/C = \log(v_1/Av_2).$$

Wright[11] suggested that m be termed the viscosity-temperature function (VTF). This last equation becomes

$$2.25*v_{210°F}^{(m+1)} = v_{100°F}.$$

Therefore, for two oils we have

$$v_{210°F}^{(m+1)}/v_{210°F}^{(m+1)} = v_{100°F}/v_{100°F},$$

and where we have common viscosities at 210°F make

$$(m + 1) - (m + 1) = N$$

$$v_{210°F}^N = v_{100°F}/v_{100°F}$$

and the relationship between N and the VI scale above 100 was found experimentally to be

$$VI = \{(\text{antilogarithm } N) - 1\}/0.0075 + 100.$$

This is the current form for calculation of the VI, and together with the original D567 part for base stocks with VIs of less than 100, is incorporated in ASTM D2270 and is currently in use.[14,15] Some publications have used the term viscosity index extension (VI$_E$) to refer to VIs of greater than 100 to indicate that the method is different.

It is worth noting that when base stocks of different VIs are combined, the VIs do not blend in proportion to their amounts and VIs. In actuality, a phenomenon termed "VI hop" occurs in which the VI of the blend is higher than the arithmetic result.

3.1.5 VISCOSITY INDEX ISSUES: VISCOSITY EFFECT

A major issue remains that has not been corrected to date. It is that D2270 is widely seen as undervaluing the VIs of base stocks with viscosities of less than

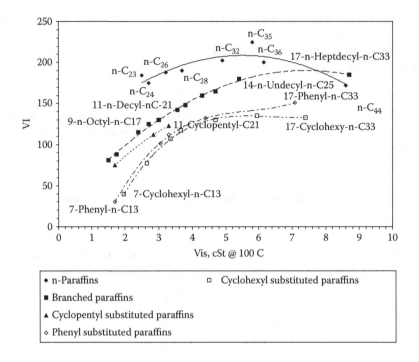

FIGURE 3.9 VI variation of pure compounds as a function of viscosity and structure. *Source:* J. Denis, "The Relationship Between Structure and Rheological Properties of Hydrocarbons and Oxygenated Compounds Used as Base Stocks," *Journal of Synthetic Lubricants* 1(1–3):201–238 (1984). With permission.

5.5 cSt at 210°F. This can be clearly seen in the pure compound VI data assembled by Denis[1] for several homologous series (Figure 3.9)—n-paraffins, branched paraffins, and cyclopentyl-, cyclohexyl-, and phenyl-substituted paraffins in which the chain lengthens but the substituent positions remain the same. The VIs for each of these decrease below about 5.5 cSt at 210°F.

Gillespie and Smith[16] did show that there is a discontinuity in VI in the region between 4 and 7.3 cSt at 210°F for some hydrocracked stocks. The basis in the method for this is illustrated in Figure 3.10 using data generated from the current ASTM D2270 method for a range of viscosities, all of VI =130, where at a viscosity at 100°C of 5 cSt the slope of the "line" changes from 0.70 above this point to 0.647 below it.

Gillespie and Smith were intrigued when they examined the behavior of the VI of dewaxed 650°F-plus hydrocrackate ("hydrocrackate" is the product from a lube hydrocracker) as the reactor severity producing it increased (Table 3.2). As expected, VI increased and viscosity decreased as cracking decreased molecular weight, but the final temperature increase caused VI to decrease rather than increase further. Many people might have ignored this discrepancy, but these individuals pursued this point further.

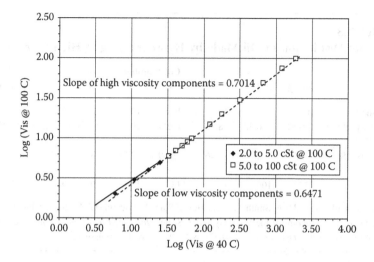

FIGURE 3.10 Illustration of the differences between the high and low viscosity regions in ASTM D2270 using a base stock a VI of 130.

Associating this behavior with the low viscosity range now measured for the bottoms sample, the authors fractionated this final product produced at 811°F (Table 3.3) into seven cuts. Inspections of these cuts showed that for those with viscosities below 5.5 cSt, low VI results predominated. In the case of the final bottoms fraction (the residue), whose viscosity was greater than 5.5 cSt at 210°F, the expected high VI materialized.

This was one of the first observations of what is now called "VI droop." VI droop refers to the decrease in dewaxed VIs with decreasing viscosities for hydrocracked base stocks.[17] It begins at around 5 to 6 cSt at 100°C (or 210°F for

TABLE 3.2
Loss of VI During High-Severity Hydrocracking for Lube Oil Manufacture

Reactor temperature, °F	738	762	782	804	811
VI of dewaxed 650°F+	88	112	125	133	127
Viscosity, cSt at 100°F	154.10	72.63	42.38	23.74	16.51
Viscosity, cSt at 210°F	13.42	9.21	6.75	4.75	3.74

Source: B. Gillespie and F. A. Smith, "Explanation of Some Lubricating Oil Hydrocracking Results in Terms of the Structure of the Viscosity Index Scale," *Industrial and Engineering Chemistry Product Research and Development* 9(4):535–540 (1970). With permission.

TABLE 3.3

Vacuum Distillation of Oil Made by Hydrocracking at High Severity

| | Charge | Cut Number | | | | | | Bottoms |
		1	2	3	4	5	6	
Volume %	100	10	10	10	10	10	10	40
Viscosity, cSt at 100°F	16.51	6.32	8.87	10.21	10.52	13.43	16.31	31.92
Viscosity, cSt at 210°F	3.74	2.00	2.46	2.69	2.75	3.24	3.69	5.86
VI	127	106	113	113	115	119	125	141

Source: B. Gillespie and F. A. Smith, "Explanation of Some Lubricating Oil Hydrocracking Results in Terms of the Structure of the Viscosity Index Scale," *Industrial and Engineering Chemistry Product Research and Development* 9(4):535–540 (1970). With permission.

that matter). Figure 3.11 demonstrates the typical VI-viscosity relationship displayed when dewaxed 600°F-plus lube hydrocrackates are fractionated. This will be discussed more in a later chapter.

One of the basic tenets of the VI system's development was that VI is essentially constant through the boiling range of a lube distillate. Figure 3.11 shows that for hydrocrackates this is not true and that VI begins to drop off at a viscosity of around 5 cSt at 100°C. This drop-off occurs in the viscosity region that includes the important 100N base stock product. The impact with the current VI system is that to make a 95 VI specification 100N, then higher viscosity fractions will have VIs

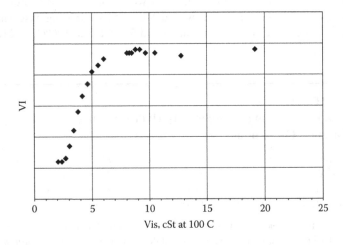

FIGURE 3.11 VI versus viscosity behavior for a dewaxed lube hydrocrackate: VIs of fractions versus viscosity at 100°C.

greater than 95 (i.e., they will be overtreated). The converse will leave 100N "undertreated" (i.e., the 100N will have a VI of less than 95). This obviously has substantial commercial implications, as will be discussed elsewhere. The point here is that this is probably due to a failure of the VI system. Zakarian[17,18] has argued that the VI system should be reassessed, and with the increasing share of lube base stocks being produced by hydrocrackers, perhaps this will finally occur.

Other criticisms[19,20] of the VI are that it does not adequately predict or reflect oxidation stability and performance of engine oils or "quality." These criticisms merely indicate that at one time expectations of the method far exceeded its real scope. The VI is indeed dependent on hydrocarbon composition (on which all behavior depends), but it should best be regarded as a measure of solely rheological properties, whatever other desirable properties these may subsequently coincide with. The term "quality" has to be considered carefully as well. Many performance factors do correlate with VI, but not necessarily so.

3.1.6 Alternative Proposals to the Viscosity Index

A number of alternative proposals to ASTM D2270 have been made but have not been able to meet both commercial and scientific criteria in the way that D2270 has. These proposals have been summarized[21] and include

- the viscosity modulus,[22] where the logarithms of the viscosities are used in the same equation as Dean and Davis;
- the independent viscosity index,[23] where viscosities in the Dean and Davis equation are replaced by loglog(viscosity + 0.8);
- the ASTM slope;[24]
- the viscosity-temperature coefficient;[25]
- the rational viscosity index;[26]
- the viscosity-temperature rating (VTR);[27]
- the viscosity index;[28] and
- the viscosity-temperature number;[29]

but none have been widely used.

3.1.7 Viscosity Calculation: The Walther Equation—ASTM D341

The viscosity of a sample at some temperature can be calculated using the Walther equation if viscosities in centistokes at two other temperatures are available (note: this equation holds true only for centistokes measurements). This linear equation is

$$\log\log(+ 0.7) = A - B*\log T,$$

where is the kinematic viscosity at temperature T (in degrees Rankine or Kelvin) and A and B are constants and log is to base 10. Viscosities at two temperatures are required to calculate that at a third temperature.

This form of the equation has a viscosity scope of 2.0 to 2.0×10^7 and the scope can be expanded down to 0.21 cSt by use of additional coefficients. Details of the method are provided in ASTM D341.[30]

For waxy samples, which may be either solid or contain wax crystals at 40°C, it is conventional to measure their viscosities at 65°C (or higher if required) and 100°C and calculate the viscosity at 40°C using ASTM D341 prior to calculating the VI.

3.2 VISCOSITY INDEX AND COMPOSITION

3.2.1 PARAFFINS AND RELATED MOLECULES

The VI of a base stock depends on its chemical composition; the value for a sample reflects the components present and proportions and boiling range, but there is no published procedure for calculating VI based on composition. The converse, calculating composition from VI, will someday materialize, but will have to solve the issue that there generally are multiple compositions that will give the same VI.

Basically, long molecules have high VIs (e.g., n-paraffins have VIs in the 180 to 200 range) while compact rigid molecules such as multiring fused naphthenes and aromatics have low or even negative VIs. Most lube molecules have intermediate shapes. Ushio et al.'s[31] summary of average VIs of lube oil components is shown in Table 3.4. The current North American trend toward high VI base stocks means lube hydrocarbon composition trends are in the direction of increased content of branched paraffinic types together with naphthenes containing just one to two fused rings, and aromatics, if present, are largely monoaromatics.

TABLE 3.4

VIs of Various Hydrocarbon Types from Multiple Regressions and Compositional Analyses on Hydrocracked Oils

Type of Hydrocarbon	VI
n-Paraffins	175
Isoparaffins	155
Mononaphthenes	142
Dinaphthenes +	70
Aromatics	50

Source: M. Ushio, K. Kamiya, Y. Yoshida, and I. Honjou, "Production of High VI Base Oil by VGO Deep Hydrocracking," presented at the Symposium on Processing, Characterization and Application of Lubricant Base Oils, American Chemical Society annual meeting, Washington, DC, *Preprints of the Division of Petroleum Chemistry* 37(4):1293–1302 (1992). With permission.

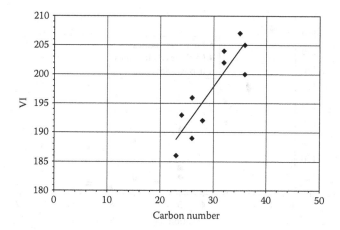

FIGURE 3.12 n-Paraffins: VI versus carbon number.
Source: "Properties of Hydrocarbons of High Molecular Weight," Research Project 42, 1940–1966, American Petroleum Institute, New York. With permission.

Figure 3.12 using data from Project 42 of the American Petroleum Institute (API) and other data show n-paraffins to have VIs in excess of 180. The limited data available suggest that VI increases with chain length and therefore individual components can be substantially higher than the least squares numbers reported by Ushio et al. Because of their high melting point and low solubility, n-paraffins (wax) are largely or completely removed in the dewaxing step during manufacturing and therefore play a minor compositional role in finished base stocks.

Branching of the n-paraffin chain reduces the VI and the more substituents there are, the lower the VI. Table 3.5 shows this effect for C_{26} and C_{30} paraffin isomers. (Bear in mind that branching improves the pour point and is the sought-after objective of hydroisomerization.)

TABLE 3.5
VIs of C_{26} and C_{30} Alkane Isomers

Carbon Number	Name	VI
26	n-Hexacosane	190[1]
26	5-n-Butyl-docosane	136[29]
26	11-n-Butyl-docosane	128[3]
26	5,14-Di-n-Butyl-octadecane	83[3]
30	n-Tricosane	~195
30	9-n-Octyldocosane	144
30	2,6,10,15,19,23-Hexamethyltetracosane	117

Source: R. T. Sanderson, "Viscosity-Temperature Characteristics of Hydrocarbons," *Industrial and Engineering Chemistry* 41:368–374 (1949). With permission.

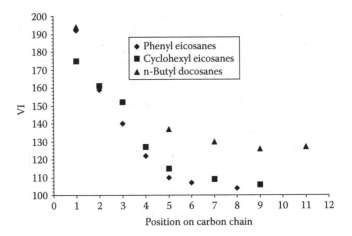

FIGURE 3.13 Effect of the position of phenyl and cyclohexyl substituents on VIs of alkanes. *Source:* "Properties of Hydrocarbons of High Molecular Weight" Research Project 42, 1948–1966. America Petroleum Institute, New York. With permission.

When there is just a single substituent, its position on the chain is important in determining VI (Figure 3.13). As the substituent, in these cases phenyl, cyclohexyl, and n-butyl, moves toward the center of the molecule from the end, the VI decreases. This VI decrease is quite significant as well, since moving from

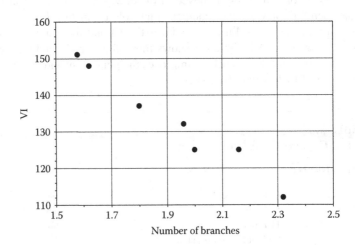

FIGURE 3.14 Hydroisomerization of n-tetradecosane (n-C_{24}): effect of the degree of chain branching on VI.
Source: M. Daage, "Base Oil Production and Processing," ExxonMobil presentation, July 2001, available at http://www.prod.exxonmobil.com/refiningtechnologies/pdf/base_ oil_ refining_ lubes_daage_france070601.pdf. With permission.

TABLE 3.6
Monoaromatic Structures and VIs for C_{20} and C_{26} Phenyl-Substituted Alkanes

Formula	Name	Number of Naphthene Rings	Number of Aromatic Rings	Viscosity, cSt at 40°C	Viscosity, cSt at 100°C	VI
$C_{26}H_{46}$	1-Phenyleicosane	0	1	10.82	3.276	192
$C_{26}H_{46}$	1,4-di-n-Decyl benzene	0	1	11.386	3.282	171
$C_{26}H_{46}$	1,3-di-n-Decyl benzene	0	1	10.734	3.061	153
$C_{26}H_{38}$	1,1-Diphenyl tetradecane	0	2	17.066	3.631	91
$C_{20}H_{32}$	2[ar]-6-dimethyl-3-octyl tetralin	1	1	12.039	2.688	32
$C_{20}H_{32}$	1[ar]-4-dimethyl-5-octyldecalin	1	1	14.803	3.029	28

Source: "Properties of Hydrocarbons of High Molecular Weight," Research Project 42, 1940–1966, American Petroleum Institute, New York. With permission.

the end of the chain to the next carbon leads to a loss of up to 30 VI units and further migration of the substituent can lead to an additional loss of 20 to 30 VI units. This is important when paraffins are catalytically isomerized, and obviously migration of the methyl group toward the center of the molecule is an unfavorable outcome from a VI standpoint.

When multiple branches can occur, as they do when n-paraffins are isomerized, VI decreases sharply as the number of methyl groups or degree of branching increases,[32] as in Figure 3.14. Monoaromatics with long chains attached generally have high VIs (much greater than 100), but VI decreases when these benzene rings are fused with others, either naphthenic or aromatic (Table 3.6).

3.2.2 POLYCYCLIC MOLECULES

Fused polycyclic naphthenes have much more compact structures than the substituted paraffins above and have even lower viscosity indexes although some can be redeemed by having n-decyl straight chains attached as in Table 3.7.

Incorporation of several aromatic rings within the structure of these polycyclic aromatics to give flatter structures reduces VI's even further into the negative regime (Table 3.8). We'll see later in some examples that components with even more negative VI's can be isolated from lube fractions.

Therefore from the study of pure compounds, the VIs of base stock components decrease in the order: isoparaffins with few branches > multiply branched iso-paraffins = mononaphthenes with long chains attached = monoaromatics with

TABLE 3.7
VIs for Fused Polycyclic Naphthenes

Carbon Number and Number of Rings	Name	Viscosity, cSt at 40°C	Viscosity, cSt at 100°C	VI	Source
16,4	Perhydropyrene	11.517	2.9	100	API 42
16,4	Perhydrofluoranthene	8.972	2.51	106	API 42
16,4	Perhydroindeno indene	9.181	2.467	87	API 42
18,4	Perhydrochrysene	23.843	3.978	21	API 42
21,5	Perhydrodibenzo[a,i] fluorene	355.18	10.66	−258	API 42
25,4	6-n-Octyl perhydro benzanthracene	94.354	7.913	28	API 42
26,4	2-Octyl Perhydro triphenylene	103.274	8.65	25	API 42
26,4	2-Decyl Perhydro indenoindene	48.071	6.958	100	API 42
26,4	n-Octyl Perhydrochrysene	156.73	9.899	−18	API 42
27,4	Cholestane	701.84	20.299	−73	API 42

Source: "Properties of Hydrocarbons of High Molecular Weight," Research Project 42, 1940–1966, American Petroleum Institute, New York. With permission.

long chains >diaromatics> polyaromatics = polycyclic naphthenes with multiple short chains attached. This ordering is consistent with lube processing experience. Two examples are (a) the solvent refining process removes polyaromatic molecules of low VI and leaves a raffinate with increased VI and (b) base stocks of

TABLE 3.8
VIs for Fused Polycyclic Aromatic and Naphthene Ring Structures

Carbon Number	Name	Number of Naphthene Rings	Number of Aromatic Rings	Viscosity, cSt at 40°C	Viscosity, cSt at 100°C	VI
16	Hexhydropyrene	2	2	35.562	4.352	−79
18	9-n-Buyl anthracene	0	3	38.987	4.364	−123
26	3-n-Hexyl perylene	0	4	867.007	23.314	−60
26	3-n-Octylchrysene	0	4	375.301	14.475	−71
26	9-n-Octyltetrahydro naphthacene	1	3	667.669	18.039	−121

Source: "Properties of Hydrocarbons of High Molecular Weight," Research Project 42, 1940–1966, American Petroleum Institute, New York. With permission.

very high VI are composed largely of iso-paraffins and monocyclic naphthenes, with few if any, low VI polycyclic naphthenes and aromatics.

3.2.3 Viscosity Index Distributions in Base Stocks: Use of Thermal Diffusion

These compositional studies raise questions about VIs within a base stock—do they contain molecules that all have approximately the same VI or is there a distribution of VIs? The answer is that there can be a substantial distribution of VIs in a petroleum-derived base stock, even greater than that previously seen across boiling range, and the overall VI measured for a particular stock is the "average." This information has been obtained by a molecular separation technique called "thermal diffusion." Whereas distillation separates predominantly by molecular weight and liquid chromatography separates by molecular structure (e.g., saturates, aromatics, polars), thermal diffusion[33] separates by a combination of density and molecular shape, which in practice means by VI. This separation method is neither a high resolution one nor is it fast, but it is effective as the following background will illustrate. Relatively few companies appear (or admit) to make use of this method; those that stand out are ExxonMobil and IFP. However, it does appear to be very valuable in obtaining otherwise inaccessible information.

Thermal diffusion of petroleum samples is carried out in the annular space defined by two coaxial cylinders whose surfaces are separated by distances of approximately 0.2 mm. These surfaces are maintained at different temperatures. Figure 3.15 demonstrates a laboratory setup[34] used to practice this technique. Separation is performed by filling the annular space with the sample, then allowing the system to equilibrate for a period up to several days. In one configuration, the cylinder diameters are about 0.5 in., the annular spacing is about 0.0115 in., and the vertical length is 6 ft. The sample is injected at the center position and the product samples are taken off at a number of sample points, frequently ten of these, on the vertical axis after an appropriate time (3 to 10 days in practice) for the diffusional separation to take place. The samples, a few milliliters or less in size, are then analyzed by modern instrumental methods.

The basic physics of the separation process has been described as follows:[34] referring to Figure 3.16, two molecules, A and B, with different structures move at different rates due to different mobilities to the hot and cold walls of the container (I), and once there will migrate with the convective streams to the upper and lower zones, again at different rates (II), with low density molecules congregating at the top and higher density ones at the bottom. Some separations achieved with model compounds are illustrated in Table 3.9.

A 34 ft. long prototype of a commercial-scale unit was built by Standard Oil (Ohio) in the late 1950s to produce 120 VI base stock on a multiple gallon per day basis,[35] but does not appear to have had commercial success. No further industrial-scale applications have been reported since then and all other published

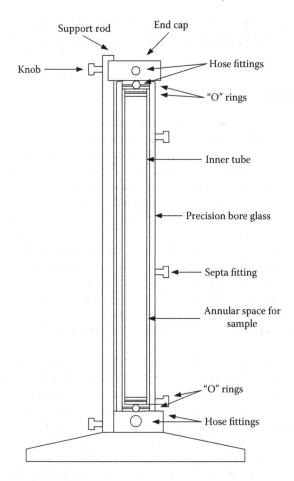

FIGURE 3.15 Schematic of the apparatus employed for thermal diffusion studies.
Source: H. A. Harner III and M. M. Bellamy, "Applications for Liquid Thermal Diffusion," *American Laboratory* January:41–44 (1972). With permission.

uses have been on an analytical scale. Some examples of the use of this technique shed light on VI and base stock composition.

Denis and Parc[2] (IFP) investigated the heterogeneity of a variety of base stocks whose origins included solvent refined, hydrocracked, wax hydroisomerized, and synthetic, in an attempt to provide an upper limit for the VI of components within these products. They employed a 1.84 m long thermal diffusion unit with an interior volume of 24 ml and ten takeoff ports. Two sets of their results are reproduced in Table 3.10, where it can be seen that in the pair of solvent refined base stocks chosen (a 150N and 400N with VIs of 105 and 97, respectively), the VIs of fractions varied from a high of 175 to a low of 15, with the low viscosity, high VI components evidently being highly paraffinic. The high

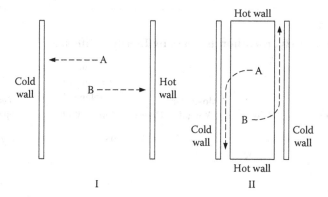

FIGURE 3.16 Behavior of two different molecules (A and B) in thermal diffusion. *Source:* H. A. Harner III and M. M. Bellamy, *American Laboratory* January:41–44 (1972). With permission.

viscosity, low VI ones are much more naphthenic and aromatic from the $\%C_A$ and $\%C_N$ numbers. Viscosities increase sharply from fractions 1 through 10 due to the compositional changes involved. Pour points of the fractions decreased with decreasing paraffinic character. The highest pour points thus belong to isoparaffins with few side chains. As one might expect, densities increased with naphthenic/aromatic character.

In other examples explored by these authors, when a hydrocracked stock with a VI of 100 was subjected to this separation method, a very similar VI span (176 to 45) was observed. Hydrocracked base stock with an even higher VI showed little change in VI at the top end, but the VI span was much less, from which one could conclude that increased hydrocracking severity has the effect of destroying low VI components rather than increasing the VI of the high-end ones. An isomerized wax base stock of 144 VI was found to have a VI range of only 30 units, with the best at 165. Indeed, the maximum VI that seems attainable by hydrocracking lube cuts from crudes or by isomerizing wax appears to be approximately 165 to 170. This corresponds to about the highest number mentioned by ExxonMobil in gas to liquids (GTL) patents.

From n-d-M analysis and nuclear magnetic resonance (NMR) spectroscopy of the fractions, the authors concluded that the highest VI structures contained an average of 0.5 to 1.5 naphthene rings per molecule substituted by long paraffin or isoparaffinic chains of 30 to 40 carbon atoms. They also found that for the lighter, high VI lube components, the ratio of CH_2 to CH_3 groups was relatively low, indicating a high proportion of branching.

In another example of this technique, D. E. Cranton,[36] at Imperial Oil, as part of a study on base stock oxidation stability, separated a midcontinent 150N solvent refined oil by thermal analysis and the ten fractions obtained were then further separated into saturates and aromatics by silica gel chromatography. Mass spectral analyses (Tables 3.11 and 3.12) show the results of the separation obtained in

TABLE 3.9

Separation of Hydrocarbon Isomers by Thermal Diffusion

Components	Volume %	Molecular Weight	Density	Final Composition, Volume %		Percent Separation
				Top	Bottom	
n-Heptane	50	100	0.6837	95	10	75
Triptane [2,2,3-trimethyl-Butane]	50	100	0.6900	5	90	
iso-Octane	50	114	0.6919	58	40	11.4
n-Octane	50	114	0.7029	42	60	
2-Methyl naphthalene	50	142	0.9905	55.5	42.5	13.1
1-Methyl naphthalene	50	142	1.0163	44.5	57.5	
trans-1,2-Dimethyl cyclohexane	40	112	0.7756	100	0	100
cis-1,2-Dimethyl cyclohexane	60	112	0.7963	0	100	
p-Xylene	50	106	0.8609	92	0	92
o-Xylene	50	106	0.8799	8	100	
m-Xylene	50	106	0.8639	100	0	80
o-Xylene	50	106	0.8799	0	81	
p-Xylene	50	106	0.8609	50	50	0
m-Xylene	50	106	0.8639	50	50	

Source: A. Letcher-Jones, "Separation of Organic Liquid Mixtures by Thermal Diffusion" *Industrial and Engineering Chemistry,* 47: 212–215 (1995). With permission.

terms of composition. Table 3.11 demonstrates that for the saturates, multiring naphthenes and polycyclic aromatics were concentrated in the later (4 and higher) fractions and were clearly separated from those with few rings (and presumably of high VI) and no rings at all in the case of the isoparaffins. A similar picture of the aromatics comes from Table 3.12, in which again there is fairly effective separation of aromatics by the number of rings, with the high (presumably) VI aromatics in fraction 1 largely being alkyl-substituted benzenes. One might also expect that the average structure of the alkylbenzenes would be different in fractions 1 and 7, for example.

TABLE 3.10
Thermal Diffusion of Solvent Refined 150N
and 400N Base Stocks

Fraction Number	Viscosity, cSt At 37.8°C	At 98.9°C	VI D2270	Density at 20°C	Pour point, °C	C_A %	C_N %	C_P %
Solvent Refined 150N								
Feed	31.2	5.18	105	0.867	−18	3.3	33.2	63.5
1	11.8	3.28	171	0.810	+12	0.0	4.5	95.5
2	13.8	3.59	162	0.817	+2	1.0	6.7	92.3
3	16.1	3.83	146	0.827	−5	0.6	17.0	82.4
4	17.4	4.00	143	0.833	−15	0.0	22.7	77.3
5	23.7	4.47	110	0.862	−26	0.0	38.7	62.2
6	41.6	5.74	82	0.882	−24	3.9	39.4	56.7
7	82.2	7.77	59	0.900	n/a	7.9	39.5	52.6
8	245	12.1	15	0.931	n/a	5.2	57.0	37.8
Solvent Refined 400N								
Feed	98.3	10.5	97	0.872	−12	9.8	16.8	73.4
1	22.4	5.05	175	0.824	+6	0.6	9.1	90.3
2	29.4	5.93	166	0.833	+1	1.0	14.2	84.8
3	36.8	6.59	147	0.841	−7	2.4	17.2	80.4
4	47.7	7.47	131	0.851	−18	3.3	19.6	77.1
5	77.8	9.38	106	0.874	−18	2.7	30.9	66.4
6	121.0	11.50	88	0.886	−18	5.8	30.7	63.5
7	221.0	15.30	69	0.899	−10	8.3	31.4	60.3
8	451.0	24.50	76	0.908	−6	15	20.7	64.3

Source: J. Denis and G. Parc, "Rheological Limits of Mineral and Synthetic Hydrocarbon Base Stocks," *Journal of the Institute of Petroleum* Vol. 56(556):70-83 (1973). With permission.

More recently, ExxonMobil has used this separation method to demonstrate that base stocks produced by their raffinate hydroconversion (RHC) process have greater homogeneity in terms of VI than solvent refined ones (Figure 3.17).[37] More specifically, their results indicate that these RHC stocks do not contain high viscosity, very low VI components which make achieving low volatility more difficult. It can be seen in Figure 3.17 that the solvent refined 150N contains components with VIs as low as −250, whereas the "worst" of the RHC 250 product only goes down to −25. The significance of this is probably more on the structures implied by the very low VI and that these are the more easily oxidized polynaphthenes and polyaromatic structures. Other examples of the use of this technique are available as well.[38,39]

A final example returns to a point raised earlier, that is, the VI changes that occur upon solvent extraction of the (largely) polyaromatics from a raw lube feed.

TABLE 3.11
Molecular Composition of Thermal Diffusion Fractions of Saturates from a Midcontinent Lubricating Oil

Fraction Number	1	2	3	4	7
Wt. % saturates in fraction	94.3	92.3	89.8	87.5	74.2
Analysis of saturates, volume %					
Isoparaffins	61.0	51.8	42.1	34.7	3.0
1-ring naphthenes	36.2	42.4	46.3	46.7	33.6
2-ring naphthenes	1.9	5.0	9.6	14.9	31.8
3-ring naphthenes	0.5	0.5	1.7	3.4	21.4
4+-ring naphthenes	0.4	0.3	0.2	0.3	10.3

Source: G. E. Cranton, "Composition and Oxidation of Petroleum Fractions," *Thermochimica Acta* 14:201–208 (1976). With permission.

Thermal diffusion[38] illustrates this very clearly in Figure 3.18, where the VI distributions of the feedstock, raffinate, and aromatic extract are highlighted remarkably well.

Finally, it should be possible to calculate VIs in the relatively near future from composition, although it will always be quicker to simply measure viscosities and calculate. Gatto et al.[40] employed compositions obtained from mass spectral analyses on some 15 hydrocracked and polyolefin base stocks to develop a correlation that predicted VI. While this one used only one component, the

TABLE 3.12
Molecular Composition of Thermal Diffusion Fractions of Aromatics from a Midcontinent Lubricating Oil

Fraction Number	1	4	5	6	7
Wt. % aromatics in fraction	5.0	9.8	12.7	15.2	18.6
Analysis of aromatics, volume %					
Alkylbenzenes	86.7	69.2	57.6	47.5	34.1
Naphthenobenzenes	13.1	20.4	23.7	26.3	25.9
Naphthalenes	0.0	0.8	9.4	14.5	20.0
Polynuclear aromatics	0.1	0.2	0.3	3.0	7.2
Thiophenes	0.1	5.4	8.3	10.7	12.8

Source: G. E. Cranton, "Composition and Oxidation of Petroleum Fractions," *Thermochimica Acta* 14:201–208 (1976). With permission.

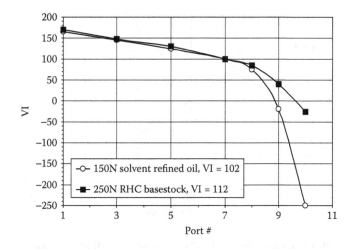

FIGURE 3.17 Thermal diffusion of an ExxonMobil RHC base stock and a conventional solvent refined base stock.
Source: I. A. Cody, D. R. Boate, S. J. Linek, W. J. Murphy, J. E. Gallagher, and G. L. Harting, "Raffinate hydroconversion process," U.S. Patent 6,325,918, December 4, 2001.

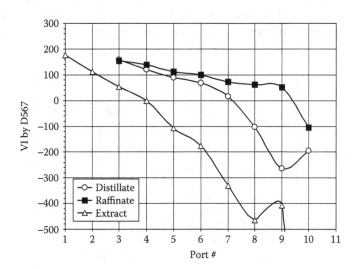

FIGURE 3.18 VIs of fractions obtained by thermal diffusion from a paraffinic distillate and its raffinate and extract.
Source: A. Letcher Jones, "Lubricating Oil Fractions Produced by Thermal Diffusion," *Industrial and Engineering Chemistry* 47(2):212–215 (1955). With permission.

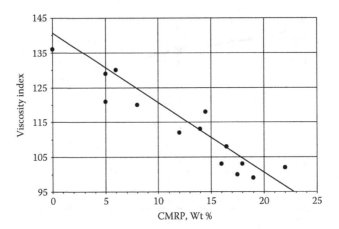

FIGURE 3.19 VI as a function of CMRP content for group II and III base stocks.
Source: V. J. Gatto M.A. Grina, T. L. Tat, H. T. Ryan. "The Influence of Chemical Structure on the Physical and Performance Properties of Hydrocracked Basestocks and Polyalphole-fins, Journal of Synthetic Lubrication 19: 13–18 (2002). With permission.

condensed multiring paraffins (CMRPs) (naphthenes with three or more rings), the correlation coefficient (R^2) was a fairly decent 0.9301. As expected, VI decreased as the CMRP content increased (Figure 3.19). One might expect that multicomponent correlations are eminently feasible, particularly if they are set

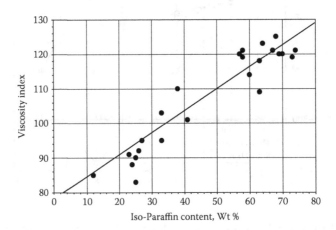

FIGURE 3.20 VI versus wt. % isoparaffin content for 4 cSt at 100°C base stocks.
Source: H. E. Henderson, B. Swinney, and W. M. Steckle, "Delivering Synthetic Performance with VHVI Specialty Base Fluids," Paper AM-00-29 presented at the annual meeting of the National Petrochemical and Refiners Association, March 26–28, 2000, San Antonio, TX. With permission.

up in commonly available viscosity ranges. In fact, it would be surprising if many companies do not already possess this capability.

This type of correlation has also been identified by Henderson et al.[41] (Petro-Canada), except that in their case, isoparaffin content was used as the single variable, and naturally a positive slope was found, with an R^2 of 0.90 (Figure 3.20).

REFERENCES

1. J. Denis, "The Relationship Between Structure and Rheological Properties of Hydrocarbons and Oxygenated Compounds Used as Base Stocks," *Journal of Synthetic Lubricants* 1(1–3):201–238 (1984).
2. J. Denis and G. Parc, "Rheological Limits of Mineral and Synthetic Hydrocarbon Base Stocks," *Journal of the Institute of Petroleum* 56(556):70–83 (1973).
3. "Properties of Hydrocarbons of High Molecular Weight," Research Project 42, 1940–1966, American Petroleum Institute, New York.
4. E. W. Dean and G. H. B. Davis, "Viscosity Variations of Oils with Temperature," *Chemical and Metallurgical Engineering* 36:618–619 (1929).
5. E. W. Dean, A. D. Bauer, and J. H. Berglund, "Viscosity Index of Lubricating Oils," *Industrial and Engineering Chemistry* 32:102–107 (1940).
6. R. E. Hersch, E. K. Fisher, and M. R. Fenske, "Viscosity of Petroleum Products. Viscosity-Temperature Characteristics of Pennsylvania Lubricating Oils," *Industrial and Engineering Chemistry* 27:1442–1446 (1935).
7. ASTM D567, "Standard Method for Calculating Viscosity Index," *ASTM Annual Book of Standards* (West Conshohocken, PA: American Society for Testing and Materials).
8. ASTM D2270, "Standard Practice for Calculating Viscosity Index from Kinematic Viscosity at 40 and 100°C," *ASTM Annual Book of Standards* (West Conshohocken, PA: American Society for Testing and Materials).
9. "Resolution 10," Proceedings of the 3rd World Petroleum Congress, The Hague, Gen. Vol., 74 (1951).
10. "Section V-Motion No. 3," Proceedings of the 4th World Petroleum Congress, Rome, Gen. Vol. 114 (1955).
11. J. C. Geniesse, "A Study of the ASTM Viscosity-Index Problems," Proceedings of the 5th World Petroleum Congress, Section V-Paper 30A, pp. 407–409 (1959).
12. W. A. Wright, "A Proposed Modification of the ASTM Viscosity Index," Proceedings of the American Petroleum Institute annual meeting, Section III—Refining, pp. 535–541 (1964).
13. W. A. Wright, "The Viscosity-Temperature Function," *ASTM Bulletin* July:84–86 (1956).
14. F. A. Litt, "Viscosity Index Calculations," *Lubrication Engineering* 42:752–753 (1986).
15. F. A. Litt, "Viscosity-Temperature Relations," *Lubrication Engineering* 42:287–289 (1986).
16. B. Gillespie and F. A. Smith, "Explanation of Some Lubricating Oil Hydrocracking Results in Terms of the Structure of the Viscosity Index Scale," *Industrial and Engineering Chemistry Product Research and Development* 9(4):535–540 (1970).

17. J. A. Zakarian, R. J. Robson, and T. R. Farrell, "All-Hydroprocessing Route for High Viscosity Index Lubes," *Energy Progress* 7(1):59–63 (1987).

18. J. A. Zakarian, "The ASTM Viscosity Index and Other Systems for Classifying Lubricating Oils," Paper FL-82-85 presented at the Fuels and Lubricants Meeting of the National Petrochemical Refiners Association, November 4–5, 1982, Houston, TX.

19. J. H. Roberts, "Impact of Quality of Future Crude Stocks on Lube Oils (Significance of VI in Measuring Quality)," Paper AM-85-21C presented at the annual meeting of the National Petrochemical Refiners Association, March 24–26, 1985, San Antonio, TX.

20. D. W. Murray, J. M. MacDonald, and P. G. Wright, "The Effect of Basestock Composition on Lubricant Oxidation Performance," *Petroleum Review* February:36–40 (1982).

21. J. C. Geniesse, "A Comparison of Viscosity-Index Proposals," *ASTM Bulletin* July:81–84 (1956).

22. J. F. T. Blott and C. G. Verver, "Methods for Expressing the Viscosity-Temperature Relationship of Lubricating Oils," *Journal of the Institute of Petroleum* 38(340):192–249 (1952).

23. F. P. Malschaert, "Un Index de Viscosite Independent des Temperatures de Mesure," Proceedings of the 2nd World Petroleum Congress, Section 2, pp. 905–910 (1937).

24. E. E. Klaus and M. R. Fenske, "The Use of ASTM Slope for Predicting Viscosities," *ASTM Bulletin* July:87–94 (1956).

25. D. F. Wilcox, "Viscosity-Temperature Coefficient," *Mechanical Engineering* 66:739 (1944).

26. E. W. Hardiman and A. H. Nissan, "A Rational Basis for the Viscosity Index System. Part I," *Journal of the Institute of Petroleum* 31:255–270 (1945).

27. J. H. Ramser, "Representation of Viscosity-Temperature Characteristics of Lubricating Oils," *Industrial and Engineering Chemistry* 41:2053–2059 (1949).

28. A. Cameron, "An Index of Viscosity," *Journal of the Institute of Petroleum* 46(434): 58–60 (1960).

29. R. T. Sanderson, "Viscosity-Temperature Characteristics of Hydrocarbons," *Industrial and Engineering Chemistry* 41:368–374 (1949).

30. ASTM D341, "Standard Viscosity-Temperature Charts for Liquid Petroleum Products," *ASTM Annual Book of Standards*, vol. 05.01 (West Conshohocken, PA: American Society for Testing and Materials).

31. M. Ushio, K. Kamiya, Y. Yoshida, and I. Honjou, "Production of High VI Base Oil by VGO Deep Hydrocracking," presented at the Symposium on Processing, Characterization and Application of Lubricant Base Oils, American Chemical Society annual meeting, Washington, DC, Preprints of the Division of Petroleum Chemistry 37(4):1293–1302 (1992).

32. M. Daage, "Base Oil Production and Processing," ExxonMobil presentation, July 2001, available at http://www.prod.exxonmobil.com/refiningtechnologies/pdf/base_oil_refining_lubes_daage_france070601.pdf.

33. A. L. Jones and E. C. Milberger, "Separation of Organic Liquid Mixtures by Thermal Diffusion," *Industrial and Engineering Chemistry* 45:2689–2696 (1953).

34. H. A. Harner III and M. M. Bellamy, Applications for Liquid Thermal Diffusion, *American Laboratory* January:41–44 (1972).

35. R. Grasselm, G. R. Brown, and C. E. Plymale, "Full-Scale Thermal Diffusion Equipment," *Chemical Engineering Progress* 67(5):59–64 (1961).
36. G. E. Cranton, "Composition and Oxidation of Petroleum Fractions," *Thermochimica Acta* 14:201–208 (1976).
37. For example, I. A. Cody, D. R. Boate, S. J. Linek, W. J. Murphy, J. E. Gallagher, and G. L. Harting, "Raffinate hydroconversion process," U.S. Patent 6,325,918, December 4, 2001.
38. A. Letcher Jones, "Lubricating Oil Fractions Produced by Thermal Diffusion," *Industrial and Engineering Chemistry* 47:212–215 (1955).
39. B. J. Mair and F. D. Rossini, "Composition of Lubricating Oil Portion of Petroleum," *Industrial and Engineering Chemistry* 47:1062–1068 (1955).
40. V. J. Gatto, M. A. Grina, and H. T. Ryan, "The Influence of Chemical Structure on the Physical and Performance Properties of Hydrocracked Basestocks and Polyalpholefins," Proceedings of the 12th International Colloquium: Tribology 2000-Plus, January 12–13, 2000, *Technische Akademie Esslingen*, pp. 295–304 (2000).
41. H. E. Henderson, B. Swinney, and W. M. Steckle, "Delivering Synthetic Performance with VHVI Specialty Base Fluids," Paper AM-00-29 presented at the annual meeting of the National Petrochemical and Refiners Association, March 26–28, 2000, San Antonio, TX.

4 Compositional Methods

4.1 INTRODUCTION

Base stock specifications, as defined by the producer or the purchaser, largely enumerate the physical properties required for the fluid—typically density, viscosity at two temperatures, viscosity index (VI), low temperature performance measures, flash and volatility properties, and solubility information from aniline point or viscosity-gravity constant (VGC)—the latter two are usually for naphthenic base stocks. While chemical composition is responsible for physical properties, it usually only surfaces as measurements of heteroatom content—sulfur and nitrogen—and aromatics content (or conversely that of saturates). Sulfur and aromatics levels in paraffinic base stocks are now criteria for American Petroleum Institute (API) classifications. However, detailed chemical compositional information is needed to understand the chemistry of the unit processes, the effects of changes in feeds, catalysts, and operating conditions, and behaviors of finished lubricant products.

This chapter discusses two methods available for obtaining compositional information on lube streams, provides some indication as to how they have been developed, and provides some examples of how the information has been employed. These two are the n-d-M method for determining the content of paraffinic, naphthenic, and aromatic carbons, and ^{13}C nuclear magnetic resonance (NMR) spectroscopy, which can provide detailed information on the environment of the three types of carbons (and their hydrogens). These two methods represent old and new technology, respectively, however, the n-d-M procedure can still play a useful role since the solvent refining technology for which it was developed is still widespread and ^{13}C NMR spectroscopy still has its most interesting days ahead. Mass spectroscopy is the standard present-day tool for petroleum hydrocarbon analyses and this method has been adequately described elsewhere.[1]

Early compositional analyses on petroleum (and lubricating oil base stocks) were focused on quantifying the three major hydrocarbon types present, namely paraffins, naphthenes or cycloparaffins, and aromatics. In that period (the 1920s to the 1950s), the availability of instrumental techniques was essentially nil in terms of our viewpoint today, since spectroscopic methods were in their infancy, as was electronics technology. Accordingly, research workers used the limited tools available at that time—density, refractive index, molecular weight, and elemental analyses. Based on work with model compounds, these led to compositional relationships between structure and these measurements and development of the concepts of VGC, refractivity intercept, and the n-d-M method.

No longer the tools of today, these methods and their variations are common throughout the lubricant literature and familiarizing oneself with these methods is worthwhile. Results from these methods are still quoted today, but mainly for naphthenic oils.

While mass spectroscopy measures the composition in terms of paraffinic, cycloparaffinic, and aromatic types, specifically the number of rings in each type and degree of unsaturation, these early methods focused on the environment of the carbon atoms and measured the percentages that are paraffinic ($\%C_P$), naphthenic ($\%C_N$), and aromatic ($\%C_A$):

$$100 = \%C_P + \%C_N + \%C_A,$$

where $\%C_N$ is the percentage of carbon atoms in naphthenic rings; carbons designated as C_P are in aliphatic chains, which may be branched or straight and can be substituents on naphthenic or aromatic rings; and C_A are those carbons that are part of aromatic rings.

4.2   n-d-M METHOD

The n-d-M method is an empirical method for determining the carbon type distribution ($\%C_P$, $\%C_N$, $\%C_A$) by simple measurement of the refractive index (n), density (d), and molecular weight (M) of the sample. It also provides the mean number of naphthenic (R_N) and aromatic (R_A) rings per molecule. The method was developed by researchers at Koninklijke/Shell in Holland after World War II. Its application includes lube feedstocks and raffinates.[2] Nearly all applications have been to solvent refined stocks. The current American Society for Testing and Materials (ASTM) method is D3238. ASTM D2140 is applicable to insulating oils.

The procedure is provided in Table 4.1, where measurements can be made either at 20°C or 70°C (for waxy fractions) and where $\%C_R$ is the percentage of cyclic carbon (aromatic + naphthenic), M is the average molecular weight, and S is the sulfur content (in weight percent [wt %]).

An application of the n-d-M method is shown in Table 4.2,[3] which provides compositional analyses on products from a Venezuelan distillate cut (Charge) that was sequentially extracted five times with furfural to produce raffinates and finally treated with silica gel to give raffinate 6. Since furfural selectively extracts the polyaromatic and aromatic components, the compositional changes that can be seen are in agreement with expectations, with a reduction of the percentage of aromatic carbon as the furfural selectively removes the polyaromatics, then presumably the di-s and finally significant amounts of monoaromatics. With this trend, the percentage naphthenic and paraffinic carbon increases, and the percentage of aromatic carbons decreases. The total number of rings (R_T) remains essentially constant; the decrease in aromatic rings presumably being offset by the increase in naphthenic ring carbon in the raffinates.

TABLE 4.1
n-d-M Procedure for Compositional Analyses

At 20°C	At 70°C

Calculate

$v = 2.51(n^{20} - 1.4750) \; d^{20} - 0.8510)$

Calculate

$w = (d^{20} - 0.8510) - 1.11(n^{20} - 1.4750)$

Calculate

$x = 2.42(n^{70} - 1.4600) - (d^{70} - 0.8280)$

Calculate

$y = (d^{70} - 0.8280) - 1.11(n^{70} - 1.4600)$

If v is positive, calculate $\%C_A$ as
$\%C_A = 430v + 3660/M$.

If x is positive, calculate $\%C_A$ as
$\%C_A = 410x + 3660/M$

If v is negative, calculate $\%C_A$ as
$\%C_A = 670v + 3660/M$

If x is negative, calculate $\%C_A$ as
$\%C_A = 720x + 3660/M$

If w is positive, calculate $\%C_R$ as
$\%C_R = 820w - 3S + 10,000/M$

If y is positive, calculate $\%C_R$ as
$\%C_R = 775y - 3S + 11,500/M$

If w is negative, calculate $\%C_R$ as
$\%C_R = 1440w - 3S + 10,600/M$

If y is negative, calculate $\%C_R$ as
$\%C_R = 1400y - 3S + 12,100/M$

$\%C_N = \%C_R - \%C_A$
$\%C_P = 100 - \%C_R$

$\%C_N = \%C_R - \%C_A$
$\%C_P = 100 - \%C_R$

R_A, **mean number of aromatic rings per molecule**

R_A, **mean number of aromatic rings per molecule**

If v is positive, calculate R_A as
$R_A = 0.44 + 0.055Mv$

If x is positive, calculate R_A as
$R_A = 0.41 + 0.055Mx$

If v is negative, calculate R_A as
$R_A = 0.44 + 0.080Mv$

If x is negative, calculate R_A as
$R_A = 0.41 + 0.080Mx$

R_T, **mean total number of rings [aromatic + naphthenic] per molecule**

R_T, **mean total number of rings [aromatic + naphthenic] per molecule**

If w is positive, calculate R_T as
$R_T = 1.33 + 0.146M(w - 0.005S)$

If y is positive, calculate R_T as
$R_T = 1.55 + 0.146M(y - 0.005S)$

If w is negative, calculate R_T as
$R_T = 1.33 + 0.180M(w - 0.005S)$

If y is negative, calculate R_T as
$R_T = 1.55 + 0.180M(y - 0.005S)$

R_N, **mean total number of naphthene rings per molecule**

R_N, **mean total number of naphthene rings per molecule**

$R_N = R_T - R_A$

$R_N = R_T - R_A$

Source: K. van Nes and H. H. van Westen, *Aspects of the Constitution of Mineral Oils* (New York: Elsevier, 1951), chap. 6.

TABLE 4.2
Solvent Extraction of Venezuelan Lubricating Oil Distillate with Furfural

Oil Sample	Percent of Charge	VI	R_T	R_A	R_N	%C_A	%C_N	%C_P
					n-d-M Method Results			
Charge	100	−17	2.80	1.21	1.59	25	24	51
Raffinate 1	78.2	14	2.80	0.89	1.91	18	29	53
Raffinate 2	64.8	36	2.79	0.65	2.14	12.5	32	55.5
Raffinate 3	59.3	43	2.74	0.50	2.24	10	33.5	56.5
Raffinate 4	54.1	48	2.76	0.39	2.37	7.5	35	57.5
Raffinate 5	53.2	57	2.74	0.29	2.35	5.5	36	58.5
Raffinate 6	40.0	68	2.57	0.18	2.39	3	36	61

Source: K. van Nes and H. H. van Westen, *Aspects of the Constitution of Mineral Oils* (New York: Elsevier, 1951), chap. 6.

n-d-M results have also been reported for hydrocracked and severely hydrocracked products.[4] In Table 4.3, the compositional properties of a 150N (approximately 6 cSt at 100°C) oil from SKs fuels hydrocracker (YU-6 base stock) are contrasted with those of paraffinic 150Ns from solvent refining, "conventional" lube hydrocracking, and a naphthenic 150N. The differences are highlighted by the complete absence of aromatic carbons in the highly paraffinic YU-6 and the very high level (79%) of paraffinic carbons. It should be noted that the n-d-M method was developed well prior to there being any hydrocracked stocks available,

TABLE 4.3
Product Quality of Products from SK Corporation's UCO Process
(Severe Hydrocracking and Hydroisomerization)

Dissolving Ability Comparison of 150N Grades

	Solvent Refined	Lube Hydrocracking	VHVI YU-6	Naphthenic 150N
Aniline point, °C	100	116	126	79
VGC	0.82	0.80	0.78	0.87
Carbon type distribution				
C_P, %	57	64	79	47
C_N, %	38	35	21	42
C_A, %	5	1	0	11

Source: W.-S. Moon, Y.-R. Cho, and J. S. Chun, "Application of High Quality (Group II, III) Base Oils to Specialty Lubricants," Paper presented at the 6th annual Fuels and Lubes Asia Conference, Singapore, January 28, 2000. With permission.

TABLE 4.4
Accuracy of the n-d-M Method Based
on Comparison with Direct Method Data

Percent Carbon	n-d-M Method	R	n-d-M Method
%C$_A$	±1.04	R$_A$	±0.042
%C$_N$	±1.81	R$_N$	±0.099
%C$_P$	±1.46	R$_T$	±0.090

Source: K. van Nes and H. H. van Westen, *Aspects of the Constitution of Mineral Oils* (New York: Elsevier, 1951), chap. 6.

so in this type of application, the method is being employed beyond its original compositional band.

The accuracy[5] of this method for %C was concluded to be better than 2% and substantially less than that for the number of rings (Table 4.4)—this is in comparison to results from a prior method, the direct method, which is based on changes in hydrogen content before and after aromatics hydrogenation.

Development of the n-d-M method was the consequence of much preceding work relating composition to density, refractive index, and molecular weight. One of the intermediate steps involved development of the VGC.

4.3 DENSITY AND VISCOSITY RELATIONSHIPS: THE VGC

The VGC is a measure of petroleum composition that connects two physical properties—specific gravity and viscosity—for distillation fractions. It was developed by Hill and Coates in 1928[6] to be an index of the paraffinic or naphthenic character. The objective was to fill in the "gap" between clearly paraffinic and clearly naphthenic samples. It is still reported for base stocks and ranges from approximately 0.78 (paraffinic base stocks) to 1.0 (highly aromatic base stocks) and its value provides some guidance for the solvency properties of the oil. Like the results of the n-d-M method, the VGC is usually reported for naphthenic products, but not for paraffinic ones.

For the development of this tool, it was expected that paraffinic cuts will have lower densities (and specific gravities) than naphthenic ones of about the same distillation range. The VGC concept arose from semilog plots of Saybolt viscosities at 100°F versus specific gravities (Figure 4.1) for a series of distillate cuts from different crude sources. Similar patterns were evidently present for fractions from different crudes.

In this figure, Pennsylvanian crude is the most paraffinic, followed by mid-Continent, and the three Gulf Coast oils in numerical order. For fractions with

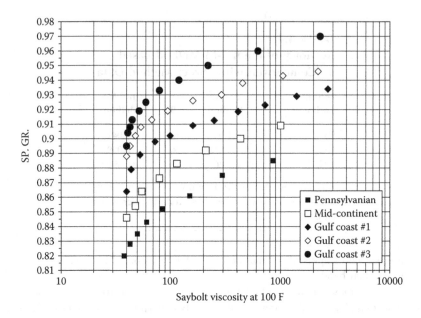

FIGURE 4.1 Specific gravity versus Saybolt viscosity at 100°C.
Source: J. B. Hill and H. B. Coats, "The Viscosity-Gravity Constant of Petroleum Lubricating Oils," *Ind. & Eng. Chem.*, vol. 20, pp. 641–644, (1928). With permission.

the same position on the viscosity scale, specific gravity decreases as paraffinicity increases, which is what one would expect given the well-known lower specific gravities of paraffins relative to naphthenes and aromatics. Alternatively, for the same specific gravity, viscosity increased from Pennsylvanian crude through the other crudes, which we can assume were increasingly aromatic/naphthenic.

The points in Figure 4.1 were found to fit by equations with the general formula

$$\text{Specific gravity} = a + b*\log(V + c), \qquad (4.1)$$

where V is the viscosity and a, b, and c are constants. The values of the constants for these five crude sources (Table 4.5) show that c is a constant (with a value of 38) and b can be expressed as a function of a, where

$$b = (1.0752 - a)/10.$$

The constant a thus has the ability to discriminate between paraffinic and naphthenic base stocks—it has an appropriate lower value for the paraffinic Pennsylvanian cuts and higher values through the mid-Continent to the Gulf Coast naphthenic cases. Replotting the data, again using semilog plots, using the revised

TABLE 4.5
Values of Constants *a*, *b*, and *c* in Equation 1

Crude	a	b	c
Pennsylvania	0.8067	0.02685	−38
Mid-Continent	0.8340	0.025	−38
Gulf Coast 1	0.8661	0.020	−38
Gulf Coast 2	0.8832	0.020	−38
Gulf Coast 3	0.8885	0.02685	−38

Source: J. B. Hill and H. B. Coats, "The Viscosity-Gravity Constant of Petroleum Lubricating Oils," *Ind. & Eng. Chem.*, vol. 20, pp. 641–644, (1928). With permission.

equation (Equation 4.2) gives straight-line plots whose slopes were taken to be identical and whose intercepts gave the value of *a* (Figure 4.2 and Table 4.6):

$$\text{Specific gravity} = a + ([1.0752 - a]/10)*\log(V - 38). \tag{4.2}$$

The VGC is simple to calculate (ASTM D2501)[7] from measurements of viscosity and specific gravity. From the kinematic viscosity at 40°C and density at 15°C, the equation is

$$\text{VGC} = (\text{density} - 0.0664 - 0.1154*\log[\text{viscosity} - 5.5])/(0.94 - 0.109*\log[\text{viscosity} - 5.5]) \tag{4.3}$$

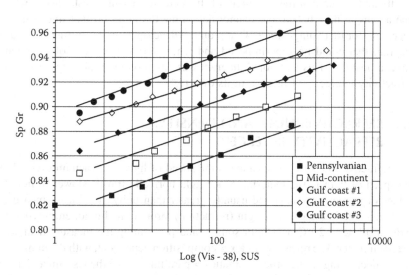

FIGURE 4.2 Semilog plot of specific gravity versus log (Saybolt viscosity at 100°F − 38). *Source:* J. B. Hill and H. B. Coats, "The Viscosity-Gravity Constant of Petroleum Lubricating Oils," *Ind. & Eng. Chem.*, vol. 20, pp. 641–644, (1928). With permission.

TABLE 4.6
Values of *a* in Equation 2

Crude	Value of *a*, VGC
Pennsylvanian	0.8067
Mid-Continent	0.8367
Gulf Coast 1	0.8635
Gulf Coast 2	0.8845
Gulf Coast 3	0.9025

Source: J. B. Hill and H. B. Coats, "The Viscosity-Gravity Constant of Petroleum Lubricating Oils," *Ind. & Eng. Chem.*, Vol. 20, pp. 641–644, (1928). With permission.

Other equations are available using Saybolt universal second (SUS) viscosities at 100°F and 210°F, kinematic viscosity at 20°C, and densities at 4°C and 20°C.[8]

Referring back to the four samples in Table 4.4, it can be seen that the severely hydrocracked SK sample has the lowest VGC value at 0.78, as it should have, being the most paraffinic, while the naphthenic 150N has the highest VGC of 0.87.

A relationship between VGC and composition ($\%C_A$, $\%C_N$, $\%C_P$) was developed by Kurtz et al.[8] using a number of samples for which all these data points were known. The format was a triangular coordinate graph (Figure 4.3) whose base was determined from fully hydrogenated samples.

It can be seen that measurement of the VGC for a sample establishes which line a sample may lie on, but cannot determine the point on that line which is necessary for that composition. That required development of a further triangular diagram that could be superimposed and where the intersections of lines gave the composition. This additional diagram was based on the following concept of refractivity intercept, of interest in its own right.

4.4 REFRACTIVE INDEX AND DENSITY: REFRACTIVITY INTERCEPT

The refractive index of a substance is the speed of light in a vacuum divided by the speed of light in the material. Since light normally travels slower in liquids, values of refractive indices are usually greater than unity. It is easily measured by the change in direction of light (refraction) through the liquid, employing the sodium d-line at 20°C as a light source, and performing the measurement with a refractometer. The refractive index is composition dependent, with the numerical values decreasing in the order aromatics > paraffins > naphthenes. Since density is also composition dependent, a relationship between the two was anticipated to be of use in compositional analyses.

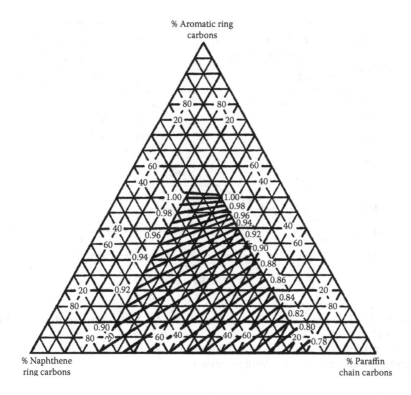

FIGURE 4.3 Viscosity-gravity constant in relation to carbon-type distribution.
Source: S. S. Kurtz, Jr., R. W. King, W. J. Stout, D. G. Parkikian, and E. A. Skrabek, "Relationship Between Carbon-Type Distribution, Viscosity-Gravity Constant, and Refractivity Intercept of Viscous Fractions of Petroleum," *Anal. Chem.*, vol. 28, pp. 1928–1936 (1956). With permission.

Indeed, when density and refractive index are plotted for related compounds, linear relationships were found to exist. The data in Figure 4.4 for C_{12} to C_{25} pure compounds are from API Project 42.[9]

The equations were of the general type[10]

$$n = d/2 + \text{refractivity intercept,}$$

where the lines are parallel and the intercept values when $d = 0$ (Table 4.7) are characteristic of the compound types.[11]

This concept, together with results on a large number of compounds for which $\%C_A$, $\%C_N$, and $\%C_P$ were known, enabled the additional set of lines superimposed on Figure 4.3. This aspect will not be discussed further here.

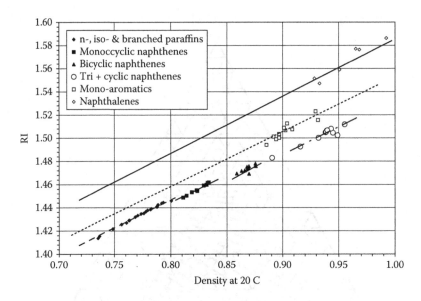

FIGURE 4.4 Refractive indices versus densities for C_{12} to C_{25} compounds from API Project 42.
Source: American Petroleum Institute, "Properties of Hydrocarbons of High Molecular Weight," API Project 42, 1960. With permission.

TABLE 4.7
Refractivity Intercept for Hydrocarbon Types

Type	Intercept	Number of Compounds
Noncyclic paraffins	1.0462	63
Monocyclic naphthenes	1.0396	81
Bicyclic naphthenes	1.0298	3
Tricyclic naphthenes	1.0200 (estimate)	0
Aromatics	1.0629	49

Source: S. S. Kurtz, Jr., and W. A. Ward, "The Refractivity Intercept and the Specific Refraction Equation of Newton. I. Development of the Refractivity Intercept and Comparison with Specific Refraction Equations," *Journal of the Franklin Institute* 222:563–592 (1936). With permission.

4.5 REFRACTIVE INDEX AND RECIPROCAL OF CARBON NUMBER

A final piece of information needed for n-d-M method development was a relationship between the refractive index and molecular weight for various compositional types. Smittenberger and Mulder[12,13] developed plots of refractive index and density at 20°C versus $1/(c + z)$, where c is the total number of carbon atoms and z is a constant similar to the hydrogen deficiency concept. They found the results (Figure 4.5 and Figure 4.6) to be a series of straight lines for homologous series of paraffins, n-alkylcyclopentanes, n-alkylcyclohexanes, and n-alkylbenzenes from gasoline through the lube oil range.[13] These lines all converged at zero for $1/(c + z)$, that is, at a single point that corresponds to infinite length carbon chains in each homologous series (e.g., an n-paraffin of infinite length), thus the "functional group," whether it be cyclopentyl, cyclohexyl, or benzene, becomes of no importance. This is not altogether surprising, since when the number of carbon atoms is infinite, the importance of the functional group becomes very small.

These plots follow equations of the type

$$d = d_\infty + q/(c + z)$$

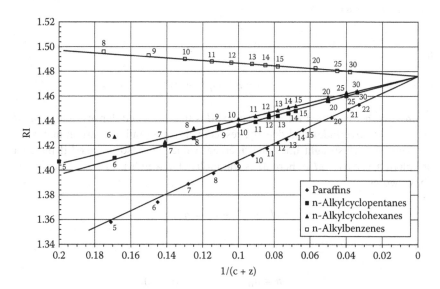

FIGURE 4.5 Refractive index versus $1/(c = z)$ for homologous series of hydrocarbons. *Source:* J. Smittenberg and D. Mulder, "Relations Between Refraction, Density and Structure of Series of Homologous Hydrocarbons. II. Refraction and Density at 20°C of n-Alkyl-cyclopentanes, -cyclohexanes and –benzenes," *Recueil des travaux chimiques*, vol. 67, pp. 826–838 (1948).

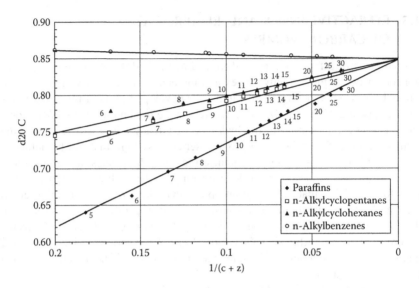

FIGURE 4.6 Density versus $1/(c + z)$ for four homologous series of hydrocarbons.
Source: J. Smittenberg and D. Mulder, "Relations Between Refraction, Density and Structure of Series of Homologous Hydrocarbons. II. Refraction and Density at 20°C of n-Alkyl-cyclopentanes, -cyclohexanes and –benzenes," *Recueil des travaux chimiques*, vol. 67, pp. 826–838 (1948).

and

$$n = n_\infty + q/(c + z),$$

where n is the measured refractive index, d is the measured density, and n and d are the refractive index and density, respectively, at an infinite carbon number.

Very similar straight lines and intercept values are also obtained if the data are plotted versus $1/M$ (where M is the molecular weight). From the equations above we get

$$\Delta d = d^{20°C} - d_\infty = q/(c + z)$$

and

$$\Delta n = n_D{}^{20°C} - n_\infty = q/(c + z),$$

which are approximately

$$\Delta d = q/M$$

and

$$\Delta n = q/M,$$

and these can be altered such that

$$\Delta d = -14k/M$$

and

$$\Delta n = -14k/M,$$

where k and k are constants, which are k_a for alkylbenzenes, k_n for n-alkylcyclo-hexanes, and k_p, k_a, k_b, and k_c for n-paraffins.

4.6 n-d-M METHOD: DEVELOPMENT

H. J. Tadema of Koninklijke/Shell found[2] that distillate composition could be expressed as linear relationships of the density, the refractive index, and the inverse of molecular weight. The equations were of the form

$$\%C = a/M + b\Delta d + c\Delta n,$$

where %C is either $\%C_A$, $\%C_N$, or $\%C_P$; M is molecular weight; $\Delta d = d^{20°C} - d_\infty$; $\Delta n = n_D^{20°C} - n_\infty$; and d_∞ and n_∞ are the limiting values for paraffins of infinite chain length determined in Smittenberg's work, as already discussed.

Using Tadema's form and Smittenberg's results:

$$\Delta d = -14k/M$$

and

$$\Delta n = -14k/M,$$

where k and k are constants and will have the values k_a, k_b, k_c, k_a', k_b', and k_c' for n-alkylbenzenes, n-alkylcyclohexanes, and n-paraffins (and assuming n-alkylcy-clopentanes can be approximated by the n-alkylcyclohexanes).

For a blend of these three components, the weight percent composition being %AR, %NA, and %PA:

$$100\Delta dM = -\%AR \cdot 14k_a - \%NA \cdot 14k_n - \%PA \cdot 14k_p$$

$$100\Delta dM = -\%AR \cdot 14k_a - \%NA \cdot 14k_n - 14k_p(100 - \%AR - \%NA).$$

and assuming there are two hydrogens per carbon atom, we get

$$\%C_A = 6(\%AR)14/M$$

$$\%C_N = 6(\%NA)14/M,$$

thus

$$\Delta d = \%C_A(k_p - k_a)/600 + \%C_N(k_p - k_n)/600 - 14k_p/M$$

and

$$\Delta n = \%C_A(k_p - k_a)/600 + \%C_N(k_p - k_n)/600 - 14k_p/M.$$

These equations thus became the form

$$\%C = a/M + b\Delta d + c\Delta n$$

used in the n-d-M method. For the complete solution of these equations, the reader is directed to Altgelt and Boduszynski.[1] The values of the constants incorporated in Table 4.1 were determined from correlations using a significant amount of compositional data available.

4.7 NMR SPECTROSCOPY: BACKGROUND

Nuclear magnetic resonance (NMR) is a spectroscopic method that can give quantitative information on the environment of the carbon and hydrogen atoms in petroleum samples. The method distinguishes between, for example, aromatic and aliphatic carbons and hydrogens, and further discriminates between atoms in certain positions within these two broad groups and others. Coupled with good separation methods, information well beyond "average" structures can be obtained. As in the case of most analytical methods, simpler structures afford more information, so the technique will be particularly useful for isomerized paraffins where mass spectroscopy has more limitations. NMR is currently by no means a routine tool in petroleum chemistry, but it is being increasingly applied in process and product development. The discussion that follows concentrates more on applications in the lubricants area than the techniques themselves.

In a more general sense, NMR is applicable to those nuclei having spin-angular momentum numbers of $I = 1/2$, and those of main interest to petroleum chemists are the naturally occurring isotopes ^1H and ^{13}C (other isotopes to which NMR is applicable include ^{15}N, ^{19}F, and ^{31}P). ^1H NMR spectroscopy (sometimes referred to as proton magnetic resonance [PMR] spectroscopy) was the first developed and showed itself to be of extraordinary use in identifying the structure of pure compounds since it provides strong signals, discriminates between hydrogen atoms in a wide variety of chemical locations, and permits one to "count" hydrogens.

^{13}C NMR is more recent, in part because ^{13}C is a less common isotope, with an abundance of only 1.1%. It does, however, employ a wider "spectroscopic" range, and is quite capable of being employed quantitatively.

Nuclear magnetic resonance spectra are obtained when the sample is placed in a strong magnetic field and exposed to electromagnetic radiation. Nuclei with spin $I = 1/2$ have two energy levels. When electromagnetic energy is applied, absorption occurs and some of the atoms in the lower energy level move to the higher level. The applied electromagnetic frequency required gives the energy difference between these levels and, indirectly, information on their chemical environment. This information is translated into the "chemical shift" relative to the protons and carbon atoms of tetramethylsilane as a reference, expressed as parts per million (ppm). Sample sizes for either form of NMR spectra are very small (less than 1 g) and are usually obtained with the sample dissolved in deuterated chloroform ($CDCl_3$). The practical chemical shift range for ^{1}H NMR is approximately 10 ppm, whereas for ^{13}C it is approximately 220 ppm. ^{13}C NMR spectra are usually obtained by repetitive scans. NMR spectroscopy provides average structural information on the molecules present. Application of separation methods to a sample prior to NMR spectroscopy can greatly assist in the interpretation.

Of the two types, the later-arriving ^{13}C NMR has been found to provide more information because of its superior ability to distinguish carbon atoms (and therefore, in most cases, hydrogens bonded to those carbons) by their positions in hydrocarbons. Techniques have been developed to portray the spectra as a series of sharp lines corresponding to the carbon types. Applications to date have been the determination of the degree of branching of alkanes, the measurement of average chain lengths, the estimation of average structures of group II and III base stocks, and the development of correlations between structure and base stock properties.

4.8 ^{1}H AND ^{13}C APPLICATIONS

A useful introductory (but very nonlubes) ^{1}H NMR spectrum[14] is that of a petroleum pitch (a highly aromatic fraction from petroleum pyrolysis) sample, whose spectrum in Figure 4.7 exhibits aliphatic hydrogens in the 0.5 to 3 ppm region and those directly attached to aromatic rings between 6.5 and 9 ppm. The hydrogens on carbons α to the aromatic rings show a distinct peak at approximately 2.5 ppm and are clearly separated from other aliphatic hydrogens on β and γ and higher carbons which are lumped together at between 1 and 1.5 ppm. Hydrogens directly attached to aromatic carbons appear between 6.5 and 9 ppm. In this particular case, integration shows that 45% of the hydrogens are aromatic, 31% are benzylic (hydrogens α to an aromatic ring), and the remaining 24% are aliphatic. Obviously this must be a highly aromatic blend with relatively few aliphatic substituent groups, and they in turn must have short side chains. The poor resolution of hydrogen types each resulting in broad "bands" is characteristic of ^{1}H NMR and is one of its limitations. Part of this is due to coupling of spins

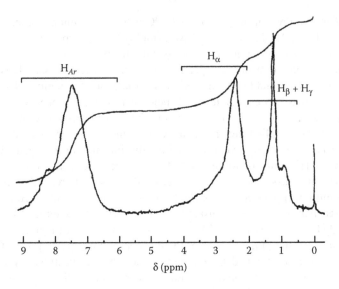

FIGURE 4.7 ^{1}H NMR spectrum of a petroleum pitch sample.
Source: E. M. Dickinson, "Structural Comparison of Petroleum Fractions Using Proton and 13C n.m.r. Spectroscopy," *Fuel*, vol. 59, pp. 290–294 (1980). With permission.

between adjacent hydrogen atoms, which leads to signal splitting and therefore an overall broadened appearance. Part is also due to the relatively limited chemical shift range exhibited by hydrogen.

The wider spectrum range (0 to 220 ppm) for ^{13}C NMR has also been accompanied by the development of means to obtain full peak intensities, which results in sharp spectra. Like ^{1}H NMR spectra, these spectra can be integrated to give quantitative results.

4.9 WAX ANALYSES

An early application of ^{1}H NMR was as one of several methods to estimate the degree of chain branching in paraffin waxes.[15] This information was sought for compounding purposes where the degree of branching was considered important for wax properties (e.g., in automobile tire manufacturing). The authors employed two NMR approaches. The first used the assignment of the upfield end (Figure 4.8) of the overall methyl peak (as previously mentioned, these aliphatic hydrogens appear between 1 and 1.5 ppm) to the branched methyl groups, while the terminal methyl groups were assigned to the large triplet. The branched methyls appear as a doublet (twin peaks; insert in Figure 4.8) because of spin coupling with the tertiary hydrogen (CH–CH$_3$), and the visible area was doubled since half the signal was obscured. Since there were two end group methyls for each chain, the percent branching was calculated by

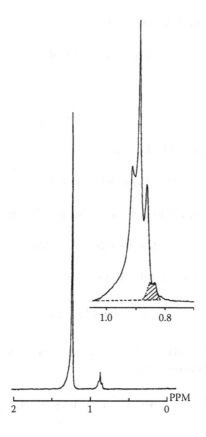

FIGURE 4.8 ^{1}H NMR Spectrum of a Wax Sample.
Source: C.H. Turner, G Blunden, F.N. Dowling and B. G. Carpenter, "Estimation of Chain Branching in Paraffin Waxes using Proton Magnetic Resonance Spectroscopy and Gas-Liquid Chromatography", *Journal of Chromatography,* 287: 305–312 (1984). With permission.

% branching = (4 × area of most upfield peak × 100/(total methyl area 2 × area of most upfield peak).

The second method used gas chromatography to measure the average chain length (n) and then NMR was used for the ratio of methyl (CH_3) to methylene (CH_2) plus methine (CH) hydrogens (which increases as branching increases) and the ratio of all methyl hydrogens to the total number of hydrogens. The ratios of these were derived as follows for an average chain length of n, using b for the fraction of total molecules that are branched, with t being the number of branches per molecule, thus the number of methylene hydrogens is

$$2(1 - b)(n - 2)$$

and the number of methine hydrogens is

$$2b(n - 2 - 2t) + bt,$$

which sum to

$$2n - 4 - 3bt.$$

The total number of methyl hydrogens is

$$3(1 - b)2 + 3(2 + t)b = 6 + 3bt,$$

and the ratio of total methyl to methylene plus methine hydrogens is

$$(6 + 3bt)/(2n - 4 - 3bt)$$

and the ratio of methyl to all hydrogens is

$$(6 + 3bt)/(2n + 2).$$

These equations are solvable for b if we make the assumption that $t = 1$ (i.e., one branch per molecule on average):

$$CH_3/(CH_2 + CH) = (6 + 3b)/(2n - 4 - 3b),$$

and the ratio of methyl to all protons is

$$(6 + 3bt)/(2n + 2),$$

where b is the fraction of all molecules that are branched, n is the average chain length determined by gas chromatography, and t is the number of branches in each molecule. The solution obtained here assumes $t = 1$.

The two NMR variations give results (Table 4.8) that agree quite well. The authors also determined the branched paraffin content by urea adduction of the n-paraffins (see Chapter 9); however, this method gave much higher results for the branched isomer total. They attributed the high adduct number (low value for the n-paraffins) to incomplete adduct formation caused by experiments carried out at 50°C rather than 25°C, a temperature which they judged in retrospect would have given more accurate results (adduction of n-paraffins is favored by low temperatures and adducts are decomposed by increasing the temperature).

Sperber et al.,[16] of Hamburg University, made structural assignments to the carbons in two macro- and two microcrystalline paraffin waxes, each deoiled and

TABLE 4.8
Wax Analyses

Product	Average Chain Length by Gas Chromatography	Estimated Percent Branching		
		By Gas Chromatography and NMR	By NMR	By Urea Adduct
Rubber compounding wax	26.5	6.3	7.9	21
Fully refined wax	26.2	5.1	5.4	20

Source: C. H. Turner, G. Blunden, F. N. Dowling, and B. G. Carpenter, "Estimation of Chain Branching in Paraffin Waxes Using Proton Magnetic Resonance Spectroscopy and Gas-Liquid Chromatography," *J. Chromatography*, vol. 287, pp, 305–312 (1984). With permission.

hydrotreated, by ^{13}C NMR. These results are summarized in Table 4.9. They were able to identify four types of methyl groups, ten methylenes, three methine (tertiary) carbons, and two types of naphthenic carbons without any prior physical separation, which totaled more than 95% of the carbon signals.

It is clear, as expected, that the macrocrystalline waxes OFM and CFA, and particularly the latter, have simpler structures than their microcrystalline counterparts. CFA has almost zero branched methyl groups, the lowest levels of tertiary carbons and methylene groups near branch positions, and no naphthenics. Wax OFM has only 0.21% branched methyl carbons. The microcrystalline waxes possess higher levels of branched methyl groups and CH_2 groups close to branches. Branching in both types of wax was mostly methyl branching, with only traces of longer branches such as ethyl- and isopropyl. It is surprising that their naphthenic carbon levels are so low.

4.10 SOME ^{13}C NMR APPLICATIONS

Chevron researchers[17] investigating catalytic wax isomerization to produce high VI base stocks employed ^{13}C NMR as a routine tool to measure the degree of branching in the base stocks during their program. The reason for this is the desire to limit the number of branches that develop, since excess branches (over those needed to achieve the target pour point) cause the VI to decline and highly branched molecules have an increased tendency to crack to smaller nonlube products. In the particular case cited, a branching index of less than 1.75 was considered desirable. Their work was based on techniques developed by Doddrell et al.[18] and Patt and Shoolery,[19] together with chemical shift data on hydrocarbons by Lindeman and Adams[20] and Netzel et al.[21]

TABLE 4.9
Characteristics and Carbon Type Distributions (%) in Waxes

	Microcrystalline		Macrocrystalline	
Designations	AW 034	AW 050	OFM	CFA
Congealing point, °C	66.0	50.5	64.5	62.5
Pen (1/10 mm)	25	51	18	15
Oil content	0.7	3.1	0.3	0.4
n-Alkanes by gas chromatography, %	12.3	19.3	9.6	76.5
Methyl carbons[a]				
α-Me	3.16	4.47	4.07	5.01
$1B_1$	1.12	0.94	0.16	0.00
$1B_2$	0.08	0.08	0.03	0.01
$1B_{3-6}$	0.17	0.17	0.02	0.00
Total branched methyls	1.37	1.19	0.21	0.01
Total methyls	4.53	5.66	4.28	5.02
Methylene carbons[b]				
β-CH_2	4.56	5.80	3.72	6.97
γ-CH_2	2.52	3.34	3.65	1.98
δ-CH_2	80.0	73.0	84.41	84.25
Total CH_2 in chains	87.07	83.13	91.79	93.20
$\alpha\delta^+$-B_1	1.32	0.82	0.34	0.09
$\alpha\delta^+$-B_2	0.04	0.19	0.12	0.12
$\alpha\delta^+$-B_{3-6}	0.17	0.66	0.00	0.00
$\beta\delta^+$-B_1	1.51	1.37	0.51	0.05
$\beta\delta^+$-B_{2-6}	0.89	1.71	0.32	0.00
$\gamma\delta^+$-B_1	1.31	1.58	0.05	0.72
$\gamma\delta^+$-B_{2-6}	0.15	0.38	0.00	0.00
$2B_2$c	0.07	0.25	0.06	0.00
CH_2 near branches	5.46	5.97	1.41	0.98
Total CH_2 overall	92.53	89.10	93.20	94.18
Methine carbons				
CHB_1	0.89	0.70	0.47	0.31
CHB_2	0.06	0.02	0.07	0.08
CHB_{3-6}	0.10	0.17	0.13	0.06
Total methine carbons	1.05	0.89	0.67	0.45
Naphthenic carbons				
CN1	0.12	0.20	0.14	0.00
CN2	0.01	0.02	0.11	0.00
Total naphthenic	0.13	0.22	0.25	0.00
Percent carbon identified	98.23	95.87	98.40	99.65

[a] These are methyl groups of methyl($1B_1$), ethyl($1B_2$), and propyl- to hexyl($1B_{3-6}$) branches.
[b] Greek letter gives position in straight chain.
[c] Methylene group in a branch.

Source: O. Sperber, W. Kaminsky, and A. Geissler, "Structure Analysis of Paraffin Waxes by 13C NMR Spectroscopy," *Petroleum Science and Technology,* vol. 23, pp. 47–54 (2005). With permission.

A number of papers have looked at the development of relationships between base stock composition as measured by NMR and either physical/chemical properties or their performance.[22–27] Most of this work has been focused on group II and III base stocks, with less or little attention paid to solvent extracted ones. These have all relied on various techniques to simplify the spectra and the assignments of peaks and make peak integration more reliable. These have many acronyms,[23] for example, GASPE (gates spin echo), PCSE (proton coupled spin echo), INEPT (insensitive nuclei enhancement by polarization transfer), DEPT (distortionless enhancement by polarization), QUAT (quaternary-only carbon spectra), 2D COSY (two-dimensional homonuclear spectroscopy), and HETCOR (heteronuclear shift correlated spectroscopy)]. Table 4.10 provides an example of some of the chemical shift data generated[26] and employed in this type of work, and Adhvaryu et al.[25] were able to develop the correlations between base stock properties and carbon types in Table 4.11, whose main features correspond to intuition (e.g., the values of API and aniline points are both decreased by aromatic carbon and increased by the

TABLE 4.10
Nomenclature of Carbon Atoms

ppm	Symbol	Assignment
10–15	CH_3	Terminal methyl groups in aliphatic chain (except the case where two methyl groups are terminal) and methyl groups branched to an aliphatic chain (except where they are branched in an α or β position to an alkyl chain from a ring).
18–23	CH_3	Methyl group branched to an aromatic or naphthenic ring. Case where two methyl groups are terminal. Methyl groups branched in an α or β position on an alkyl chain from a ring.
23–33	CH_2	Methylene groups of alkyl chains (except where they are branched in an α or β position from an aromatic ring or in an α position from a naphthenic rings).
33–36	CH	Methine groups in aliphatic chains.
36–43	CH_2	Methylene groups branched in an α or β position from an aromatic ring or in an α position from a naphthenic ring.
43+	CH	Methine groups of naphthenic rings.
	C_{ar}	Aromatic carbon atoms.
	CH_{ar}	Aromatic protonated carbon atoms.
	C_{qp}	Aromatic bridgehead quaternary carbon atoms.
	C_{qs}	Aromatic substituted quaternary carbon atoms.

Source: A. Adhvaryu, J. M. Perez, and J. L. Duda, "Quantitative NMR Spectroscopy for the Prediction of Base Oil Properties," *Tribology Transactions,* vol. 43(2), pp. 245–250 (2000). With permission.

TABLE 4.11
Equations Developed by Adhvaryu et al. for the Property Prediction of Base Stocks

Equations	Range	R^2	Error of Estimate	Repeatability of Standard Test Method
API = 35 − 3.19x·$C_{ar}^{0.5}$ − 1.31*CH_3 + 1.74x·CH_2 − 0.987x·CH_2	20–30	0.91	0.45	1.5
Aniline point, °C = 95.4 − 13.2x $C_{ar}^{0.5}$ − 13.2x·CH_3 + 1.48x·CH_2 + 17.0x·$CH^{0.5}$	90–120	0.91	2.2	0.16
Pour point, °C = 17.3x·$CH_2^{0.5}$ − 30.6x·$CH_2^{0.5}$ 5.51	0–(−24)	0.84	2.2	+3
Viscosity, cSt at 100°C = 1.17x·CH_2 + 2.61x·CH_2 − 21.6	4–35	0.91	2.7	0.35% of mean
Viscosity, cSt at 40°C = 24.9x·CH_2 + 71.9x·CH_3 − 548	25–550	0.88	4.2	0.35% of mean

Source: A. Adhvaryu, J. M. Perez, and J. L. Duda, "Quantitative NMR Spectroscopy for the Prediction of Base Oil Properties," *Tribology Transactions* 43:245–250 (2000). With permission.

TABLE 4.12
Correlation Coefficients for Base Stock Physical Properties

Physical Property	Correlation Coefficient
Viscosity at 40°C	0.953
Viscosity at 100°C	0.949
Viscosity index	0.996
Noack volatility	0.629
Noack volatility[a]	0.998
API gravity	0.997
Aniline point	0.997
Pour point	0.995

[a] Combined with simulation distillation data at 700°F.

Source: T. M. Shea and S. Gunsel, "Modeling Base Oil Properties Using NMR Spectroscopy and Neural Networks," *Tribology Transactions,* vol. 46(3), pp. 296–302 (2003). With permission.

increased presence of chain CH_2, and pour points depend largely on counter-imposed effects of chain CH_2 and branched CH_2).

Montanari et al.[24] (Eniricerche & Agip Petroli), from ^{13}C NMR on samples that included isomerized wax and PAO base oils, developed a good ($R^2 = 0.9951$) correlation between the fraction of carbons on "long" chains (i.e., those with length greater than three) and pour point. Shea and Gunsel[27] used a set of 30 base oils, of groups II, III, and IV, to apply neural network solutions to the ^{13}C NMR data and found excellent correlation coefficients for viscosities at 40°C and 100°C, VI, API gravity, aniline, and pour points (Table 4.12).

In the long term, it is conceivable that multiple analyses may be avoided when characterizing samples by simply running a ^{13}C NMR. Some additional examples of the application of ^{13}C NMR to isomerization of Fischer-Tropsch wax are given in Chapter 12.

REFERENCES

1. K. H. Altgelt and M. M. Boduszynski, *Composition and Analysis of Heavy Petroleum Fractions* (New York: Marcel Dekker, 1993).

2. K. van Nes and H. H. van Westen, *Aspects of the Constitution of Mineral Oils* (New York: Elsevier, 1951).

3. K. van Nes and H. H. van Westen, *Aspects of the Constitution of Mineral Oils* (New York: Elsevier, 1951), 429, Table 100.

4. W.-S. Moon, Y.-R. Cho, and J. S. Chun, "Application of High Quality (Group II, III) Base Oils to Specialty Lubricants," Paper presented at the 6th annual Fuels and Lubes Asia Conference, Singapore, January 28, 2000.

5. K. van Nes and H. H. van Westen, *Aspects of the Constitution of Mineral Oils* (New York: Elsevier, 1951), 327, Table 62.

6. J. B. Hill and H. B. Coats, "The Viscosity-Gravity Constant of Petroleum Lubricating Oils," *Industrial and Engineering Chemistry* 20:641–644 (1928).

7. ASTM D2501, "Standard Test Method for Calculation of Viscosity-Gravity Constant of Petroleum," *ASTM Annual Book of Standards*, vol. 05.01 (West Conshohocken, PA: American Society for Testing and Materials).

8. S. S. Kurtz, Jr., R. W. King, W. J. Stout, D. G. Parkikian, and E. A. Skrabek, "Relationship Between Carbon-Type Distribution, Viscosity-Gravity Constant, and Refractivity Intercept of Viscous Fractions of Petroleum," *Analytical Chemistry* 28:1928–1936 (1956).

9. "Properties of Hydrocarbons of High Molecular Weight," Research Project 42, 1940–1966, American Petroleum Institute, New York.

10. S. S. Kurtz, Jr., "Physical Properties and Hydrocarbon Structure," in B. T. Brooks, C. E. Boord, S. S. Kurtz and L. Schmerling, eds., *The Chemistry of Petroleum Hydrocarbons*, vol. 1 (New York: Reinhold Publishing, 1954).

11. S. S. Kurtz, Jr. and W. A. Ward, "The Refractivity Intercept and the Specific Refraction Equation of Newton. I. Development of the Refractivity Intercept and Comparison with Specific Refraction Equations," *Journal of the Franklin Institute* 222:563–592 (1936).

12. J. Smittenberg and D. Mulder, "Relations Between Refraction, Density and Structure of Series of Homologous Hydrocarbons. I. Empirical Formulae for Refraction and Density at 20°C of n-Alkanes and n-alpha-Alkenes," *Recueil des travaux chimiques* 67:813–825 (1948).

13. J. Smittenberg and D. Mulder, "Relations Between Refraction, Density and Structure of Series of Homologous Hydrocarbons. II. Refraction and Density at 20°C of n-Alkyl-cyclopentanes, -cyclohexanes and -benzenes," *Recueil des travaux chimiques* 67:826–838 (1948).

14. E. M. Dickinson, "Structural Comparison of Petroleum Fractions Using Proton and ^{13}C NMR Spectroscopy," *Fuel* 59:290–294 (1980).

15. C. H. Turner, G. Blunden, F. N. Dowling, and B. G. Carpenter, "Estimation of Chain Branching in Paraffin Waxes Using Proton Magnetic Resonance Spectroscopy and Gas-Liquid Chromatography," *Journal of Chromatography* 287:305–312 (1984).

16. O. Sperber, W. Kaminsky, and A. Geissler, "Structure Analysis of Paraffin Waxes by ^{13}C NMR Spectroscopy," *Petroleum Science and Technology* 23:47–54 (2005).

17. S. J. Miller, "Preparing a High Viscosity Index, Low Branch Index Dewaxed Base Stock," U.S. Patent 6,663,768.

18. D. T. Doddrell, D. T. Pegg, and M. R. Bendall, "Distortionless Enhancement of NMR Signals by Polarization Transfer," *Journal of Magnetic Resonance* 48:323–327 (1982).

19. S. L. Patt and J. N. Shoolery, "Attached Proton Test for Carbon-13 NMR," *Journal of Magnetic Resonance* 46:535–539 (1982).

20. L. P. Lindeman and J. Q. Adams, "Carbon-13 Nuclear Magnetic Resonance Spectrometry," *Analytical Chemistry* 43:1245–1252 (1971).

21. D. A. Netzel, D. R. McKay, R. A. Heppner, F. D. Guffey, S. D. Cooke, D. L. Varie, and D. E. Linn, "^{1}H- and ^{13}C-NMR Studies on Naphtha and Light Distillate Saturate Hydrocarbon Fractions Obtained from *in-situ* Shale Oil," *Fuel* 60:307–320 (1981).

22. A. S. Sarpal, G. S. Kapur, A. Chopra, S. K. Jain, S. P. Srivastava, and A. K. Bhatnagar, "Hydrocarbon Characterization of Hydrocracked Base Stocks by One- and Two-Dimensional NMR Spectroscopy," *Fuel* 75:483–490 (1996).

23. A. S. Sarpal, G. S. Kapur, S, Mukherjee, and S. K. Jain, "Characterization by ^{13}C NMR Spectroscopy of Base Oils Produced by Different Processes," *Fuel* 76:931–937 (1997).

24. L. Montanari, E. Montani, C. Corno, and S. Fattori, "NMR Molecular Characterization of Lubricating Base Oils: Correlation with Their Performance," *Applied Magnetic Resonance* 14:345–356 (1998).

25. A. Adhvaryu, J. M. Perez, and J. L. Duda, "Quantitative NMR Spectroscopy for the Prediction of Base Oil Properties," *Tribology Transactions* 43:245–250 (2000).

26. S. K. Sahoo, D. C. Pandey, and I. D. Singh, "Studies on the Optimal Hydrocarbon Structure in Next Generation Mineral Base Oils," *Proceedings of the International Symposium on Fuels and Lubricants*, New Delhi, pp. 273–278 (2000).

27. T. M. Shea and S. Gunsel, "Modeling Base Oil Properties Using NMR Spectroscopy and Neural Networks," *Tribology Transactions* 46:296–302 (2003).

5 Oxidation Resistance of Base Stocks

5.1 INTRODUCTION

Formulated lubricating oils are designed to meet many requirements and one of the most important ones is that the operating life of the oil be as long as possible. This is achieved if the lubricants maintain their "as new" properties (i.e., the physical and chemical properties of the lubricant change as slowly as possible during use). The principal factor affecting the life of a lubricant is oxidation of the base stock, which is the major component of the oil. Some additives in a finished lubricant are there specifically to protect the base stock from oxidation. Oxidation causes the formation of volatile oxidation products, corrosive acids, sludges, lacquer deposits, surface active compounds, and high molecular weight oxidation products. These can cause a decrease in flash point, an increase in viscosity, poorer oil-water separation, less corrosion protection, deposit formation, and increased wear.

This chapter discusses the relationship between base stock composition and some of the conclusions that have been drawn about base stock and product stabilities in use. This composition is determined by the refinery lube plant feedstock, the process, and the chemical changes that occur in the process. Much of the early published research was performed on the base stocks themselves. This was justified on the basis that in many early lubricants there were no antioxidant packages present, since that was a developing area of technology. In addition, as inhibitors are consumed, the resistance of the base stock to oxidation is an increasingly important contributor to the life of the lubricant and therefore hydrocarbon composition inevitably plays a major role. Important progress has also been made from studies on formulated products in which the response of base stocks to inhibitors is being measured. This type of measurement has frequently employed standardized test methods developed within the industry. In many of these cases, details about the inhibitor systems have not been published, but the value of the comparative results are not challenged. These results in turn have been employed to promote new processes as they have been developed.

Changes in base stock manufacturing technology over the past 30 years have been partially to satisfy the increasingly severe oxidative stress that modern equipment imposes on lubricants that prior products could not have satisfied. These changes have also been driven by a reduction in manufacturing costs. For example, production of hydrocracked lubes became widespread when the process was recognized as dramatically widening the range of crudes that could be used, reducing feed costs, and increasing yields, but at the same time delivering base stock of quality superior to that from solvent extraction. Dewaxing by hydroisomerization is replacing the labor- and energy-expensive solvent dewaxing process with a more economical one that also creates a viscosity index (VI) and improves the oxidative performance of the base stocks.

The overall compositional trend over the last 50 years in base stock composition because of these and earlier developments has been to reduce the levels of or eliminate polyaromatics and nitrogen- and sulfur-containing compounds, reduce the levels of monoaromatics and polycyclic naphthenes, and increase the levels of isoparaffins and monocyclonaphthenes. More recently, the combination of performance demands from the marketplace (the need for base stocks of lower volatilities and therefore higher VIs) and the introduction of new technologies has accelerated these compositional changes. The forthcoming widespread commercialization of gas to liquids (GTL) technology to make base stocks from natural gas will bring these compositional changes to their ultimate conclusion, with base stocks being entirely isoparaffinic in composition, with sulfur, nitrogen, and aromatics content all being essentially zero. These have all been in the direction of formulated products with better oxidation performance.

Studies on the relationship between composition and the resistance of base stocks to oxidation started many years ago when solvent extracted stocks were the only ones available. World War II made this an area of particular importance and intensified the effort. Many of the significant early studies date from this period.

Oxidation itself is a chemical process, accelerated by increased temperatures, in which oxygen reacts with base stock hydrocarbons to form chemically altered materials by dehydrogenation and conversion to oxygen-containing compounds. The physical and chemical properties of the oxidation products are very different from those of the base stock.

Oxidation of hydrocarbons has been known for many years to involve the formation of key intermediate hydroperoxides and dialkylperoxides ("peroxides" in general) from the reaction of oxygen and hydrocarbons via free radical intermediates. At low temperatures, the peroxides formed slowly accumulate and eventually decompose either thermally or by metal-induced reactions or by ionic routes. At high temperatures, formation and thermal decomposition of the peroxides occurs rapidly. Thermal decomposition leads to the production of additional free radicals (the "propagation" step of the reaction) and the formation of oxygen-containing products (e.g., acids, alcohols, ketones, polar compounds, and polymeric materials) that can ultimately bring about lubricant failure.

The widely recognized chemical steps[1-4] involved are

Initiation: production of free radicals,

$$RH \rightarrow R^{\bullet}$$

This initiation step may occur in a number of ways: by thermal decomposition of hydrocarbons with weak C–H bonds, by oxidation of hydrocarbons by metal ions via free radical routes, by reaction with oxygen gas, etc.

Propagation: formation of peroxides and new free radicals,

$$R^{\bullet} + O_2 \rightarrow RO_2^{\bullet}$$

$$RO_2^{\bullet} + RH \rightarrow RO_2H + R^{\bullet}$$

This step first incorporates molecular oxygen into the hydrocarbon as an unstable peroxide radical. This is the start of "oxidation." In the second step, the peroxy radical, which is short lived, attacks another hydrocarbon molecule. If this reaction is fast (i.e., the C–H bond being broken is weak), oxidation can occur quickly.

Decomposition of peroxides: thermal decomposition to generate additional free radicals,

$$RO_2H \rightarrow RO^{\bullet} + {}^{\bullet}OH$$

This step doubles the number of free radicals in the pool which can then attack other hydrocarbon molecules (the propagation step). Peroxide decomposition begins at temperatures greater than 100°C, accelerates at higher temperatures, and is central to severe oxidation. At lower temperatures, hydroperoxides may simply accumulate in the absence of compounds that can destroy them by non-radical means.

Inhibition to destroy peroxides or free radicals,

$$RO_2H + Inhibitor \rightarrow Stable\ products$$

$$RO^{\bullet} + Inhibitor \rightarrow Stable\ products.$$

These two steps remove the intermediates which propagate the chain reaction and stop the chain. Inhibition chemistry is the centerpiece of the application of antioxidant packages in lubricants.

Termination: destruction of reactive free radicals,

$$R^{\bullet} + R^{\bullet} \rightarrow R - R$$

$$RO_2^{\bullet} + R^{\bullet} \rightarrow Nonradical\ products.$$

These are recombination reactions that end these particular chains.

5.2 STUDIES ON SOLVENT REFINED BASE STOCKS

The key method for studying lubricant oxidation, developed about the time of World War II and still very much in use today, has been to measure the extent and rate of oxidation from oxygen uptake measurements at elevated temperatures. The equipment traditionally employed includes an oxidation flask for the sample whose temperature can be accurately controlled, together with an oxygen delivery system that maintains constant oxygen pressure and records the rate of oxygen consumption. Means were also developed to identify and measure some of the simple oxidation products, for example, water production (formed through dehydrogenation), organic acids, carbon dioxide, and carbon monoxide. Precise experimental details have varied as equipment technology has improved. Early work focused on the oxidation stability of the base stock sample without any additives, since at that time, additive technology was still in its infancy. In later studies, the base stock was supplemented with a basket of additives to simulate the inhibitors and accelerators encountered in use in the "real" world. Metal oxidation accelerators (representing trace metal ions from the system being lubricated) are frequently part of the standard package of additives. Differential scanning calorimetry (DSC),[5,6] a rapid microtechnique using digitally controlled equipment, is now in widespread use as an alternative to oxidation uptake measurements. In this method, since oxidation is an exothermic process, the exotherms are measured as a few milligrams of a sample are heated at controlled rates under oxygen pressure.

We are not yet at the stage (and likely we will never quite get there) when any of these or newer laboratory methods are regarded as the final acceptance criteria of a new lubricant. However, those mentioned above and many others have become essential as screening tests to identify the most probable candidates for final acceptance by industry-recognized methods (e.g., engine tests in the case of automotive or diesel engine oils). Ultimately, commercial experience proves success or failure of all laboratory tests.

M. R. Fenske[7] (Pennsylvania State University) was one of the early pioneers in this area, and in 1941 he and coworkers published their results using the oxygen consumption[8] technique on fractions separated from a semirefined Pennsylvanian lubricant stock (38 cSt at 100°F, 101 VI [ASTM D2270]), and it is worth providing details of this seminal work. They started with 124 fractions from a combination of vacuum fractionation (into cuts of narrow distillation ranges) followed by solvent extractions.[9] The solvent extractions produced refined lubricant fractions (raffinates) and extracts. Distillations effected separations approximately by molecular weight and the extractions by aromaticity, which generally translates into separation by VI. The fractions produced had VIs between 120 and −270 and molecular weights estimated to be between 250 and 680. Those raffinates selected for oxidation studies[7] fell into two groups, one with relatively high VIs (111 or greater; samples Q, T, and U in Table 5.1), and therefore highly saturated, with less than 2% aromatics, while those in the second group had low VIs (5 to 83; samples B, F, G, and S) and higher aromatic levels (15% to 40%). As might be expected, the high VI fractions were largely paraffinic, with carbons estimated

TABLE 5.1
Inspection Results on a Pennsylvania Distillate Lubricant Stock and Selected Raffinate Fractions

	Whole Oil O	B	F	G	Q	S	T	U
Viscosity, cSt								
At 100°F	37.86	130.0	101.8	744.0	18.53	124.0	21.39	38.95
At 210°F	5.76	8.70	9.05	24.5	3.86	11.38	4.35	6.38
VI (D2270)	101	−5	58	12	111	83	123	125
Density, 20°C/4°C	0.8710	0.9451	0.9052	0.9435	0.8486	0.8888	0.8432	0.8476
VG constant	0.820	0.901	0.949	0.877	0.803	0.823	0.795	0.789
RI, n_D^{20}	1.4841	1.5353	1.5043	1.526	1.4691	1.4920	1.4670	1.4693
Sp. Refr. at 20°C	0.3286	0.3296	0.3273	0.3254	0.3290	0.3264	0.3291	0.3287
Aniline point, °C	99.6	47.3	78.2	62	103.2	97.3	108.8	116.7
Percent aromatic rings	9	40	25	37	2	15	0	0
Percent naphthenic rings	16	7	15	20	18	20	18	15
Percent paraffin chains	73	53	60	43	80	65	82	85

Source: M. R. Fenske, C. E. Stevenson, N. D. Lawson, G. Herbolsheimer, and E. F. Koch, "Oxidation of Lubricating Oils—Factors Controlling Oxidation Stability," *Industrial and Engineering Chemistry* 33:516–524 (1941). With permission.

to be more than 80% paraffinic chains by the Waterman method, while the low VI ones contained approximately 40% to 50% paraffinic chains.

Oxidation studies were conducted between 140°C and 180°C for a 50 hour period with pure oxygen being circulated at 1 atm pressure with a flow rate of 15 L/hr through a 250 g sample. Volatile oxidation products were removed as formed by the gas flow and trapped downstream. At the end of the experiment the base stock was analyzed for acid content, amount of precipitates, and changes in viscosity. Oxygen consumption was measured throughout the 50 hour time period.

The two groups showed different behavior when oxygen uptake was plotted versus time (Figure 5.1 illustrates these two behaviors). The highly paraffinic samples exhibited an initial induction period during which there was little oxygen uptake and whose length depended on the sample and temperature, the induction periods being halved by a 10°C increase in temperature. This induction period was followed by rapid reaction (i.e., oxygen uptake)—about the same slope for all paraffinic samples studied—with the rate curve eventually decreasing with time. This type of oxidation, featuring an induction period then rapid oxidation and a later slowing oxidation rate, was termed autocatalytic and is characteristic of paraffinic lubes. Subsequent research has shown that the induction period corresponds to a buildup of peroxides, whose rapid free radical decomposition

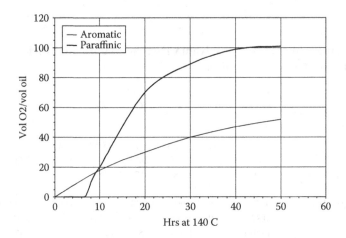

FIGURE 5.1 Oxygen consumption curve types for base stocks at 140°C.
Source: M. R. Fenske, C. E. Stevenson, N. D. Lawson, G. Herbolsheimer, and E. F. Koch, "Oxidation of Lubricating Oils—Factors Controlling Oxidation Stability," *Industrial and Engineering Chemistry* 33:516–524 (1941). With permission.

eventually ends the induction period and causes an initial high rate of oxygen uptake. The formation of antioxidants as by-products eventually slows this rate.

More aromatic fractions under the same conditions demonstrated immediate oxidation (no induction pause) and relatively slow oxygen uptake rates that were approximately constant throughout the oxidation period, unlike the paraffinic samples. These, too, eventually slowed down with increased oxygen consumption. This type of oxidation in which the rate slows as the reaction progresses is termed autoretardant. In these cases, oxidation begins immediately due to the presence of easily oxidizable materials (e.g., aromatics with alkyl side chains) and eventually slows again due to the presence or generation of inhibitors/antioxidants.

As oxidation temperatures increased, Fenske et al.[7] found that the type of oxidation changed; for example, using one of the more aromatic fractions, "carbon oxidation" with the formation of CO_2, CO, volatile acids, and insolubles became more significant than that of "hydrogen oxidation" to form water (see Table 5.2). Accompanying this was an increase in insolubles as well as lacquer formation at higher temperatures. The addition of antioxidants severely reduced oxygen uptake, but with the particular ones chosen, insolubles and lacquer formation increased (Table 5.3).[7]

Clay treating was widely used at that time to improve base stock quality by removing polar compounds or those polyaromatics that would adsorb on clay sites; it has largely been replaced now by hydrofinishing. The benefits of removing easily oxidizable polar molecules[7] can be seen somewhat in Table 5.4.

Clay treating has a marked effect on the oxidation stability of oil #7 (Table 5.4), described as a moderately aromatic residual-type base stock of intermediate VI.

TABLE 5.2
Effect of Temperature on Oil Oxidation (Oil = Whole Oil O from Table 1)

Oxidation Temperature, °C	150	160	170	180
Average partial pressure of O_2, mm	700	698	685	698
O_2 absorbed, mmol	34	58	150	364
Distribution of absorbed O_2, %				
To H_2O	65.5	59.5	53.2	48.5
To CO_2	3.8	5.7	5.6	9.1
To CO	0.5	2.8	2.2	2.8
To volatile acids	1.0	3.3	3.5	3.9
To fixed acids	2.1	2.2	2.0	2.1
To isopentane insolubles	3.1	6.7	5.8	7.1
Neutralization number of oxidized oils	0.2	0.5	0.7	2.6
Milliequivalents volatile acids	0.3	2.2	5.3	14.3
Precipitatable oxidation products	0.09	0.38	0.74	2.22
A. Total isopentane insolubles, wt %	0.09	0.26	0.21	0.88
B. Oil-sol, isopentane insolubles, wt %	0.00	0.12	0.53	1.54
C. Milligrams lacquer on 3 in. × 1 in. slide	0.9	1.5	3.2	9.4

Source: M. R. Fenske, C. E. Stevenson, N. D. Lawson, G. Herbolsheimer, and E. F. Koch, "Oxidation of Lubricating Oils—Factors Controlling Oxidation Stability," *Industrial and Engineering Chemistry* 33:516–524 (1941). With permission.

The oxygen consumption greatly decreased after clay treating as acids and insolubles are formed. This is in spite of the fact that the treated material was oxidized at a 10°C higher temperature. This temperature increase normally doubled the rate of these reactions. In addition, the rate of oxygen uptake was greatly decreased after clay treating.

In contrast, in the case of oil #3 (Table 5.4), which is a high VI paraffinic type with low aromatics content, clay treating actually increased the susceptibility of the base stock to oxidation, reducing the induction period from more than 5 hours to less than 1 hour. We saw previously that paraffinic-type base stocks frequently exhibit induction periods. The authors speculated that in this case the clay removed naturally occurring inhibitors, a conclusion we would still agree with.

Larsen et al.[10] (Shell Development Company) used the same technique to study the oxidation of pure compounds. All the saturated hydrocarbons they tested—n- and branched paraffins and cycloparaffins—were found to react rapidly after a preliminary induction period and in an autocatalytic manner. Figure 5.2 shows their oxygen uptake and relative rates—these are not adjusted for the number of hydrogens. It can be seen that perhydroanthracene, a tricyclic condensed naphthene, oxidizes most quickly of this group, and more recent studies

TABLE 5.3
Effect of Antioxidants on Oxygen Uptake and Products at 170°C

Antioxidant:	None	M	N	P
Wt %		3.0	0.2	0.2
O_2 partial pressure, mm	650	666	700	699
Millimoles O_2 absorbed	993	236	268	163
Distribution of absorbed O_2, %				
To H_2O	44.3	61.2	54.0	58.9
To CO_2	11.2	9.2	7.9	7.0
To CO	3.2	1.9	2.4	2.5
To volatile acids	7.5	2.5	4.5	4.5
To fixed acids	2.5	1.7	1.9	2.1
Neutralization number of oil	7.6	1.6	1.6	0.8
Milliequivalents volatile acid	74.6	5.7	12.0	7.4
Precipitatable oxidation products				
A. Total isopentane insolubles, wt %	1.89	1.24	0.26	0.47
B. Oil-sol, isopentane insolubles, wt %	1.89	0.02	0.02	0.00
C. Milligrams lacquer on 3 in. × 1 in. slide	0.1	6.3	1.9	8.8
Percent increase in viscosity at 100°F	133	8.2	19.3	10.2

Source: M. R. Fenske, C. E. Stevenson, N. D. Lawson, G. Herbolsheimer, and E. F. Koch, "Oxidation of Lubricating Oils—Factors Controlling Oxidation Stability," *Industrial and Engineering Chemistry* 33:516–524 (1941). With permission.

have confirmed that polycyclic naphthenes are not good components to have for excellence in base stock oxidation stability. Larsen et al. also found that branched paraffins (e.g., polyisobutylene) reacted faster than n-paraffins such as decane (due to the greater reactivity of tertiary hydrogens) and that increased molecular weight increased reaction rates. Over the course of their experiments with saturated hydrocarbons, there was no development of retardation (i.e., the oxidation rates did not slow down with reaction time), unlike the "saturate" fractions obtained by Fenske. Table 5.5 shows the results of later work by Walling and Thaler on the relative reactivity of t-butoxy radical (similar to those expected from decomposition of peroxides or hydroperoxides in base stocks) with different types of carbon–hydrogen bonds.[11] It can be seen that tertiary hydrogens and the hydrogens on a naphthene ring (cyclohexane) are much more reactive to alkoxy radicals than the secondary or primary hydrogens found in n-paraffins.

Condensed monoaromatic naphthenes (i.e., monoaromatics such as decalin) with fused cycloparaffinic rings showed no induction period and reacted rapidly from the start, with inhibition developing in the later stages of the oxidation, as is typical of the autoretardant mechanism. In contrast, alkylated benzenes and

TABLE 5.4
Effect of Clay Treatment

Oil	3	3	7	7
Clay treatment	No	Yes	No	Yes
Oxidation temperature, °C	150	150	140	150
Run length, hr	20	20	50	20
O_2 partial pressure, mm	662	679	622	693
Millimoles O_2 absorbed	855	1052	729	105
Distribution of absorbed O_2, %				
To H_2O	39.3	40.4	56.0	68.0
To CO_2	9.8	10.7	8.4	4.1
To CO	2.6	2.6	2.7	2.2
To volatile acids	6.4	6.6	3.4	1.4
To fixed acids	4.0	5.0	0.5	3.6
Neutralization number of oil	11.7	12.7	8.9	0.9
Milliequivalents volatile acid	55	70	24.5	1.5
Precipitatable oxidation products				
A. Total isopentane insolubles, wt %	2.52	4.33	11.3	0.08
B. Oil-sol, isopentane insolubles, wt %	2.35	4.25	10.1	0.06
C. Milligrams lacquer on 3 in. × 1 in. slide	0.7	1.4	0.0	0.0
Percent increase in viscosity at 100°F	108	158	508	22.1

Source: M. R. Fenske, C. E. Stevenson, N. D. Lawson, G. Herbolsheimer, and E. F. Koch, "Oxidation of Lubricating Oils—Factors Controlling Oxidation Stability," *Industrial and Engineering Chemistry* 33:516–524 (1941). With permission.

naphthalenes oxidized relatively slowly with essentially no induction periods. Some examples of these different behaviors are shown in Figure 5.3, where tetralin and two anthracenes—one an octahydro case, the other with the central ring "saturated"—react very quickly with no development of inhibition, whereas the alkyl benzenes and naphthalenes (not shown) react very slowly.

The rapid initial reaction of the naphthenoaromatics is attributed to the ease of removal of a hydrogen on the carbons alpha to the aromatic ring to form a relatively stable (resonance-stabilized) alkyl radical in the propagation stage of the oxidation reaction (this will subsequently react with oxygen to produce the first oxidized peroxy intermediate radical). The high rate of attack on tetralin is significant, as this compound is the prototype for all condensed naphthenoaromatics. The slow rates of the alkylbenzenes must mean that their free radicals are less stable. Later studies by Williams et al.[12] measured the rates of attack by t-butoxy radicals on pure compounds and confirmed the high reactivity of the tetralin hydrogens on a per hydrogen basis (Table 5.6).

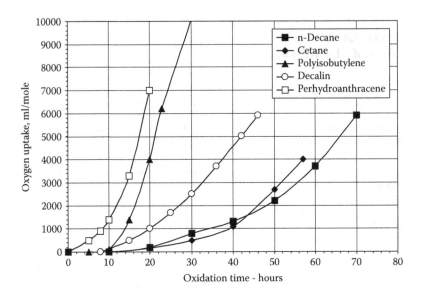

FIGURE 5.2 Oxygen uptake curves at 110°C for saturated hydrocarbons.
Source: R. G. Larsen, R. E. Thorpe, and F. A. Armfield, "Oxidation Characteristics of Pure Hydrocarbons," *Industrial and Engineering Chemistry* 34(2):183–193 (1942). With permission.

In contrast to the examples cited above, alkylated naphthalenes oxidized considerably slower (approximately 1/30 the rate) compared to other hydrocarbons and this effect is believed to be due to formation of phenolic antioxidants. Unsubstituted positions on aromatic rings are not liable to hydrogen abstraction by free radicals due to high C–H bond strengths. Such positions in naphthalene

TABLE 5.5
Relative Reactivity per Hydrogen of Aliphatic Hydrogens Toward t-Butoxy Radicals

$$X^{\bullet} + RH \rightarrow XH + R^{\bullet}$$

Type of Hydrogen	Relative Rates at 40°C
Primary	1
Secondary	8
Tertiary	44
Cyclohexane	15

Source: C. Walling and W. Thaler, "Positive Halogen Compounds. III. Allylic Chlorination with t-Butyl Hypochlorite. The Stereochemistry of Allylic Radicals," *Journal of the American Chemical Society* 83:3877–3884 (1961). With permission.

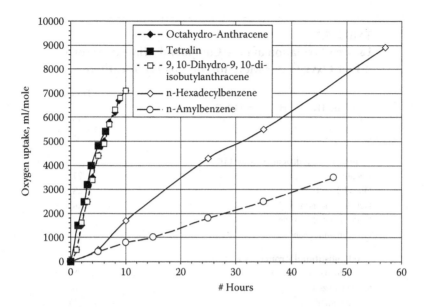

FIGURE 5.3 Oxygen uptake curves at 110°C for alkyl benzenes and naphthenoaromatics. *Source:* Source: R. G. Larsen, R. E. Thorpe, and F. A. Armfield, "Oxidation Characteristics of Pure Hydrocarbons," *Industrial and Engineering Chemistry* 34(2):183–193 (1942). With permission.

TABLE 5.6
Relative Rates of Hydrogen Abstraction by t-Butoxy Radicals from t-Butyl Peroxide

RH + t-BuO$^\bullet$ → R$^\bullet$+ t-BuOH

RH	Relative Rate per Hydrogen
Toluene	1.0
Ethylbenzene	3.2
Iso-propylbenzene	5.1, 6.4
m-Xylene	1.2
p-Xylene	1.5
t-Butylbenzene	0.1
Cyclohexane	2.0
Tetralin	7.6
Diphenylmethane	4.2

Source: W. A. Pryor, *Free Radicals* (New York: McGraw Hill, 1966), Table 12.7. A. L. Williams, E. A. Oberright and J. W. Books, "The Abstraction of Hydrogen atoms from Liquid Hydrocarbons by t-Butoxy Radicals;" *Journal of the American Chemical Society*, 78: 1190–1193 (1956). With permission.

TABLE 5.7
Larsen's Oxidation of Pure Compounds: Time to Absorb 2 L of Oxygen per Mole for Various Hydrocarbons

	Hours
Paraffinic Hydrocarbons at 110°C	
n-Decane	47
Cetane	45
Naphthenic Hydrocarbons at 110°C	
Decalin	27
Dicyclohexyl	28
Octadecyldecalin	24
Octadecylcyclohexane	37
Perhydroanthracene	12
Alkyl Substituted Aromatics at 110°C	
n-Amylbenzene	28
t-Amylbenzene	80
Diphenylmethane	>70
Hexaethylbenzene	23
Hexadecylbenzene	12
Polycyclic Aromatics at 150°C	
Naphthalene	>150
α-Methylnaphthalene	62
β-Methylnaphthalene	>150
Iso-amylnaphthalene	55
Fluorene	26
Di-iso-butyl anthracene	90
Phenanthrene	>50
Condensed Cycloparaffin Monoaromatics at 110°C	
Tetralin	2
Octadecyltetralin	4
Octahydroanthracene	2
5-iso-Butylacenaphthene	8

Source: R. G. Larsen, R. E. Thorpe, and F. A. Armfield, "Oxidation Characteristics of Pure Hydrocarbons," *Industrial and Engineering Chemistry* 34(2):183–193 (1942). With permission.

and higher systems are subject to addition, with the formation of more stable free radicals which can act as inhibitors or chain termination agents.

Table 5.7 summarizes part of Larsen et al.'s results in terms of time (hours) to absorb 2 L of oxygen per mole and organizes them by chemical type, where the distinctions between them become much clearer.[13,14] The stability at 110°C of the paraffinic hydrocarbons versus both naphthenic and monocyclic ones is immediately apparent. The oxidation rates of the polycyclic aromatics had to be measured at 150°C to get numbers of the same magnitude as the others. Standouts for their instability were tetralin and related compounds, which have one or more fused cycloparaffin rings attached to a benzene ring—these showed exceptional reactivity to oxidation and are therefore vulnerable components in base stocks.

5.3 IMPACT OF AROMATICS AND SULFUR LEVELS

Von Fuchs and Diamond (Shell Oil) pursued these results further by examining the effects of increasing the content of the aromatics on base stock oxidation rates.[15] They undertook this study because of the increasing realization that while solvent extraction technology improved lubricant performance, overextraction of the base stock could make it less resistant to oxidation because of removal of the "autoretardant" components. Since extraction removed aromatics and nitrogen and sulfur compounds, a relationship between these levels and oxidation stability seemed likely.

For this work they chose several types of base stocks—some were raffinates from solvent extraction and therefore contained different percentages of aromatics and some were blends of white oils (of near-zero aromatic content) with added quantities of aromatic fractions separated from bright stocks. Oxidations were performed at a number of temperatures with and without added metals (metallic iron and copper to simulate wear metals) and were followed by the oxygen uptake method. Their overall conclusion was that aromatics did indeed inhibit oxidation, whether metals were present or not, and that the effect with their samples exhibited a maximum at about 5% aromatics content. Figure 5.4 illustrates this phenomenon, with measurements of the time for uptake of 1800 ml of oxygen per 100 ml of sample. It can be seen that the reaction rate at zero and 8% aromatics is twice as fast as at 5% (times vary by a factor of two). It was clear that aromatics contained these "naturally occurring" inhibitors, but this work was unable to shine any light on just what these inhibitors were.

When synthetic oxidation inhibitors such as phenyl-alpha-naphthylamine were used, the lengths of the induction periods were proportional to the inhibitor concentration. The effect of inhibitors (i.e., the length of the induction period) was greatest for those samples which had been most severely extracted by the furfural, namely, those with the fewest aromatics, a good augury for this era of low aromatic base stocks.

Larsen et al.[10] concluded that the stability of lubricating oils was due to the presence of natural inhibitors and that these were not hydrocarbons. Denison[16] (Standard Oil of California) addressed this issue and deduced that it was the sulfur-containing components of the aromatics fraction that inhibited oxidation. The presence of sulfur compounds in solvent refined base stocks was therefore deemed critical for their performance. In his studies he found that desulfurized lubricating oils behaved like white oils under oxidation conditions (i.e., they were very unstable to oxygen and they oxidized very rapidly in an autocatalytic manner), whereas for the original undesulfurized oils, oxidation rates were slow

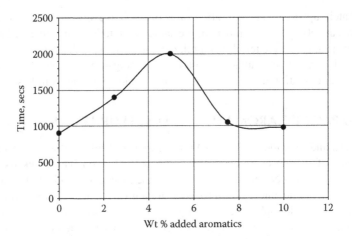

FIGURE 5.4 Optimum aromaticity: time to consume a fixed quantity of oxygen versus weight percent added aromatics.

Source: G. H. von Fuchs and H. Diamond, "Oxidation Characteristics of Lubricating Oils, Relation Between Stability and Chemical Composition," *Industrial and Engineering Chemistry* 34(8):927–937 (1942). With permission.

(Figure 5.5). To obtain these desulfurized base stocks, he carried out reductions by sodium metal and hydrogen at 200 psi partial pressure. Analyses by the methods available at that time indicated that the levels of aromatic, naphthenic, and paraffinic content were unaffected, while sulfur levels were reduced by more than 75% (Table 5.8).

Denison also measured the kinetics of peroxide formation and found that their formation was proportional to oxygen consumption (Equation 5.1), signaling these as crucial preliminary products in the oxidation mechanism:

$$-d(O_2)/dT = k(\text{peroxide}).   \quad\quad (5.1)$$

Even more interesting is that addition of a lubricating oil (percent sulfur = 0.53) to an oxidized white oil caused rapid disappearance of the peroxides, and the kinetics were proportional both to the peroxide and lubricating oil concentrations (Figure 5.6). He concluded that the sulfur compounds were the "natural inhibitors" being discussed at the time and that they acted by destroying the peroxide oxidation intermediates. Therefore progress of oxidation of solvent refined stocks is in part a competition between peroxides formation and their destruction by sulfur compounds—if the latter were in high concentration, peroxides levels remained very low for the most part and the base stock was seen as resisting oxidation.

S. Korcek (Ford Motor Company) found that the hydroperoxides/peroxides formed during the oxidation of a simple n-paraffin of molecular weight close to that of a lube base stock, namely, hexadecane, were extremely complex and

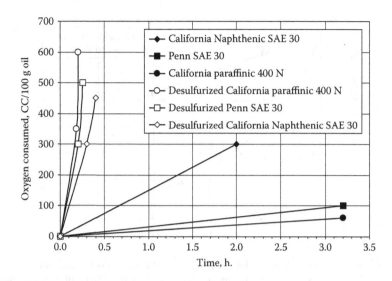

FIGURE 5.5 Oxygen uptake by lubricating oil base stocks and their desulfurized analogs. *Source:* G. H. Denison, Jr., "Oxidation of Lubricating Oils, Effect of Natural Sulfur Compounds and of Peroxides," *Industrial and Engineering Chemistry* 36(5):477–482 (1945). With permission.

TABLE 5.8
Effect of Sodium Treatment on the Composition of a Lubricating Oil

Oil	Percent Sulfur		Percent Aromatic Rings		Percent Naphthene Rings		Percent Paraffin Side Chains	
	Original	Final	Original	Final	Original	Final	Original	Final
California naphthenic SAE 30	0.53	0.07	15	14	33	33	52	53
California paraffinic 400N	0.22	0.08	1	1	27	27	72	72
Gulf Coast SAE 30	0.20	0.04	10	7	32	35	58	58
Pennsylvania SAE 30	0.10	0.01	7	4	16	20	77	75

Source: G. H. Denison, Jr., "Oxidation of Lubricating Oils, Effect of Natural Sulfur Compounds and of Peroxides," *Industrial and Engineering Chemistry* 36(5):477–482 (1945). With permission.

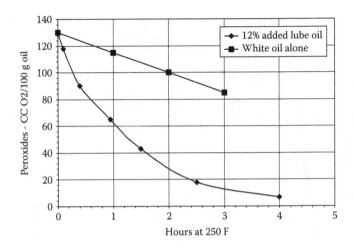

FIGURE 5.6 Decomposition of white oil peroxides with and without added lubricating oil. *Source:* G. H. Denison, Jr., "Oxidation of Lubricating Oils, Effect of Natural Sulfur Compounds and of Peroxides," *Industrial and Engineering Chemistry* 36(5): 477–482 (1945). With permission.

included isomeric monohydroperoxides, dihydroperoxides, and cyclic peroxides.[17] Not surprisingly, the decomposition products will be even more complex, and in the case of actual solvent refined base stocks, virtually impossible to identify on an individual basis.

This concept of "optimum aromaticity" and the role of sulfur compounds as inhibitors were further established by a study by Burn and Greig (British Petroleum) of the oxidation of solvent extracted base stocks.[18] They chose samples from a North African (Sahara) and three Middle East (Iran, Abu Dhabi, and Kuwait) crudes. The aromatic + heterocyclic $(A + H)$ and paraffin + naphthene $(P + N)$ components were separated by alumina chromatography from each base stock (Table 5.9 includes their composition and sulfur contents) and recombined in several ratios and the resistance of the blends to oxidation measured by the oxygen uptake method.

Oxidation stability of the original base stocks and the new ones created were assessed by determining the times (t_5) for oxygen uptake of five times the liquid volume. High values for t_5 obviously correspond to good resistance to oxidation. The results shown in Figure 5.7 demonstrate that not only are stabilities (t_5) different, as expected, but their t_5 maxima are attained at different values for $(A + H)$, ranging from 10% to 20% aromatics, and that all oils investigated exhibit optimum aromaticity. The main feature of these blends is the very steep increase in stability that takes place when initial quantities of the aromatics are blended back in. The exception is the case of the low sulfur Sahara base stock, where very low stabilities persist until aromatic levels are greater than 10%. Maximum stabilities depend on the crude source. The order of maximum stabilities at the peaks is Iran > Sahara > Kuwait > Abu Dhabi.

TABLE 5.9
Lubricating Oil Compositional Data

Source	(A + H) wt %	Wt % Sulfur in Oil	Wt % Sulfur in (A + H)	Wt % Nonthiophenic Sulfur in (A + H)
Sahara	11.0	0.05	0.46	0.18
Iran	14.6	0.25	1.68	1.05
Abu Dhabi	27.2	0.67	2.47	0.94
Kuwait	29.4	1.0	3.43	1.72

Source: A. J. Burn and G. Greig, "Optimum Aromaticity in Lubricating Oil Oxidation," *Journal of the Institute of Petroleum* 58(564):346–350 (1972). With permission.

In addition, the magnitude of the responses to (A + H) addition varies, being greatest for the high sulfur oil (Kuwait), followed in order by Abu Dhabi, Iran, and Sahara. This order corresponds to their sulfur content (Table 5.9). When the threshold on-set concentrations for the effects of the (A + H) fractions are measured and converted to on-set sulfur levels, these turn out to be virtually identical (Table 5.10). Thus chemically bound sulfur, either in alkyl sulfide structures or thiophene types or both, must be key "naturally occurring" inhibitors.

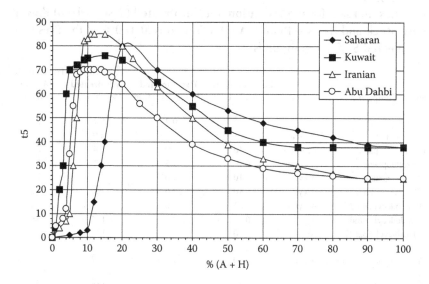

FIGURE 5.7 Oxygen uptake times (t_s) versus percent (A + H) content for base stocks of North African and Middle East origin.
Source: A. J. Burn and G. Greig, "Optimum Aromaticity in Lubricating Oil Oxidation," *Journal of the Institute of Petroleum* 58(564):346–350 (1972). With permission.

TABLE 5.10
Minimum Sulfur Content for Response

Oil	Threshold (A + H) Content, wt %	Threshold Sulfur Content, wt %
Sahara	12	0.06
Iran	4	0.07
Abu Dhabi	2.5	0.06
Kuwait	1.5	0.05

Source: A. J. Burn and G. Greig, "Optimum Aromaticity in Lubricating Oil Oxidation," *Journal of the Institute of Petroleum* 58(564):346–350 (1972). With permission.

These authors also constructed blends involving the $(N + P)$ and $(A + H)$ fractions from a hydrogenated Kuwait sample from which the sulfur had been removed (remaining sulfur was 23 ppm) and model aliphatic and aromatic sulfur compounds. Measurement of t_5 (Table 5.11) showed that aliphatic sulfides generally had superior oxidation inhibitory effects on all fractions but

TABLE 5.11
t_5 Values (in ksec) for the Oxidation of Hydrotreated (desulfurized) Kuwait Base Stock and Its (N + P) and (A + H) Fractions at 160°C in the Presence of Selected Sulfur Compounds

Sulfur Compound	Type	Concentration, Mole/L	Hydrotreated (N + P) Fraction	Hydrotreated Whole Oil	Hydrotreated (A + H) Fraction
None			0.7	5.1	22
Dibenzothiophene	Aromatic	0.01	—	4.9	—
Dibenzothiophene	Aromatic	0.10	0.9	5.4	25
Benzothiophene	Aromatic	0.10	1.3	67	38
Thiaadamantane	Alkyl	0.10	1.0	45	27
Dibenzyl sulfide	Alkyl	0.10	27.0	68	38
Di-n-dodecyl sulfide	Alkyl	0.01	1.0	115	36
Di-n-dodecyl sulfide	Alkyl	0.10	103	113	38
S-Thiabicyclo[3,2,1] octane	Alkyl	0.10	1.5	129	49
Trans-8-thiabicyclo[4.3.0] nonane	Alkyl	0.10	17	147	68
Trans-7-thiabicyclo [4.3.0]nonane	Alkyl	0.10	—	147	—

Source: A. J. Burn and G. Greig, "Optimum Aromaticity in Lubricating Oil Oxidation," *Journal of the Institute of Petroleum* 58(564):346–350 (1972). With permission.

were most noticeable in the $(N + P)$ fraction. What was surprising was that the benzothiophenes did have some inhibiting effect on the hydrotreated base stock itself and its aromatic fraction. Any inhibition by sulfur will inevitably require availability of the sulfur free electron pair, and that pair in thiophene derivatives will undoubtedly be partly involved in aromaticity of the five-membered ring.

Overall the results show that the $(A + H)$ fraction was more stable to oxidation than the $(N + P)$ fraction, and addition of sulfur-based inhibitors, whether aromatic or alkyl, brought about only modest improvement in oxidation by the method used here. For the whole oil, alkyl sulfide inhibitors were clearly superior to aromatic ones and the effect of the alkyl ones on the $(N + P)$ fraction was clearly dependent on the structure of the sulfide.

When the $(A + H)$ fraction from the desulfurized oil was recombined with the $(N + P)$ fraction, t_5 slowly increased, but when di-n-dodecyl sulfide was added to the $(A + H)$ fraction before recombining, the typical curve illustrating "optimum aromaticity" developed (Figure 5.8), thus both sulfur and aromatic compounds were necessary for this feature to develop. Note that in these cases, the ratio of sulfur to aromatic compounds remained constant. On the other hand, when di-n-dodecyl sulfide was added separately over a range of concentrations to the desulfurized base stock, its $(N + P)$ fraction, and its $(A + H)$ fraction, it had markedly different effects on the components (Figure 5.9). In the case of the $(A + H)$ fraction, the sulfide did improve stability to about 40 ksec under those conditions. For the saturate $(P + N)$ fraction and the mostly saturated (85% saturates) desulfurized base stock, the sulfide was very effective in inhibiting oxidations with lifetimes in the range 110 to 155 ksec, which was better than for the original base stocks themselves.

Denison and Condit[19] previously postulated how sulfur compounds inhibit base stock oxidation. In this mechanism the hydroperoxide/peroxides intermediates reacted with sulfur compounds to give products that could not propagate further by the free radical mechanism. In their oxidation preventative role, the peroxides oxidized the sulfur compounds first to sulfoxides and these in turn could be further oxidized to sulfones:

$$R–S–R + R–O–O–R \rightarrow R–SO–R$$

$$R–SO–R + R–O–O–R \rightarrow R–SO_2–R.$$

Alkyl sulfides turned out to be good inhibitors because they react readily with peroxides to form sulfoxides and sulfones, whereas thiophenes and diarylsulfides, which form sulfoxides and sulfones more slowly (the sulfur electron pair is more involved in resonance stabilization), were not such good inhibitors. Experiments with the sulfides themselves showed them to react slowly with oxygen, therefore they were not acting in a sacrificial manner.

When the rates of disappearance of a dialkyl sulfide and the corresponding sulfoxide were measured during their reaction with peroxides in white oils, their

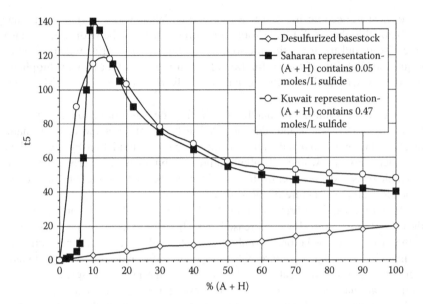

FIGURE 5.8 Effect on oxidation stability of di-n-octyl sulfide added to the desulfurized (A + H) fraction.
Source: A. J. Burn and G. Greig, "Optimum Aromaticity in Lubricating Oil Oxidation," *Journal of the Institute of Petroleum* 58(564):346–350 (1972). With permission.

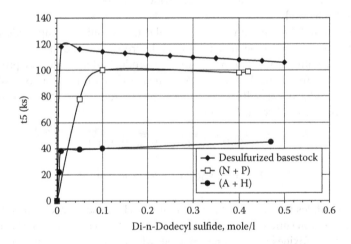

FIGURE 5.9 Effect of di-n-dodecyl sulfide on the oxidation stability of a desulfurized base stock and its (A + H) and (N + P) fractions.
Source: A. J. Burn and G. Greig, "Optimum Aromaticity in Lubricating Oil Oxidation," *Journal of the Institute of Petroleum* 58(564):346–350 (1972). With permission.

experiments showed that sulfide initially underwent oxidation very rapidly with consumption of the peroxide to produce the corresponding sulfoxide. The latter then reacted at a slower rate with more peroxide to form sulfone. These findings clarified the inhibitory action of natural sulfur compounds by their destruction of peroxides as they are formed to give initially sulfoxide and then sulfone. By this mechanism, diaryl sulfides and thiophene derivatives would be expected to oxidize much less easily and therefore be poorer inhibitors.

Barnard et al.[20] presented a somewhat different picture of the involvement of sulfur compounds in oxidation inhibition involving olefinic hydrocarbons. They studied the oxidation of squalene (an olefin) in the presence of sulfur compounds and concluded by careful measurement of oxygen uptake that it was not the sulfide that inhibited oxidation but the initially formed sulfoxide, and that inhibition was very dependent on the chemical structure of the sulfide. However, they did not suggest any specific mechanism for the inhibition.

In their second paper, Denison and Condit tracked the destruction of peroxides by dialkyl sulfides and sulfoxides and found that the former caused the rapid disappearance of peroxides until the sulfide had been converted to sulfoxide, and thereafter the slower rate of disappearance was identical to that caused by the sulfoxides.[19] These authors also found that oxidation of the sulfoxide produced strong acids, which they conjectured were likely sulfonic acids. These acids can cause ionic decomposition of peroxides and this suggests a second route for interrupting the oxidation chain. In contrast, the dialkyl sulfones oxidized by a different route to produce weak—probably carboxylic—acids.

Similar results were obtained by Berry et al.[21] (Texaco), who also measured oxygen uptake for blends of aromatic and saturate base stock components from low and high viscosity solvent refined oils. They found that minimum oxygen consumption occurred at approximately 18% aromatics with the samples they were using. Using model aromatic compounds added to white oils, alkyl substituted benzenes did not reduce oxidation rates, whereas alkyl naphthalenes and anthracenes definitely did so (as expected due to easier formation of phenol inhibitors). Some sulfides and naphthalene-type molecules showed substantial synergistic effects in inhibiting oxidation. Figure 5.10 compares oxygen consumptions at 140°C of three white oil blends with (a) a low concentration (0.1%) of mercaptan, (b) a high concentration (15%) of β-methyl-naphthalene, and (c) both antioxidants, but at reduced (0.05% and 10%, respectively) levels. It can be seen that the presence of both a sulfur compound and an aromatic brings about a very substantial improvement in stability, and the synergism of the two compound types is very real.

Lenoardi[22] (Socony Mobil) worked with saturate and aromatic fractions from both raw distillate and the corresponding solvent extracted lube cut and measured oxygen uptake at 149°C. They found in both cases that saturates alone oxidized rapidly and were stabilized by addition of the aromatics fraction. An optimum aromatics level of about 5% to 10% was found for the raw distillate. Working with oxidized samples, they concluded that when oxidation took place, paradoxically

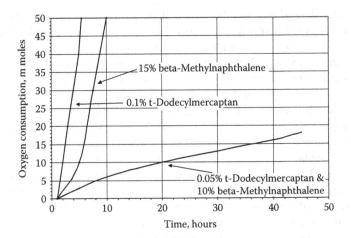

FIGURE 5.10 Oxidation stability of white oil formulations with added β-methyl naphthalene and t-dodecyl mercaptan.

Source: R. E. Berry, F. C. Toettcher, and E. C. Knowles, "Composition and Stability Studies on Lubricating Oils," presented at the Symposium of Additives in Lubricants, Division of Petroleum Chemistry, meeting of the American Chemical Society, Atlantic City, New Jersey, September 17–21, 1956. Figure copyrighted by Chevron Corporation and used with permission.

the saturates were unaffected and all oxidation that occurred was in the aromatic molecules, presumed to be at carbon atoms alpha to the aromatic rings. When the sulfur in the aromatic fraction was removed by reaction with Raney nickel, these aromatics no longer inhibited oxidation. More importantly, they found that addition of 5% by weight of calcium hydroxide to either whole oil to neutralize any strong acids caused oxidation rates to accelerate, since acids were no longer available to decompose peroxides.

Cranton[5] (Imperial Oil) took these results further, examining the relationships between saturate and aromatic types and oxidation inhibition. In this work, a solvent refined 150N base stock from a U.S. midcontinent crude was initially separated by thermal diffusion (Chapter 3) into 10 fractions. Each fraction was then separated by column chromatography into saturates, aromatics, and polars. Polar components were not employed further as part of this study. Mass spectra analyses of the aromatic and saturate fractions showed, as expected, that the thermal analyses technique fractionated the material by molecular shape, with the top fractions (lower fraction numbers) favoring more linear higher VI structures and the bottom fractions containing higher proportions of polycyclic naphthenes in the saturates case and polynuclear aromatics (and thiophene derivatives) in the aromatics (Table 5.12 and Table 5.13). VIs were not reported, but we would expect that fractions with low numbers would have substantially higher VIs than those with high numbers.

TABLE 5.12

Composition of Saturates from Thermal Diffusion Fractions of Solvent Extracted U.S. Midcontinent 150N Base Stock

Fraction Number	1	2	3	4	7
Wt % saturates in fraction	94.3	92.3	89.8	87.5	74.2
Isoparaffins	61.0	51.8	42.1	34.7	3.0
1-ring naphthenes	36.2	42.4	46.3	46.7	33.6
2-ring naphthenes	1.9	5.0	9.6	14.9	31.8
3-ring naphthenes	0.5	0.5	1.7	3.4	21.4
4+-ring naphthenes	0.4	0.3	0.2	0.3	10.3

Source: G. E. Cranton, "Composition and Oxidation of Petroleum Fractions," *Thermochimica Acta* 14:201–208 (1976). With permission.

The oxidative stability of the fractions and some blends (no antioxidants or accelerators present) were measured from induction times at 170°C and 190°C using DSC.[5] These times, in Table 5.14, show that for the saturates alone, oxidative stability at 170°C decreases with increasing complexity (more naphthenes, fewer isoparaffins); that is, polycyclic naphthenes do not have good oxidation resistance, as we have seen already. Polynuclear aromatic fractions added to the saturates at 16% stabilize the blends (induction times increase), with the more polynuclear and more thiophenic fractions having the greatest effect. Monoaromatics had no effect or actually decreased the stability of the more isoparaffinic fractions. We have seen this effect before, where naphthalenes and higher polycyclic aromatics due to sulfur

TABLE 5.13

Composition of Aromatics from Thermal Diffusion Fractions of Solvent Extracted U.S. Midcontinent 150N Base Stock

Fraction Number	1	4	5	6	7
Wt % aromatics in fraction	5.0	9.8	12.7	15.2	18.6
Alkyl benzenes	86.7	69.2	57.6	47.5	34.1
Naphtheno benzenes	13.1	20.4	23.7	26.3	25.9
2-ring aromatics	0.0	0.8	9.4	14.5	20.0
Polynuclear aromatics	0.1	0.2	0.3	3.0	7.2
Thiophenes	0.1	5.4	8.3	10.7	12.8

Source: G. E. Cranton, "Composition and Oxidation of Petroleum Fractions," *Thermochimica Acta* 14:201–208 (1976). With permission.

TABLE 5.14
Induction Times (minutes) at 170°C Measured on
Saturate Fractions and Their Blends with 16% Aromatics

Saturate Fraction Number	Aromatic Fraction Number	Induction Time, Minutes
1	—	8.0
1	1	5.0
1	7	3.0
2	—	5.8
2	4	7.5
2	6	7.5
3	—	7.5
3	5	12.0
4	—	3.2
4	4	5.0
4	6	17.5
7	—	3.0
7	1	4.0
7	7	>20

All fractions obtained by thermal diffusion and chromatography.

Source: G. E. Cranton, "Composition and Oxidation of Petroleum Fractions," *Thermochimica Acta* 14:201–208 (1976). With permission.

content and phenol formation inhibit oxidation and monoaromatics are not so effective.[20] The naphthenoaromatics in aromatic fraction #1 destabilize saturates fraction #1 because of their easily attacked C–H bonds.

Ford Motor Company researchers, in evaluating solvent refined base stocks for automatic transmission fluid (ATF) use, found that at a high degree of oxidation, times to a fixed oxygen uptake gave the best smooth curves when plotted against parameter 1:[23]

$$(C_A + S) \times C_P/C_N, \quad \text{Parameter 1} \tag{5.2}$$

where S is the total sulfur percentage, C_A is the percentage of aromatic carbon measured by ^1H nuclear magnetic resonance (NMR), and C_P and C_N are the percentage of paraffinic and naphthenic carbon atoms, with C_P being measured by infrared and C_N by difference.

These gave a set of curves (Figure 5.11) similar to those generated in the case of "optimum aromaticity." At lower degrees of oxidation, the C_P/C_N term was not necessary. Interesting as well is that the rate of insolubles formation was lowest for those samples found to exhibit optimal oxidation stability.

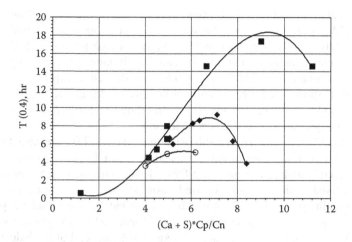

FIGURE 5.11 Base stock oxidation times related to carbon types and sulfur content. *Source:* S. Korcek and R. K. Jensen, "Relation Between Base Oil Composition and Oxidation Stability at Increased Temperatures," *Lubrication Engineering* 19:83–94 (1976). With permission.

5.4 LUBRICANT PERFORMANCE, COMPOSITION, AND THE TREND TO HYDROCRACKED BASE STOCKS

The most valuable assessment of a base stock is how it performs in use as a formulated product and how well it meets the expectations of the formulator and the customer. The answer to this in the real world is complex, since many factors (application, specific conditions of use, price, availability, etc.) are involved. There are a number of standardized tests that provide comparative information under controlled conditions. For the most part, these tests employ inhibited base stock samples (i.e., those containing antioxidants). Perhaps most importantly, these tests are evaluations of the base stock's response to antioxidants, which is the result the formulator really wants and needs. These tests have also been used successfully to provide scientific information on the effects of base stock composition. This approach, coupled with advances in analytical technology, has been particularly successful when applied to the developing field of hydrocracked base stocks, which began in the 1970s. These stocks differ from solvent refined ones in having only very low levels of aromatics, sulfur, and nitrogen, so the focus was much more on hydrocarbon composition itself.

Some of these tests are

- ASTM D943[24] (oxidation characteristics of inhibited mineral oils, sometimes referred to as the turbine oil test [TOST]) is used to compare inhibited steam-turbine oils. The test measures the time required to oxidize the oil to a fixed extent of acid formation (total acid number [TAN]) of 2.0. The oil, plus inhibitors, water, and a copper oxidation catalyst, is oxidized by bubbling oxygen through the oil at 95°C with regular measurement of acid formation by titration with aqueous potassium hydroxide (KOH). The result from the test is expressed as the time (hours) required to an acid number of 2.0 mg KOH. Time to completion of this test may be several thousand hours (2000 hours is 83 days, so this test requires patience!).

- ASTM D2272 (rotating bomb oxidation test [RBOT])[25] was also developed to test steam-turbine oils and is much faster (approximately several hundred minutes) than D943. It measures the time to consume a fixed quantity of oxygen (induction period). This is done using a rotating pressure vessel in which the inhibited oil, water, and a copper metal catalyst are maintained at 150°C under 90 psi oxygen pressure. When all inhibitor is consumed (indicated by the end of the induction period), oxygen pressure drops quickly and the time is recorded. Obviously a long induction period indicates a base stock/inhibitor combination stable to oxidation. Figure 5.12 illustrates the difference in RBOT times between uninhibited (about 40 minutes) and inhibited (about 200 minutes) oils and also demonstrates how DSC can be used as an alternative research method.[6]
- ASTM D4742[26] (thin film oxygen uptake test [TFOUT]) was developed for automotive engine oils. The test sample is mixed with standard components to simulate engine conditions and then tested at 90 psig

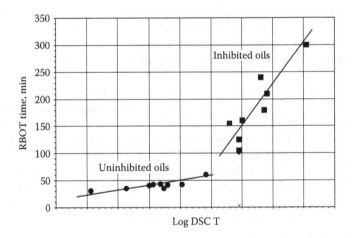

FIGURE 5.12 RBOT times in minutes versus DSC onset temperatures, log(*T*), for uninhibited and inhibited base stocks.
Source: F. Noel, "The Characterization of Lube Oils and Fuel Oils by DSC Analysis," *Journal of the Institute of Petroleum* 57(558):354–358 (1971). With permission.

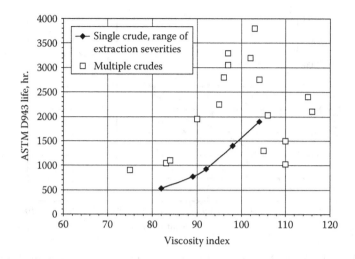

FIGURE 5.13 Base stock VI and ASTM D943 life.
Source: D. W. Murray, J. M. MacDonald, A. M. White, and P. G. Wright, "The Effect of Base Stock Composition on Lubricant Oxidation Performance," *Petroleum Review* February:36–40 (1982). With permission.

oxygen pressure at 160°C in a modified RBOT bomb. The time is recorded when a sharp pressure reduction occurs; longer times correspond to more oxidatively stable formulations.
• ASTM engine tests (e.g., sequence IIIC, IIID, etc.) are used to test fully formulated engine oils under high temperature oxidizing conditions to determine performance against specific parameters.

Murray et al.[27,28] (Imperial Oil) used a number of these tests to develop methods for predicting performance from composition. They found that ASTM D943 lives for solvent refined base stocks from a single crude correlated well with VI and therefore with severity of solvent extraction. However, when applied to base stocks from different crudes and different processes, the correlation with VI no longer held (Figure 5.13).

However, after plotting the saturates content of base stocks against either D943 or D2272 lives, encouraging straight-line correlations were obtained. Figure 5.14 illustrates the results obtained for 150Ns obtained from a range of crudes. In the case of the more severe ASTM sequence IIIC and IIID engine tests, good correlations were obtained against saturates content and these correlations were found to be improved when the sulfur content was included in the correlation. Figure 5.15 shows the agreement between observed times to 375% viscosity increase versus those predicted, where the latter was calculated using both saturates and sulfur content. These results clearly point out the advantages of high saturate base stocks. The authors indeed concluded that "the dominant effect on inhibitor response is the saturate content ... the higher the saturate content, the better the inhibitor response."

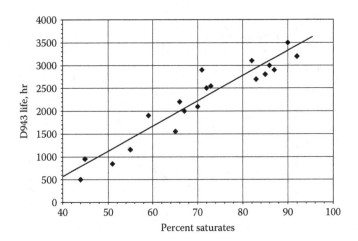

FIGURE 5.14 ASTM D943 inhibited lives versus percent saturates in the base stock samples.
Source: D. W. Murray, J. M. MacDonald, A. M. White, and P. G. Wright, "The Effect of Base Stock Composition on Lubricant Oxidation Performance," *Petroleum Review* February:36–40 (1982). With permission.

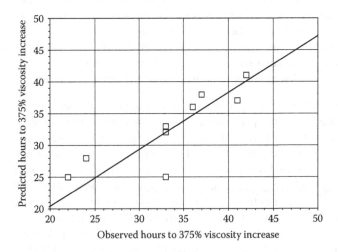

FIGURE 5.15 Predicted versus observed ASTM sequence IIID results: hours to failure at 375% viscosity increase for solvent refined oils.
Source: D. W. Murray, J. M. MacDonald, A. M. White, and P. G. Wright, "The Effect of Base Stock Composition on Lubricant Oxidation Performance," *Petroleum Review* February:36–40 (1982). With permission.

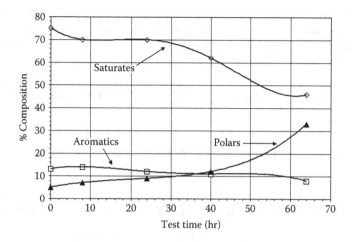

FIGURE 5.16 Compositional changes during engine test operation: SAE 10W30 in ASTM sequence IIIC.
Source: D. W. Murray, C. T. Clarke, G. A. McAlpine, and P. G. Wright, "The Effect of Base Stock Composition on Lubricant Performance," SAE Paper 821236 (Warrendale, PA: Society of Automotive Engineers, 1982). With permission.

These conclusions were very important not only because of their general application, but also because the tests employed were "real world" ones in everyday use for evaluation of base stocks and formulated products. The results thus carried great weight.

This was also a very useful step forward, since crude assessments for (solvent refined) lubes had employed VIs of the lube cuts as a convenient measure of quality and quality implied oxidative resistance. Users had experienced the fallibility of this methodology[29] and were in search of better. It is worth mentioning that VI is still frequently misused today as a proxy for quality. Probably the best response when the word "quality" is used is to ask "quality for what?" and proceed carefully from there.

Investigation by these authors as to what components in the base stock undergo oxidation and when, led to the results in Figure 5.16, which shows that under the severe operating conditions of modern engines, both saturates and aromatics are attacked by oxygen and converted to compounds falling under the "polars" rubric. Mass spectrum analysis of the base stock after engine testing found (rather surprisingly) that paraffins, and 1- through 4+-ring naphthenes all degraded at the same rates. Less surprisingly from what we have seen earlier, among the monoaromatics, alkylbenzene content increased while that of naphthenoaromatics decreased.

Similar compositional relationships to those of Murray et al. have also been reported by Mookken et al.[30] (Indian Oil Corporation) for turbine oils formulated from base stocks produced by a variety of processes and covering a range of viscosities from 30 to 103 cSt at 40°C and VIs from 92 to 113. Their multiple regressions yielded equations 5.3 through 5.5, all of which emphasize that saturates content plays a major role in oxidation stability, and while sulfur does emerge as a positive factor, its significance is less.

$$D943 \text{ life, hr} = 52.2*\%\text{Saturates} + 147.7*\%\text{S} - 26.5*\%\text{Aromatics} - 3.7*\text{N, ppm}$$

$$(5.3)$$

$$RBOT \text{ life, min} = 4.6*\%\text{Saturates} + 36.2*\%\text{S} - 2.9*\%\text{Aromatics} - 0.4*\text{N, ppm}$$

$$(5.4)$$

$$\text{Total oxidation products, IP 280 (\%)} = 1.6*\%\text{Saturates} - 301*\%\text{S}$$

$$+ 17*\%\text{Aromatics} + 3.7*\text{N, ppm}/1000 \qquad (5.5)$$

The three equations see both aromatics and nitrogen as having negative influences on these standard test results (no sign of optimum aromaticity in these samples!). The correlation coefficients were remarkably good, 0.99, 0.98, and 0.87, respectively.

More specific relationships between oxidation stability and hydrocarbon structures were obtained by Gatto et al.[31] in their study of 15 group II, III, and IV base stocks. These authors found that condensed double-ring paraffins (CDRPs) and condensed multiring paraffins (CMRPs) (three or more condensed ring naphthenes) reduced additized base stock oxidation stabilities measured by both RBOT and TFOUT procedures. Correlation coefficients (R^2) were 0.9 and 0.82, respectively. The negative effects of the di+-condensed cycloparaffins were logically attributed to the presence of weak tertiary hydrogen–carbon bonds.

Nitrogen, measured as basic nitrogen, is a minor factor in each of these three equations, but in every instance plays a negative role, shortening D943 and D2272 lives and increasing the amounts of oxidation products formed in the IP 280 test. A group from Texaco (now part of Chevron) and Nippon Oil examined this aspect using the same laboratory tests as above, supplemented by high-pressure differential scanning calorimetry (HPDSC), as well as an open beaker test and engine tests.[32] Their interest was in automotive and industrial oils. Statistical analyses of results for 12 base stocks, produced by a range of processes (solvent refining, hydrocracking, and isomerization), indicated that basic nitrogen was a major contributor to sludge formation in their open beaker test (Figure 5.17). It can be seen that tetracyclic naphthenes and triaromatics were also identified as contributing in a significant way to sludge.

Several of these base stocks were clay treated, which brought basic nitrogen levels below 1 ppm, but changed composition only slightly (Table 5.15). RBOT results were improved substantially (Figure 5.18) by this removal of nitrogen compounds and presumably also by the removal of the other resin-type and polycyclic aromatic molecules associated with them that would adhere to a clay surface.

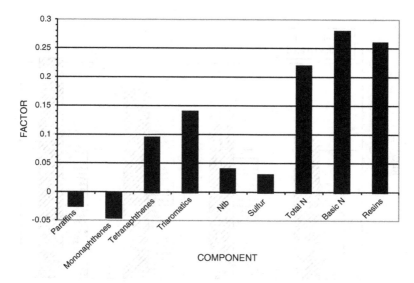

FIGURE 5.17 Thermal sludge weighting factors for base stock components.
Source: Y. Yoshida, J. Igarashi, H. Watanabe, A. J. Stipanovic, C. Y. Thiel, and G. P. Firmstone, "The Impact of Basic Nitrogen Compounds on the Oxidative and Thermal Stability of Base Oils in Automotive and Industrial Applications," SAE Paper 981405 (Warrendale, PA: Society of Automotive Engineers, 1998). With permission.

A wide-ranging study by French researchers (IFP and Elf) on oxidation mechanisms in solvent refined, isomerized wax, and poly-alpha-olefin (PAO) base stocks oxidation was published in 1995.[33] Included was the finding that in modified TFOUT tests, insolubles formation was related to aromatics content, becoming significant at about 6% C_A, but surprisingly not increasing further with C_A (Figure 5.19).

TABLE 5.15
The Effect of Clay Treatment on Composition

Base Oil	Before Clay Treatment				After Clay Treatment			
	Nitrogen, ppm	Aromatics	Saturates	Sulfur	Nitrogen, ppm	Aromatics	Saturates	Sulfur
BO5	22	27.8	71.9	0.19	<1	27.3	72.2	0.17
BO5B	17	30.4	69.3	0.14	<1	30.0	69.7	0.13
BO7	1	0.9	98.9	0.0008	<1	0.9	98.8	0.0009
BO10	32	5.8	93.6	0.018	<1	n/a	n/a	0.02

Source: Y. Yoshida, J. Igarashi, H. Watanabe, A. J. Stipanovic, C. Y. Thiel, and G. P. Firmstone, "The Impact of Basic Nitrogen Compounds on the Oxidative and Thermal Stability of Base Oils in Automotive and Industrial Applications," SAE Paper 981405 (Warrendale, PA: Society of Automotive Engineers,). With permission.

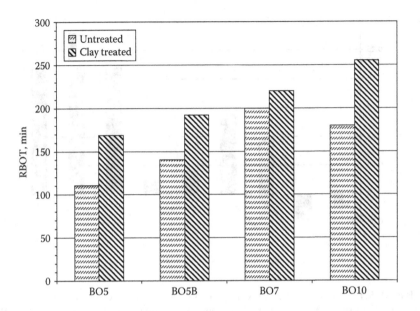

FIGURE 5.18 Effect of clay treatment on RBOT lifetimes.
Source: Y. Yoshida, J. Igarashi, H. Watanabe, A. J. Stipanovic, C. Y. Thiel, and G. P. Firmstone, "The Impact of Basic Nitrogen Compounds on the Oxidative and Thermal Stability of Base Oils in Automotive and Industrial Applications," SAE Paper 981405 (1998) (Warrendale, PA: Society of Automotive Engineers). With permission.

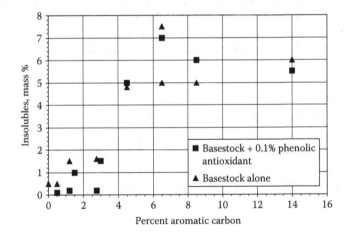

FIGURE 5.19 Dependence of insolubles formed on the percent aromatic carbon in base stocks: results from TFOUT tests.
Source: X. Maleville, D. Faure, A. Legros, and J. C. Hipeaux, "Oxydation des Huiles de Bases Minérales d'Origine Pétrolière," *Revue de l'Institute Francais du Petrole* 50(3): 405–443 (1995). With permission.

The most important conclusion is probably the near zero percent insolubles from base stocks with no aromatic content.

Smith et al.[34] (Texaco, now part of Chevron) applied statistical methods to predict performance in laboratory screening tests such as D943 and engine tests (sequence IIIE and VE) with what appears to be considerable success. The base stocks selected were from a wide variety of sources—naphthenic and paraffinic virgin oils, refined mineral oils, hydrocracked oils, wax isomerized oils, and PAOs. Twenty input variables were used, including 16 compositional ones based on mass spectral data, to develop their ability to predict oxidation stabilities. For ASTM D943, the average absolute difference between actual and predicted times was 387 minutes (in a test that with modern base stocks should run for 3000 to 4000 hours). Their results (Figure 5.20) show that for turbine oils, the best results were clearly from PAOs, followed closely by a base stock from wax isomerization, then hydrocracked base stocks, paraffinic, and naphthenic in descending order. This order indicates that isoparaffin content favors inhibitor response, while increasing quantities of aromatics + naphthenes increase reactivity to oxidation (or inhibitor consumption due to base stock oxidation). It can be seen from Figure 5.20 that of the base stocks chosen in this work, solvent refined oils (either naphthenic or paraffinic) did not fare well in comparison to highly saturated hydrocracked and hydroisomerized stocks. Similar models were also developed for crankcase oils.

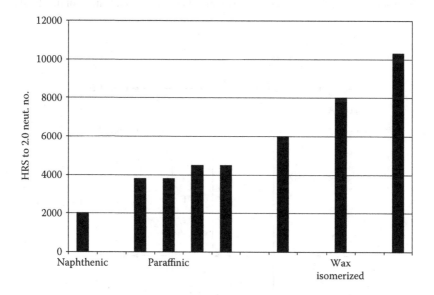

FIGURE 5.20 Predicted times to 2.0 neutralization number in ASTM D943 for turbine oils derived from base stocks of different origins.
Source: M. P. Smith, A. J. Stipanovic, G. P. Firmstone, W. M. Cates, and T. C. Li, "Comparison of Mineral and Synthetic Base Oils Using Correlations for Bench and Engine Tests," *Lubrication Engineering* 52(4):309–314 (1996). With permission.

Further evidence for the desirability of removing aromatics and polyaromatics in hydrocracked base stocks came from a Mobil[35] study in 1994 which found that ASTM D943 times increased as aromatics and polyaromatics levels were reduced by hydrotreating. In this study, which was one of the first to focus on hydrocracked stocks, ASTM D943 tests were run on a number of first- and second-stage hydrocracked oils and several solvent refined reference stocks.

In their program, they hydrocracked heavy vacuum gas oil (HVGO) (feed A) and light vacuum gas oil (LVGO) (feed B) feeds, split the product at around 650°F, and then solvent dewaxed to obtain the initial base stock, which in most cases was second staged, as outlined in Figure 5.21. Table 5.16 provides inspections on 450N base stocks produced from the HVGO, two (A1 and A4) without any second-stage hydrotreating and three with (A2, A3, and A5), and these obviously have significantly lower aromatics content and ultraviolet (UV) absorptivities at 226 and 400 nm (226 nm was judged the best wavelength at which to measure total aromatics and 400 nm for polynuclear aromatics levels). Base stocks A1 and A4 were not second staged. Base stocks A4 and A5 were produced by first-stage processing at higher (by 700 psi) hydrogen pressures than for A1, A2, and A3. This table also includes inspections on the two commercial base stock samples—Com A and Com B—they worked with, one of which is clearly a hydrocracked product from its low (about 2%) aromatics level and near-complete absence of sulfur and nitrogen. Com B is probably a solvent refined and hydrofinished base stock with a high (24.8%)

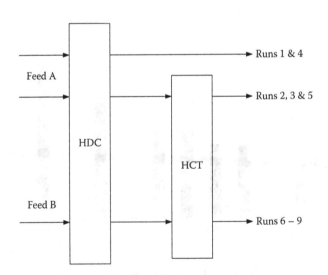

FIGURE 5.21 Hydrocracking and hydrotreating schematic for Mobil study.
Source: A. S. Galiano-Roth and N. M. Page, "Effect of Hydroprocessing on Lubricant Base Stock Composition and Product Performance," *Lubrication Engineering* 50(8): 659–664 (1993). With permission.

TABLE 5.16
Inspections and D943 Results on 450N Base Stocks from Mobil Study

	Al	A2	A3	A4	A5	Com A	Com B	SR1 Auto	SR2 Turbine
First-stage pressure	Base	Base	Base	+700	+700				
Second-stage temperature	No	Base	+60	No	+80				
Product properties									
Viscosity, SUS at 100°F	426	437	424	448	378	468	559	445	438
Viscosity index	97	98	99	97	103	103	95	96	102
Pour point, °F	5	5	5	5	5	5	5	10	25
Sulfur, ppm	32	5	2	26	8	12	170	3500	6800
Nitrogen, ppm	10	11	16	13	3	11	5	120	42
Aromatics, wt %	14.7	10.5	8.4	8.2	4.6	2.0	24.8	22.8	27.0
Composition by mass spectrometry									
Total saturates	85.3	91.8	89.5	91.6	95.4	97.6	74.2	77.2	73.0
Paraffins	15.5	14.9	14.1	14.6	16.3	20.9	14.8	23.4	26.9
Mononaphthenes	23.6	23.0	23.7	24.5	25.0	31.7	16.8	16.4	13.6
Polynaphthenes	46.2	53.9	51.7	52.5	54.1	45.0	43.6	37.4	32.5
Total aromatics	14.7	8.2	10.5	8.4	4.6	2.0	24.8	22.8	27.0
Mono	8.3	3.8	5.3	3.9	2.2		14.9	12.0	14.2
Di	2.6	1.6	2.0	1.5	1.1		4.6	2.9	2.4
Tri	0.7	0.5	0.4	0.5	0.3		1.1	1.0	1.4
Poly	2.0	1.5	2.0	1.8	0.6		3.1	5.6	6.6
Aromatic sulfur compounds	0.8	0.5	0.7	0.6	0.3		0.9	1.3	2.3
UV absorptivity									
At 226 nm	4.8	2.2	2.0	2.3	0.66	0.38	5.3	6.3	5.2
At 400 nm ($\times 10^3$)	11.1	0.99	2.5	4.8	2.7	2.7	8.4	4.0	2.2
ASTM D943, hr	3050	6775	6500	6340	6912	9000	2839	2200	3820

Source: A. S. Galiano-Roth and N. M. Page, "Effect of Hydroprocessing on Lubricant Base Stock Composition and Product Performance," *Lubrication Engineering* 50(8):659–664 (1993). With permission.

aromatics level. SR1 and SR2 are also solvent refined stocks. Inspections on the 140N products from the light feed (feed B) are shown in Table 5.17. All 140Ns were second staged and process variations investigated were increased hydrogen pressure during hydrocracking and the effect of both increasing hydrogen pressure

TABLE 5.17
Hydroprocessing of the Light Feed

Run Number	6B	7B	8B	9B
First-stage pressure, psig	Base	Base	Base	+600
Second-stage temperature	Base	+25	+80	+80
Second-stage pressure, psig	Base	Base	Base	+600
Product properties				
SUS at 100°F	141	134	134	146
VI at 0°F pour point	98	98	100	99
Sulfur, ppm	9	24	5	5
Nitrogen ppm	<1	<1	<1	<1
Aromatics, mass %	<5	<5	<5	<5
UV absorptivity				
226 nm ($\times 10^1$)	6.3	5.1	2.1	1.2
400 nm ($\times 10^4$)	3.1	2.5	1.4	1.2

Source: A. S. Galiano-Roth and N. M. Page, "Effect of Hydroprocessing on Lubricant Base Stock Composition and Product Performance," *Lubrication Engineering* 50(8):659–664 (1993). With permission.

and temperature during second staging. In both tables (Table 5.16 and Table 5.17), it can be seen that either increasing hydrogen pressure in the first stage or temperature in the second stage decreases product aromatics content and UV absorptivity.

The oxidation stability results in Figure 5.22 show that for the heavy neutrals, ASTM D943 life increases significantly as aromatics level decreases toward zero—very approximately about 1000 hours are added per unit change in 226 nm absorptivity—in the zero to 25% aromatics. A similar curve was also generated for polyaromatics as measured by UV absorbance at 400 nm, but whether this is really an independent dataset is open to debate. Products from the light feed also showed this relationship between 226 nm intensity and oxidation stability. It is very clear from this work that the hydrocracked and hydrotreated products take the D943 lives into new territory.

Chevron developed a bench-scale test (called Oxidator BN) for measuring the relative oxidation stability of base stocks when formulated as engine oils and has used this as a means to explore the relationships with composition.[36] Their method measures the time for uptake of a standard amount of oxygen (1 L of oxygen uptake by a 100 g sample) at 171°C (340°F) when the base stock is blended with a standard package that includes both oxidation accelerators (metal naphthenates) and an inhibitor. The longer the time period, the more oxidatively stable the system. The method's credibility was established by the finding that an excellent linear correlation exists between this test's results and those from ASTM's sequence IIID engine test using ASTM reference oils employed to calibrate the sequence IIID test (Figure 5.23). Subsequent work with hydrocracked and solvent refined oils found that while a reasonable correlation ($R^2 = 0.85$) could be obtained between

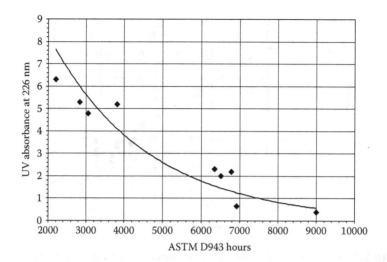

FIGURE 5.22 Variation of ASTM D943 times for heavy neutrals (450Ns) with total aromatics levels measured by UV absorbance at 226 nm.
Source: A. S. Galiano-Roth and N. M. Page, "Effect of Hydroprocessing on Lubricant Base Stock Composition and Product Performance," *Lubrication Engineering* 50(8): 659–664 (1993). With permission.

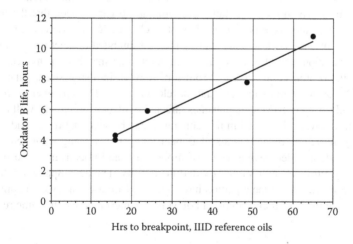

FIGURE 5.23 Comparison of Chevron's Oxidator BN results with those of ASTM sequence IIID using ASTM reference oils.
Source: R. J. Robson, "Base Oil Composition and Oxidation Stability," presented at the Symposium on Trends in Lube Base Stocks, Division of Petroleum Chemistry, meeting of the American Chemical Society, Philadelphia, August 26–31, 1984. Figure copyrighted by Chevron Corporation and used with permission.

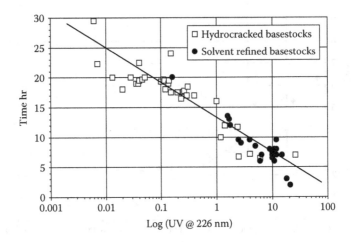

FIGURE 5.24 Chevron Oxidator BN results for hydrocracked and solvent refined stocks: Oxidator times (hr) versus log (UV absorbance at 226 nm).
Source: R. J. Robson, "Base Oil Composition and Oxidation Stability," presented at the Symposium on Trends in Lube Base Stocks, Division of Petroleum Chemistry, meeting of the American Chemical Society, Philadelphia, August 26–31, 1984. Figure copyrighted by Chevron Corporation and used with permission.

Oxidator BN life and (benzocycloparaffins + benzodicycloparaffins + diaromatics], the best ($R^2 = 0.91$) was with UV absorbance at 226 nm (Figure 5.24), a wavelength at which monoaromatics predominate. It can be seen that the longest Oxidator BN results correspond to the lowest levels for 226 nm absorbance (i.e., near elimination of base stock aromatics gives the best inhibited product performance as regards oxidation stability). It is quite remarkable that the chart spans nearly four decades in absorbance at 226 nm. From further work in a subsequent paper, other authors[15] from Chevron's research and development (R&D) department concluded that 5- and 6-ring polycyclic naphthenes affected oxidation stability as well as VI.

From the work described in this chapter, it can be seen that the thread joining several generations of base stocks has been one of increasing saturated hydrocarbon levels and decreasing levels of mono-, di-, and polyaromatics driven by the need to improve both VI and oxidation stability. The accompanying changes on the saturates side of the equation have been to decrease polycyclic naphthenes and increase those of their mono- and dicyclic counterparts, for the same reasons.

REFERENCES

1. K. U. Ingold, "Inhibition of the Autoxidation of Organic Substances in the Liquid Phase," *Chemical Reviews*, 61:563–589 (1961).
2. J. A. Howard, "Inhibition of Hydrocarbon Autoxidation by Some Sulfur Containing Transition Metal Complexes," in *Frontiers in Free Radical Chemistry*, W. Pryor, ed. (New York: Academic Press, 1980), 237–282.

3. J. L. Reyes-Gavilan and P. Odorisio, "A Review of the Mechanism of Antioxidants, Metal Deactivators and Corrosion-Inhibitors," *NLGI Spokesman,* 64(II):22–33 (2001).

4. V. J. Gatto, W. E. Moehle, T. W. Cobb, and E. R. Schneller, "Oxidation Fundamentals and Their Application to Turbine Oil Testing," *Journal of the ASTM International,* 3(4):Paper ID JAI 13498 (2006).

5. G. E. Cranton, "Composition and Oxidation of Petroleum Fractions," *Thermochimica Acta,* 14:201–208 (1976).

6. F. Noel, "The Characterization of Lube Oils and Fuel Oils by DSC Analysis," *Journal of the Institute of Petroleum,* 57(558):354–358 (1971).

7. M. R. Fenske, C. E. Stevenson, N. D. Lawson, G. Herbolsheimer, and E. F. Koch, "Oxidation of Lubricating Oils—Factors Controlling Oxidation Stability," *Industrial and Engineering Chemistry,* 33:516–524 (1941).

8. M. R. Fenske, C. E. Stevenson, R. A. Rusk, N. D. Lawson, M. R. Cannon, and E. F. Koch, "Oxidation of Lubricating Oils—Apparatus and Analytical Methods," *Industrial and Engineering Chemistry, Analytical Edition,* 13(1):51–60 (1941).

9. M. R. Fenske and R. E. Hersch, "Separation and Composition of a Lubricating Oil Distillate," *Industrial and Engineering Chemistry,* 33:331–338 (1941).

10. R. G. Larsen, R. E. Thorpe, and F. A. Armfield, "Oxidation Characteristics of Pure Hydrocarbons," *Industrial and Engineering Chemistry,* 34:183–193 (1942).

11. C. Walling and W. Thaler, "Positive Halogen Compounds. III. Allylic Chlorination with t-Butyl Hypochlorite. The Stereochemistry of Allylic Radicals," *Journal of the American Chemical Society,* 83:3877–3884 (1961).

12. A. L. Williams, E. A. Oberright, and J. W. Brooks, "The Abstraction of Hydrogen Atoms from Liquid Hydrocarbons by t-Butoxy Radicals," *Journal of the American Chemical Society,* 78:1190–1193 (1956).

13. A. C. Nixon, "Autoxidation and Antioxidants of Petroleum," in *Autoxidation and Antioxidants,* vol. II, W. O. Lundberg, ed. (New York: Interscience, 1962), Table XXVII, p. 795.

14. D. C. Kramer, J. N. Ziemer, M. T. Cheng, C. E. Fry, R. N. Reynolds, B. K. Lok, M. L. Sztenderowicz, and R. R. Krug, "Influence of Group II and III Base Oil Composition on VI and Oxidation Stability," presented at the 66th annual meeting of the National Lubricating Grease Institute, Tucson, Arizona, October 24–27, 1999.

15. G. H. von Fuchs and H. Diamond, "Oxidation Characteristics of Lubricating Oils, Relation Between Stability and Chemical Composition," *Industrial and Engineering Chemistry,* 34:927–937 (1942).

16. G. H. Denison, Jr., "Oxidation of Lubricating Oils, Effect of Natural Sulfur Compounds and of Peroxides," *Industrial and Engineering Chemistry,* 36:477–482 (1945).

17. R. K. Jensen, S. Korcek, L. R. Mahoney, and M. Zinbo, "Liquid-Phase Autoxidation of Organic Compounds at Elevated Temperatures. I. The Stirred Flow Reactor Technique and Analysis of Primary Products from n-Hexadecane Autoxidation at 120–180°C," *Journal of the American Chemical Society,* 79:1574–1579 (1979).

18. A. J. Burn and G. Greig, "Optimum Aromaticity in Lubricating Oil Oxidation," *Journal of the Institute of Petroleum,* 58(564):346–350 (1972).

19. G. H. Denison, Jr. and P. C. Condit, "Oxidation of Lubricating Oils, Mechanism of Sulfur Inhibition," *Industrial and Engineering Chemistry,* 37:1102–1108 (1945).

20. D. Barnard, L. Bateman, M. E. Cain, T. Colclough, and J. I. Cunneen, "The Oxidation of Organic Sulphides. Part X. The Co-oxidation of Sulfides and Olefins," *Journal of the Chemical Society,* :5339–5344 (1961).
21. R. E. Berry, F. C. Toettcher, and E. C. Knowles, "Composition and Stability Studies on Lubricating Oils," presented at the Symposium of Additives in Lubricants, Division of Petroleum Chemistry, meeting of the American Chemical Society, Atlantic City, New Jersey, September 17–21, 1956.
22. S. J. Lenoardi, E. A. Overright, B. A. Orkin, and R. V. White, "Autoinhibition of Mineral Oils," Division of Petroleum Chemistry, meeting of the American Chemical Society, Miami, Florida, April 7–12, 1957.
23. S. Korcek and R. K. Jensen, "Relation Between Base Oil Composition and Oxidation Stability at Increased Temperatures," *Lubrication Engineering*, 19:83–94 (1976).
24. ASTM D943, "Standard Test Method for Oxidation Characteristics of Inhibited Mineral Oils," *ASTM Annual Book of Standards*, vol. 05.01 (West Conshohocken, PA: American Society for Testing and Materials).
25. ASTM D2272, "Standard Test Method for Oxidation Stability of Steam Turbine Oils by Rotating Pressure Vessel," *ASTM Annual Book of Standards*, vol. 05.01 (West Conshohocken, PA: American Society for Testing and Materials).
26. ASTM D4742, "Standard Test Method for Oxidation Stability of Gasoline Automotive Engine Oils by Thin-Film Oxygen Uptake (TFOUT)," *ASTM Annual Book of Standards*, vol. 05.01 (West Conshohocken, PA: American Society for Testing and Materials).
27. D. W. Murray, J. M. MacDonald, A. M. White, and P. G. Wright, "The Effect of Base Stock Composition on Lubricant Oxidation Performance," *Petroleum Review,* February:36–40 (1982).
28. D. W. Murray, C. T. Clarke, G. A. McAlpine, and P. G. Wright, "The Effect of Base Stock Composition on Lubricant Performance," SAE Paper 821236 (Warrendale, PA: Society of Automotive Engineers, 1982).
29. J. H. Roberts, "Impact of Quality of Future Crude Stocks on Lube Oils (Significance of VI in Measuring Quality)," Paper AM-85-21C presented at the 1985 National Petrochemical and Refiners Association annual meeting, San Antonio, Texas, March 24–26, 1985.
30. R. T. Mookken, D. Saxena, B. Basu, S. Satapathy, S. P. Srivastava, and A. K. Bhatnagar, "Dependence of Oxidation Stability of Steam Turbine Oil on Base Oil Composition," *Lubrication Engineering*, October:19–24 (1997).
31. V. J. Gatto, M. A. Grina, and H. T Ryan, "The Influence of Chemical Structure on the Physical and Performance Properties of Hydrocracked Base Stocks and Polyalphaolefins," Proceedings of the 12th International Colloquium: Tribology 2000-Plus, January 12–13, 2000, Technische Akademie Esslingen, pp. 295–304.
32. Y. Yoshida, J. Igarashi, H. Watanabe, A. J. Stipanovic, C. Y. Thiel, and G. P. Firmstone, "The Impact of Basic Nitrogen Compounds on the Oxidative and Thermal Stability of Base Oils in Automotive and Industrial Applications," SAE Paper 981405 (1998) (Warrendale, PA: Society of Automotive Engineers).
33. X. Maleville, D. Faure, A. Legros, and J. C. Hipeaux, "Oxydation des Huiles de Bases Minérales d'Origine Pétrolière," *Revue de l'Institute Francais du Petrole,* 50:405–443 (1995).

34. M. P. Smith, A. J. Stipanovic, G. P. Firmstone, W. M. Cates, and T. C. Li, "Comparison of Mineral and Synthetic Base Oils Using Correlations for Bench and Engine Tests," *Lubrication Engineering* 52:309–314 (1996).
35. A. S. Galiano-Roth and N. M. Page, "Effect of Hydroprocessing on Lubricant Base Stock Composition and Product Performance," *Lubrication Engineering* 50:659–664 (1993).
36. R. J. Robson, "Base Oil Composition and Oxidation Stability," presented at the Symposium on Trends in Lube Base Stocks, Division of Petroleum Chemistry, meeting of the American Chemical Society, Philadelphia, August 26–31, 1984.

6 Conventional Base Stock Production: Solvent Refining, Solvent Dewaxing, and Finishing

6.1 SOLVENT REFINING

Conventional lube manufacturing employs mostly separation technology originally developed prior to World War II and significantly improved since then. The steps usually are

- Solvent refining to adjust the viscosity index (VI) and improve the base stock's response to oxidation. This solvent extraction separates the low VI and easily oxidizable fraction from the more desirable high VI and oxidation-resistant components.
- Solvent dewaxing to ensure the base stock is liquid in winter temperatures. By chilling with a solvent, wax is separated by crystallization out and is filtered off.
- "Finishing" to bring about final quality improvement. In clay finishing, some of the remaining unstable polar compounds adhere to clay surfaces and are removed from the oil. This separation step has largely been replaced by catalytic hydrofinishing.

Solvent refining as a means to separate "good" components from "bad" was developed in the early stages of the petroleum industry, specifically for improving kerosene quality by use of sulfur dioxide.[1] It was subsequently applied to lubricants, initially employing a variety of solvents to reduce the amounts of oxidation-unstable aromatics, which were recognized as being "bad" components and also, conveniently, were recognized as having low VIs. Since it was at one time the only technology available, solvent refining became the dominant process worldwide for lube oil base stock refining, accounting for approximately 92% of the world's production recently, but it has since declined to about 75%. Solvent refined or extracted base stocks all fall into the American Petroleum Institute (API) group I type unless accompanied by severe hydrotreating. In North America,

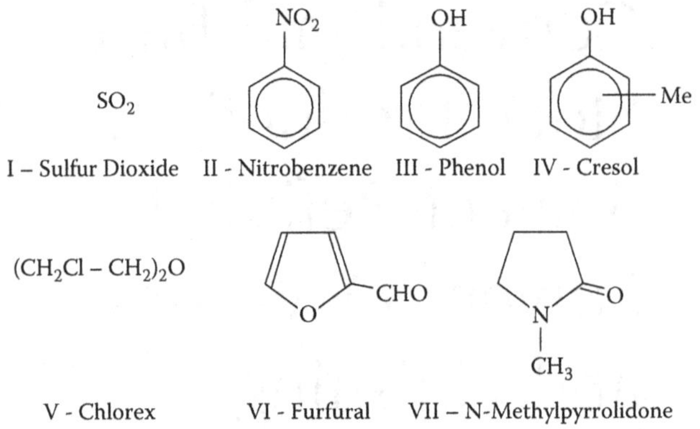

I – Sulfur Dioxide   II - Nitrobenzene   III - Phenol   IV - Cresol

V - Chlorex   VI - Furfural   VII – N-Methylpyrrolidone

FIGURE 6.1 Structures of significant solvents employed in solvent refining of base stocks.

solvent refining is rapidly being replaced by the hydroprocessing route, which converts "bad" to "good" components and leads to group II and III base stocks.

Since the solvents employed must be insoluble or nearly so in the hydrocarbon feed, process development necessarily focused on solvents with polar structures. The solvents that eventually achieved significant commercial use (Figure 6.1 and Table 6.1) are liquid sulfur dioxide (I), nitrobenzene (II), phenol (III), cresylic acid (o-, m-, p-cresol) (IV), β,β-dichloroethylether (Chlorex, V), furfural (VI), and n-methyl-2-pyrollidone (NMP, VII). The Duo-Sol process[2] employs a

TABLE 6.1
Properties of Some Solvents Employed in Solvent Refining

Solvent	Molecular Weight	Boiling Point, °C	Melting Point, °C	Density, n_4^{20}	Refractive Index
Cresylic acid	108.1	191	30.9	1.0273	1.5361
Chlorex	143.01	178	−24.5	1.2199	1.4575
Furfural	96.09	161.7	−38.7	1.1594	1.5261
MEK	72.11	79.6	−86.3	0.8054	1.3788
Nitrobenzene	123.1	210.8	5.7	1.1945	1.5562
NMP	99.1	202	−23	1.026_{25}^{25}	1.4684
Phenol	94.1	181.7	43	1.0576	1.5408
SO$_2$	64.1	−10	−72.7	1.434	—
Toluene	92.1	110.6	95	0.8669	1.4961

Source: D. R. Lide, ed., *CRC Handbook of Chemistry and Physics*, 73rd ed., Boca Raton, FL: CRC Press, 1992. With permission.

propane-phenol-cresylic acid mixture. Extraction produces the extract—soluble in the solvent—and raffinate—the high-value product insoluble in the solvent.

Criteria for a good solvent are

- Good solubility for the low VI aromatics and polyaromatics to be extracted.
- Poor solubility for high VI paraffins and naphthenes that should remain in the raffinate.
- Good thermal and oxidative stability to minimize losses and the development of contamination in use; furfural readily oxidizes if exposed to air since it contains an aldehyde functional group, but its other properties outweigh the special provisions needed for its use.
- A significant density difference between solvent and raffinate to ensure good separation of two-phase mixtures.
- A low solvent viscosity aids phase separation.
- A low solvent melting point prevents freeze up in winter.
- A low solvent boiling point reduces energy requirements and ensures successful separation of solvent from raffinate and extract.
- Nontoxic and noncorrosive.
- Low cost.

Figure 6.2 and Figure 6.3 are good illustrations of how the viscosities and VIs of the raffinate and extract differ in solvent refining. This is from quite old work in which a Texas Van Zandt distillate was extracted with acetone and the extract and raffinate were then distilled into fractions and analyzed.[3] Figure 6.2 shows that the viscosities at 100°F of the polynuclear aromatic extracts are higher than those of the more paraffinic raffinates and this difference increases almost exponentially as molecular weight increases. This is what we would expect from properties of model compounds. The raffinate viscosities do not increase as quickly since they have significant paraffinic character and higher VIs as well. Figure 6.3 demonstrates the disparities in VI between raffinate and extract; the extract is largely material with a VI less than zero, while the raffinate VI is between 80 and 100 and increases slowly with increasing boiling point.

In spite of the issues with phenol (toxicity, high melting point) and furfural (poor oxidative stability), these two became the most commonly used solvents and are applicable to the full slate of products, from a 40N to bright stock. Joining them in the 1970s was NMP (licensed by ExxonMobil as Exol N extraction process[4] and by Texaco as Texaco MP refining process[5]) when its advantages (good selectivity, lower solvent:oil ratio, low toxicity, low melting point but higher solvent cost) were recognized. ExxonMobil has licensed more than 20 NMP plants, mostly as conversions of phenol plants.[6] By 1993, half of North America's solvent refining plants had been converted to NMP. As of 2005 there were 14 such plants in North America, but the number has been steadily declining with the encroachment of hydroprocessing. There are currently about 115 solvent refining plants worldwide.[7]

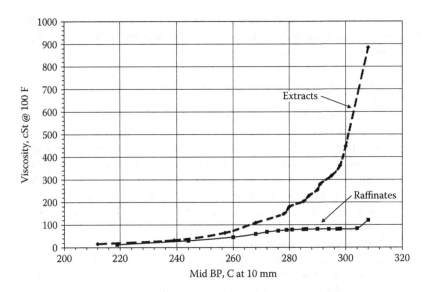

FIGURE 6.2 Viscosities at 100°F of distillation fractions from an acetone extract and the corresponding raffinate from a Texas van Zandt distillate: viscosity versus mid-boiling point at 10 mm Hg.
Source: M. R. Cannon and M. R. Fenske, "Composition of Lubricating Oil," *Ind. and Eng. Chem.*, vol. 31, pp. 643–648 (1939).

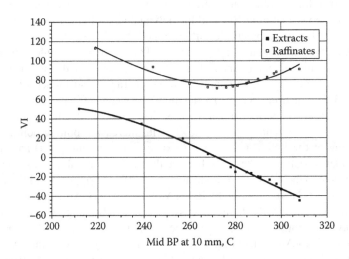

FIGURE 6.3 VIs of distillation fractions from an acetone extract and the corresponding raffinate from a Texas van Zandt crude distillate: VIs versus mid-boiling point at 10 mm Hg.
Source: M. R. Cannon and M. R. Fenske, "Composition of Lubricating Oil," *Ind. and Eng. Chem.*, vol. 31, pp. 643–648 (1939).

Figure 6.4 provides a schematic of an NMP plant.[4] The key unit is the extraction tower into which NMP is fed from the top to extract an ascending stream of lube feed. Good contact between the two phases is brought about through the use of packed towers, baffles, or rotating disks.[8] Product quality is controlled by contacting temperature, solvent:oil ratio, and charge rate. The raffinate proceeds to a solvent stripping section from which emerges the solvent-free raffinate to tankage. The extract-containing solvent from the bottom of the

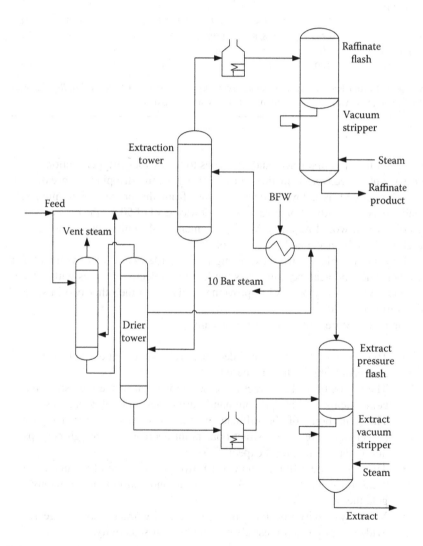

FIGURE 6.4 Schematic of NMP solvent refining plant.
Source: B. M. Sankey, D. Bushnell and D. A. Gudelis, "Exol N: New Lubricants Extraction Process," *Proceedings, 10th World Petroleum Congress,* vol. 4, pp. 407–414 (1979). With permission.

TABLE 6.2
Single Batch Extractions with Chlorex, Nitrobenzene, and Acetone

Properties	Original Oil	Chlorex		Nitrobenzene		Acetone	
		Raffinate	Extract	Raffinate	Extract	Raffinate	Extract
Viscosity, SUS at 100°F	272	233	582	216	406	236	390
VI	77.5	104	5.4	104.5	54.1	89.4	49.8
VGC	0.845	0.828	0.898	0.819	0.871	0.833	0.891
Pour point, °F	0	5	−10	5	−10	0	−15
Yield, vol. %	100	74.8	25.2	51.1	48.9	79.2	20.8

Source: V. L. Kalichevsky and K. A. Kobe, "Petroleum Refining with Chemicals," in *Refining with Adsorption* (New York: Elsevier, 1956), 244–311. With permission.

extraction tower is first dried, and then goes to the extract stripper section, where the solvent is recovered, in this illustration by steam stripping. The extract is usually regarded as a low value by-product from the process. Minimum feed quality in solvent extraction is said to be a dewaxed VI of 50 for the heavy vacuum gas oil, which would require a VI improvement of 45 units via this process[9] to reach the usually acceptable VI value of 95.

Results in Table 6.2 from extracting a single (dewaxed) feed with Chlorex, nitrobenzene, and acetone, illustrates the fact that selectivity varies with solvent, acetone giving a very poor VI improvement relative to the other two in spite of high raffinate yield.

General features of these extractions are

- The viscosity of the raffinate decreases relative to that of the feedstock since paraffinicity has increased.
- The viscosity of the extract increases above that of the feed since the components here are predominantly aromatics and polyaromatics.
- The pour point of the raffinate increases—n-paraffins are a higher percentage here—and extract pour point decreases, although perhaps not as far as one might expect.
- The VIs of the raffinate and extract diverge—the size of the delta is a measure of the selectivity of the solvent, and Chlorex here obviously gets the nod.
- Viscosity-gravity constants of raffinate and extract also diverge for evident reasons and must also reflect solvent selectivity.

In the case of NMP, while one of its major advantages is energy savings, it also offers better yields at the same raffinate quality. Table 6.3 demonstrates the superior results claimed for NMP versus furfural and phenol at the same treat rate (ratio of

TABLE 6.3

Selectivity as Expressed in Yields for NMP, Phenol, and Furfural for a 60 Naphthenic Distillate to 0.871 Density Raffinate

	NMP	Phenol	Furfural
Treat (vol. %)	160	160	160
Raffinate yield (vol. %)	87.0	83.5	83.5

Source: B. M. Sankey, "A New Lubricants Extraction Process," *Canadian Journal of Chemical Engineering* 63:3–7 (1985). With permission.

solvent to oil) and the same raffinate density.[10] Table 6.4 shows that in a 150N case, while raffinate compositions from phenol and NMP extraction are very similar, the extract has lower saturates pointing to higher NMP selectivity for the "good" components. This is accompanied by a higher refractive index (higher aromatics content) for the extract. This is in spite of a 5% increase in raffinate yield for the NMP case at a constant raffinate product quality with a VI of 104.

For a more comprehensive discussion of solvent extraction, the reader is referred to Chapter 5 in Sequeira's excellent text.[8]

TABLE 6.4

Composition of Streams from Phenol and NMP Extraction of Middle East 150N Distillate

	Feed	Phenol		NMP	
		Raffinate	Extract	Raffinate	Extract
Density, kg/dm³	0.9148	0.8639	0.9841	0.8629	0.9973
Refractive index (75°C)	1.4922	1.4571	1.5377	1.4561	1.5463
Color, ASTM		<2.0		1.0	
Dewaxed oil VI at 9°C pour point	62	104		104	
LC Analysis, wt. %					
Saturates	48.9	74.8	23.5	74.3	13.8
Aromatics	40.7	24.1	66.4	23.9	74.7
Polars + unrecovered	10.4	1.1	10.1	1.8	11.5
Refractive Index (75°C)					
Saturates	1.4459	1.4443	1.4488	1.4446	1.4502
Aromatics	1.5471	1.4956	1.5619	1.4907	1.5697

Source: B. M. Sankey, D. Bushnell and D. A. Gudelis, "Exol N: New Lubricants Extraction Process," Proceedings, 10th World Petroleum Congress, 4: 407–414 (1979). With permission.

6.2 SOLVENT DEWAXING

Dewaxing of waxy paraffinic lubricants was first performed using the only means of refrigeration then available—winter—and allowing the wax to settle in the cold tanks or barrels.[11,12] This "cold-settling" was subsequently replaced by year-round refrigeration, wax removal by filtration or centrifugation, and the use of solvents and solvent mixtures such as sulfur dioxide-benzene (Edeleanu process), propane (developed by Standard Oil of Indiana), and benzene-acetone (The Texas Company, later Texaco). Propane dewaxing is still in use and is considered best for heavy feeds such as bright stocks.[13] The benzene-acetone process eventually morphed into the toluene-methyl ethyl ketone (MEK) process, the most common solvent dewaxing process today. Toluene replaced benzene due to the latter's toxicity, higher melting point, and lower filtration rates, while acetone gave way to MEK due to the latter's higher boiling point (56.2°C versus 79.6°C) and therefore reduced solvent losses.[12] In the toluene-MEK process, toluene as a single component is capable of dissolving both wax and oil, whereas MEK is a poor solvent for the wax (at low temperatures oil becomes insoluble too). At filtration temperature, the oil is still soluble in the solvent mixture although the wax is not.

To dewax via the toluene-MEK process, the waxy feed/toluene mixture is initially heated above the cloud point to remove any microcrystals, then it is cooled in a heat exchanger with water and subsequently using scraped surface pipe chillers to about 10°F to 20°F below the target pour point (Figure 6.5). Filtration normally employs rotary filters using specially manufactured filter cloths. The wax produced usually contains 5% to 20% oil, depending on the feed, and can be deoiled in a subsequent step to produce hard wax and a by-product soft wax stream called "foots oil," which is largely isoparaffins. The solvent is stripped from the oil and recycled. This process is applicable both to solvent refined and hydrocracked waxy base stocks.

The ratio of MEK to toluene is adjusted depending on the stream being dewaxed. Higher toluene ratios are required for heavier stocks such as bright stock to avoid oil immiscibility. Figure 6.6 illustrates this for a high VI 250N and a bright stock, each at two different pour points, and for which samples were dewaxed at different MEK contents with percent oil measurements after filtration. Immiscibility temperatures are at the breaks in the curves. It can be seen that bright stock dewaxing requires significantly higher toluene content than for the lighter 250N stock, and in each case, lower oil pour points move the curves and breakpoints to the left (i.e., higher toluene levels) to avoid oil immiscibility. Light waxy base stocks generally produce macrocrystalline wax (large crystals), and as viscosity increases the wax becomes increasingly microcrystalline.

Solvent dewaxing is a chemical separation process in which no chemical reactions occur. Therefore the composition of the dewaxed oil is completely dependent on that of the waxy feed and represents the subtraction of the "wax" composition. Solvent dewaxing works because it exhibits a preference to crystallize out the highest carbon number n-paraffins, followed by isoparaffins and

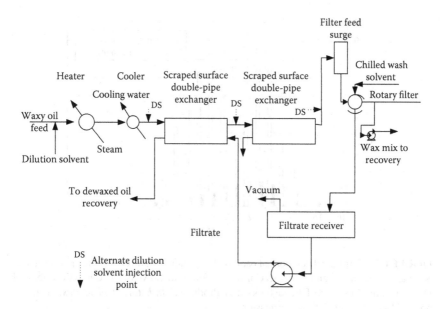

FIGURE 6.5 Simplified flow diagram: toluene/MEK dewaxing.
Source: S. Marple and L. J. Landry, "Modern Dewaxing Technology" in *Advances in Petroleum Chemistry and Refining,* vol. X, Ed., J.J. McKetta, Jr., Interscience, 1965. With permission.

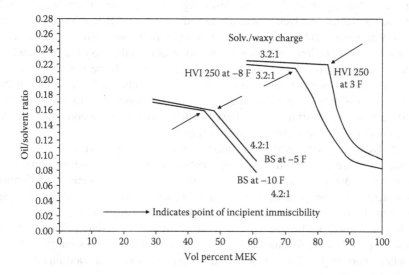

FIGURE 6.6 Dewaxing miscibility relationships for high VI 250 and bright stock.
Source: S. Marple and L. J. Landry, "Modern Dewaxing Technology" in *Advances in Petroleum Chemistry and Refining,* vol. X, Ed., J.J. McKetta, Jr., Interscience, 1965. With permission.

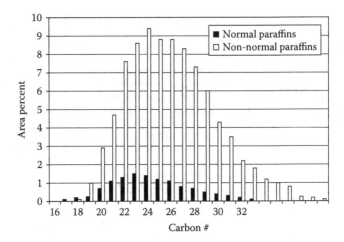

FIGURE 6.7 Normal and non-normal paraffin distributions in a waxy 100N feed to dewaxing. *Source:* R. J. Taylor and A. J. McCormack, "Study of Solvent and Catalytic Lube Oil Dewaxing by Analysis of Feedstocks and Products," *Ind. Eng. Chem. Res.*, vol. 31, pp. 1731–1738 (1992). With permission.

then other hydrocarbons that have paraffin-like structures. This is the same order in which these components crystallize when a waxy distillate is cooled, and therefore the selectivity in solvent dewaxing is needed. Taylor and McCormack (Texaco Research and Development) studied[14,15] this for MEK-toluene dewaxing of solvent refined 100N, 320N, and 800N waxy raffinates from Arabian crudes. Carbon number distributions for the n- and isoparaffins in the waxy feed (Figure 6.7) show that in the 100N sample, the isoparaffins outweigh the normal ones and have a distribution maximum at about one carbon number higher than that of the normals. This difference is due to a normal fractionation effect in producing the distillate: the isoparaffins have a lower boiling point than the corresponding n-paraffins of the same carbon number (Chapter 2).

Solvent dewaxing with MEK-toluene predominantly removes the normal paraffins (Figure 6.8), but the "gap" between the distribution maximas has increased to about two to three carbon numbers. When the dewaxed oil is further dewaxed to a lower pour point to isolate remaining "wax," its composition can be seen (Figure 6.9) to be dominated by the nonnormal paraffins, but there are still some lower molecular weight normal paraffins remaining. Thus solvent dewaxing "selectively" removes higher carbon number paraffins but the oil still contains normal paraffins, of lower carbon number and significant quantities as well of isoparaffins.

A significant advance in solvent dewaxing was announced by ExxonMobil[16–18] in 1972 in the form of the DILCHILL™ (DILution CHILLing) process, which improves the crystallization step by producing large, dense, spherical "crystals"

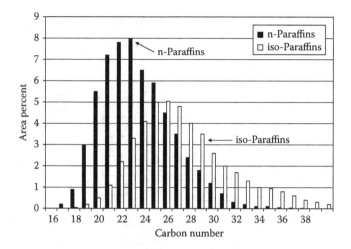

FIGURE 6.8 Normal and non-normal paraffin distributions in wax obtained by solvent dewaxing a waxy 100N.
Source: R. J. Taylor and A. J. McCormack, "Study of Solvent and Catalytic Lube Oil Dewaxing by Analysis of Feedstocks and Products," *Ind. Eng. Chem. Res.*, vol. 31, pp. 1731–1738 (1992). With permission.

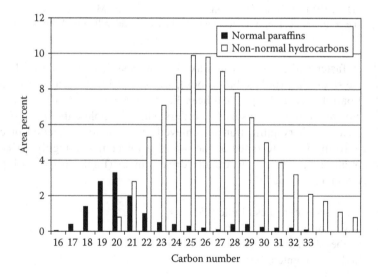

FIGURE 6.9 Normal and non-normal paraffins remaining in the dewaxed oil after solvent dewaxing a waxy 100N.
Source: R. J. Taylor and A. J. McCormack, "Study of Solvent and Catalytic Lube Oil Dewaxing by Analysis of Feedstocks and Products," *Ind. Eng. Chem. Res.*, vol. 31, pp. 1731–1738 (1992). With permission.

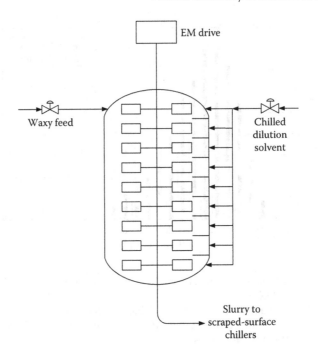

FIGURE 6.10 Schematic of DILCHILL™ crystallizer unit.
Source: "DILCHILL Solvent Dewaxing Technology," ExxonMobil, available at http://www.prod.exxonmobil.com/refiningtechnologies/lubes/mn_dilchill.html. With permission.

that filter faster and reduce oil in wax. The process is applicable to the full waxy base stock slate (60N to base stock), and as of 1999, eleven units totaling 120,000 barrels per day (bpd) had been licensed.[18] The DILCHILL™ unit is located upstream of scraped surface chillers which complete the final 10°C to 15°C of chilling. Crystallization is achieved by sequential injection of cold solvent into the waxy feed fed at the top of a tower into a highly turbulent environment created by an electrically driven mixer (Figure 6.10 and Figure 6.11). Advantages include

- Increased filter rates (Table 6.5)
- Lower oil in wax (Table 6.6)
- Higher dewaxed oil yields
- Reduced maintenance.

The process can also include a deoiling section (see Figure 6.11) to make hard wax. The soft wax that is to become foots oil is in the last few layers of the DILCHILL™ crystal and dissolves when exposed to hot solvent, leaving the low-oil wax to be filtered off. This represents a major energy savings over the conventional recrystallization process, where all the wax must be redissolved.

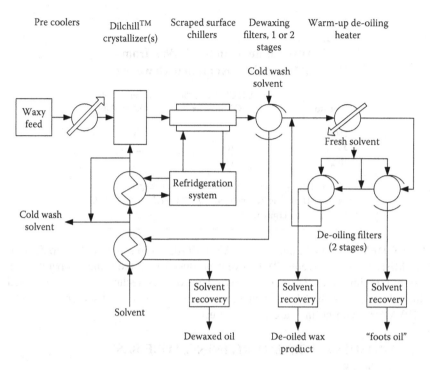

FIGURE 6.11 DILCHILL™ dewaxing/warm-up deoiling process.
Source: "DILCHILL Solvent Dewaxing Technology," ExxonMobil, available at http://www.
prod.exxonmobil.com/refiningtechnologies/lubes/mn_dilchill.html. With permission.

W. R. Grace & Co. and ExxonMobil introduced a further innovation in solvent
dewaxing in 1998 with the commercialization of membrane technology for partial
separation of oil or wax and solvent.[19,20] This was at ExxonMobil's Beaumont,
Texas, refinery and was employed on their filtrate streams from dewaxing

TABLE 6.5
DILCHILL™ Dewaxing Process: Throughput Advantage

Feedstock	DILCHILL™, bpd per 1000 ft² Filter Area	Conventional Dewax Plant, bpd per 1000 ft² Filter Area
100N	3450	2500
150N	4000	2900
450N	2150	1500
1000N	1700	1200

Source: D. A. Gudelis, J. D. Bushnell and J. F. Eagen, "Improvements in
Dewaxing Technology," *API Proceedings* 65:724–737 (1973). With permission.

TABLE 6.6
Oil in Wax Contents in Slack Wax from
DILCHILL™ and Conventional Dewaxing

Feedstock	DILCHILL™, % Oil in Wax	Conventional, % Oil in Wax
150N	<5	10
450N	<10	20
1000N	<15	40

Source: D. A. Gudelis, J. D. Bushnell, and J. F. Eagen, "Improvements in Dewaxing Technology," *API Proceedings* 65:724–737 (1973). With permission.

(MAX-DEWAX™) and deoiling (MAX-DEOIL™) to separate 25% to 50% of the MEK-toluene solvents. The solvent produced is at the same temperature as the filtrate involved. The advantages cited for this technology are increased throughput and reduced oil content of the slack wax in the case of MAX-DEWAX™, with no increase in energy use.

6.3 FINISHING SOLVENT REFINING LUBE BASE STOCKS

The purpose of a finishing step for solvent extracted base stocks is to further improve color and performance by removing some of the remaining polar compounds (generally higher molecular weight sulfur-, nitrogen- and oxygen-containing compounds) which are among the more easily oxidizable components during lubricant use and contribute to the formation of sludge, color, and oxidation products. Most of these compound types have been removed in the extraction step, but some can still remain and are dealt with in the finishing step. There are two processes which have been employed for this:

- Clay-treating, which separates them by adsorption, and
- Hydrofinishing, which converts them into acceptable lube components by hydrogenation.

6.3.1 CLAY TREATING

The clay treating process[21] percolates base stock through a heat-activated solid adsorbent, usually bauxite (a form of aluminum oxide; e.g. Porocel) or a naturally occurring clay (Fuller's earth, Attapulgis clay). Polar impurities such as nitrogen-, oxygen-, and sulfur-containing compounds are adsorbed on the solid surface and removed from the oil. Figure 6.12 provides a general schematic for this process.[22] Heated oil charge is fed to a bauxite bed and allowed to percolate through at a predetermined rate. When specifications can no longer be met, the clay is declared

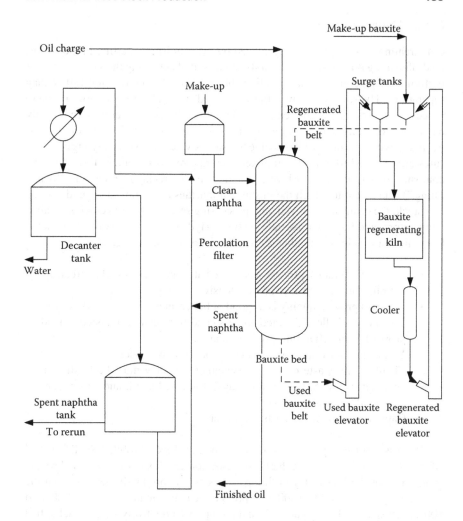

FIGURE 6.12 Schematic of a clay finishing unit.
Source: "Hydrocarbon Refining and Static Bed Percolation" published by the Porocel Corporation, Houston, Texas. With permission.

exhausted and charge feed is discontinued. A flush with naphtha removes the remaining oil and some of the adsorbed components, then a steam purge is employed to remove the naphtha and the clay can be transferred to a belt that carries the hydrocarbon-free clay to the regenerator kiln to burn off the remaining adsorbed material. This step also adjusts the water content of the clay to the correct level, which is needed since water level affects adsorption. The regenerated material is then returned to the percolation filter bed and reused. This process has also been employed as part of the traditional route for purifying waxes and as a cleanup method for hydrotreated white oils.[23]

6.3.2 HYDROFINISHING

Clay treating was progressively replaced by hydrofinishing starting in the 1950s. Hydrofinishing is a low-pressure catalytic method of achieving the same or similar performance outcome as clay treating, but by chemical conversion rather than separation. (Hydrofinishing was the term used originally for this low pressure process, and certainly as applied to solvent refined lubes. The term more recently has come to be used in a more generic sense in "hydrofinishing" or the last treatment with hydrogen of hydrocracked stocks where considerably higher pressures are required to almost completely saturate monoaromatics. To add to this confusion of terminology, IFP refers to their hydrocracking process as "hydrofining.") The extent to which the unstable impurities are removed depends on the feedstock, the catalyst employed, and processing severity (space velocity, catalyst temperature, and hydrogen partial pressure). Hydrofinishing of solvent extracted base oils (paraffinic or naphthenic) has the following advantages.[24]

- It is a continuous process versus a batch process for clay treating, therefore it is much less labor intensive.
- The final product quality is as good or better than that produced by clay.
- The process is flexible since process conditions can be adjusted to feed properties and changing product quality.
- Yields are equivalent or better than for the clay process.
- There are no waste disposal problems (used clay must be landfilled, an expensive undertaking nowadays, if allowed at all, and clay regeneration affects air quality).
- Capital and operating costs are lower.

Processes for this purpose were developed by ExxonMobil,[25] Shell,[26] IFP,[27] BP,[28] UOP,[29] and others. Bechtel[30] licenses former Texaco processes. Process conditions for the Exxon Hydrofining™ process, as originally developed in the 1950s, are given as 100 to 1200 psig pressure, catalyst temperatures of 400°F to 800°F, and space velocities up to 2, and it employs a relatively simple trickle bed reactor system. Since no cracking or significant aromatics reduction occurs, hydrogen consumption is low. From the examples of the chemical and physical changes that occur during both clay and hydrotreating given in Table 6.7,[25,31] it can be seen that the common change is color improvement attributed to polyaromatics saturation, while physical properties such as VI or pour undergo no significant changes, but viscosity may be slightly decreased. Catalysts with any cracking activity are best avoided in this application because cracking leads to yield loss. Hydrotreating does affect sulfur levels, whereas clay treating only has a slight effect on sulfur levels.

It can be seen in Table 6.8 that these changes are sufficient to improve the behavior of the oils relative to those from clay treating. In the Indiana oxidation test, hydrofinishing reduced sludge (insolubles), viscosity and acid number

TABLE 6.7
Inspections on Hydrofinished and Clay-Treated Base Stocks from Western Canadian Crude

SAE Grade		10			30	
Finishing Process	Feed	Clay-treating	Hydro-finishing	Feed	Clay-treating	Hydro-finishing
Yield, vol. %	100	99	99.5	100	98	99.5
Flash, °F	395	405	400	510	505	510
Viscosity at 100°F, SUS	157	157	157	643	624	632
VI	91	91	91	91	91	91
Color, ASTM	1.0	0.5	−5 Say	4.1	1.0	1.0
Pour point, °F	5	10	10	25	25	25
Cloud point, °F	14	14	14	28	34	34
Sulfur, wt. %	0.15	0.12	0.06	0.18	0.15	0.07
Carbon residue, wt. %	—	—	—	0.052	0.028	0.024

Source: J. B. Gilbert, R. Kartzmark, and L. W. Sproule, "Hydrogen Processing of Lube Stocks," *Journal of the Institute of Petroleum* 53(526):317–327 (1967). With permission.

TABLE 6.8
Oxidation Stability of Hydrofinished and Clay-Treated Oils from Western Canadian Crude

SAE Grade	10		30	
Finishing Process	Claytreating	Hydrofinishing	Claytreating	Hydrofinishing
Indiana oxidation[a]				
Sludge, mg/10 g	90	47	69	17
Viscosity increase ratio	1.4	1.3	1.4	1.3
Acid number	2.5	2.2	3.0	2.0
Staeger oxidation[b]				
Color, ASTM	7.5	4.0	7.5	6.4
Increase in acid number	0.49	0.04	0.33	0.08
Turbine oil oxidation, D943				
Hours to 2.0 acid number	1325	2050	1080	1250

Source: J. B. Gilbert, R. Kartzmark, and L. W. Sproule, "Hydrogen Processing of Lube Stocks," *Journal of the Institute of Petroleum* 53(526):317–327 (1967). With permission.

[a] Oil sample heated at 171.7°C per 10 L of air per hour per 300 ml sample.
[b] Oil is heated to 230°F in the presence of copper for 380 hr.

increase, and color decreases; and acid number increases in the Staeger test; and there is increased lifetime in the turbine oil oxidation test.

Table 6.9 shows application of the IFP[27] process to naphthenic and paraffinic oils, where the main changes are color improvement and a reduction in sulfur content, but accompanying these is some reduction in viscosity (due to either some slight hydrocracking or the benefit of saturating some polyaromatics of very low VI). The Conradson carbon reduction also points to fewer polyaromatics in the product.

Butler and Kartzmark[32] (Imperial Oil) published an interesting paper that provided details on the chemistry of the hydrofinishing step through compositional studies of the feed and products and is worth discussing at some length. This was from early studies on the Hydrofining™ process. Their work was on a very light Tia Juana distillate and overall they found significant reductions in both aliphatic sulfides and thiophenic compounds. The research focused on changes that occurred in components in the feedstock and its hydrofinished product. In both cases they performed analyses on narrow distillate cuts from the feed and product. They then used the n-d-M analyses method (Chapter 4) on each of these cuts to follow the changes in aromatic types during these reactions as a function of boiling range.

On the feedstock, their compositional work found that the number of aromatic rings per molecule increased with molecular weight (and boiling point), as one might intuitively expect, and in fact, two regions are discernible and are defined by boiling range (Figure 6.13). Below about 620°F, there are around 0.45 aromatic rings/molecule, whereas above 650°F, this number increases to 0.65 aromatic rings/molecule. In the product, interestingly, these are almost reversed, with approximately 0.45 rings/molecule above 600°F and generally more than 0.6 rings/molecule below 600°F. These changes are also reflected in the specific gravities and refractive indices of the fractions. As will be seen, these changes were due to benzothiophenes and similar compounds undergoing desulfurization and therefore a decrease in boiling point.

Sulfur contents in Figure 6.14 show that sulfur reduction during hydrofinishing can be substantial and that most of the remaining sulfur is in the high molecular weight end. Mass spectroscopic analyses (Figure 6.15 and Figure 6.16) showed that this sulfur decrease coincided with equally precipitous declines in the levels of benzothiophene derivatives, but hydrocarbon aromatic levels had actually increased (Figure 6.17). These authors concluded that in hydrofinishing, the benzothiophenic and dibenzothiophenic compounds were being desulfurized with the accompanying opening of the five-membered ring and converted to substituted monoaromatics so that total aromatics levels did not change. Similar results for aromatics were observed for hydrofinishing higher boiling point distillates from Tia Juana and western Canadian crudes. The authors did not study the detailed changes in aliphatic sulfides during Hydrofining™, but noted their levels fell dramatically in the product fractions, by a factor of 10 in the very light Tia Juana case. Overall, sulfur levels showed similar changes, and the remaining sulfur tended to concentrate at higher molecular weights. This work also found

TABLE 6.9
IFP Hydrofinishing Results on Solvent Refined Naphthenic and Paraffinic Base Stocks

| | Naphthenic | | | | Paraffinic | | |
| | Distillates Type V-20 | | Total Distillates, Lagunillas Crude | | SAE 30, Middle East Crude | | |
	Charge	Hydrofinished Product	Charge	Hydrofinished Product	Charge	Hydrofinished Product	Acid + Clay
Density	0.940	0.935	0.941	0.932	0.895	0.885	0.895
Viscosity at 98.9°C	18.22	17.45	7.53	6.96	9.97	9.67	—
VI	—	—	—	—	95	96	96
Color, D1500	8	3.8	8	2.4	6–7	2.5	2.5
Carbon concentration, wt. %	0.58	0.35	0.24	0.05	0.20	0.10	0.20
Acid index, mg/g	1.04	0.04	5	0.05	0.03	0.015	0.03
Yield, vol. %	—	97.5	—	97	—	98.2	96
H_2 consumption, m^3 at normal temperature and pressure/m^3 charge	—	18	—	32	—	30	—
Pour point, °C	—	—	—	—	–6	–6	–6

Source: R. Dutriau, "Hydrocracking and Hydrorefining of Lubes—I.F.P. Processes," *Chemical Age of India* 17:402–410 (1966).

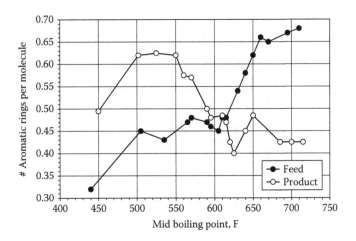

FIGURE 6.13 Hydrofinishing of very light Tia Juana lube distillate: changes in the distributions of the number of aromatic rings per molecule in feed and product by boiling range. *Source:* R. M. Butler and R. Kartzmark, "Chemical Changes in Lubricating Oil on Hydrofining," *Proceedings of the 5th World Petroleum Congress*, Section III, Paper 11, pp. 151–160 (1959). With permission.

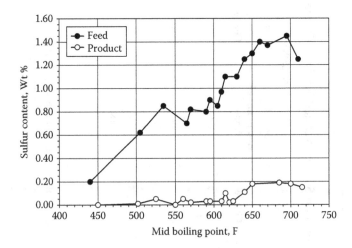

FIGURE 6.14 Hydrofinishing of very light Tia Juana distillate: changes in the sulfur distributions in the feed and product by boiling range.
Source: R. M. Butler and R. Kartzmark, "Chemical Changes in Lubricating Oil on Hydrofining," *Proceedings of the 5th World Petroleum Congress*, Section III, Paper 11, pp. 151–160 (1959). With permission.

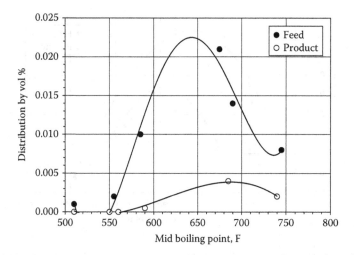

FIGURE 6.15 Hydrofinishing of very light Tia Juana distillate: changes in the benzothiophene distributions in the feed and product.
Source: R. M. Butler and R. Kartzmark, "Chemical Changes in Lubricating Oil on Hydrofining," *Proceedings of the 5th World Petroleum Congress,* Section III, Paper 11, pp. 151–160 (1959). With permission.

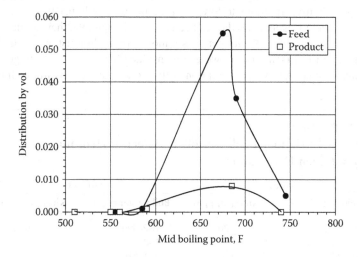

FIGURE 6.16 Hydrofinishing of very light Tia Juana distillate: changes in the dibenzothiophene and naphthothiophene distributions in the feed and product.
Source: R. M. Butler and R. Kartzmark, "Chemical Changes in Lubricating Oil on Hydrofining," *Proceedings of the 5th World Petroleum Congress,* Section III, Paper 11, pp. 151–160 (1959). With permission.

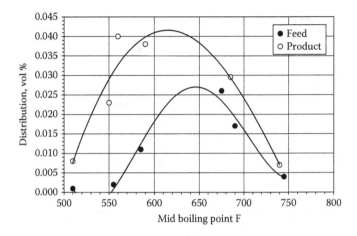

FIGURE 6.17 Hydrofinishing of very light Tia Juana distillate: changes in the distribution of substituted monoaromatics in the feed and product.
Source: R. M. Butler and R. Kartzmark, "Chemical Changes in Lubricating Oil on Hydrofining," *Proceedings of the 5th World Petroleum Congress*, Section III, Paper 11, pp. 151–160 (1959). With permission.

that there was little change in total nitrogen content through the hydrofinishing process, but there was a significant change in nitrogen distribution, with a shift from higher boiling fractions to lower boiling points (Table 6.10).

While total nitrogen levels did not decrease in this example, they did in another case where 78% to 95% of nitrogen was removed by hydrofinishing.[25] The beneficial effects of nitrogen removal were demonstrated[25] by removing nitrogen from Tia Juana base stocks by (1) adsorption with fluorosil and (2) hydrotreating, and

TABLE 6.10
Very Light Tia Juana Lube Distillate: Effect of Hydrofining on Nitrogen Distribution

Feed		Product	
Boiling Range, °F	Nitrogen, ppm	Boiling Range, °F	Nitrogen, ppm
IBP–489	20	IBP–490	76
524–550	10	530–555	40
700–730	620	712–735	330
730+	1280	735+	635
Total sample	190	Total sample	210

Source: R. M. Butler, and R. Kartzmark, "Chemical Changes in Lubricating Oil on Hydrofining," *Proceedings of the 5th World Petroleum Congress*, Section III, Paper 11, pp. 151–160 (1959). With permission.

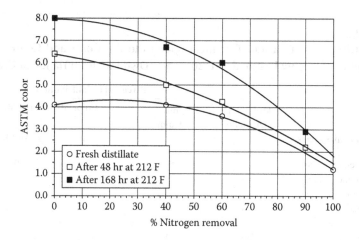

FIGURE 6.18 Effect of nitrogen removal by fluorosil on the color of fresh and oxidized Tia Juana distillate samples.
Source: J. B. Gilbert, R. Kartzmark, and L. W. Sproule, "Hydrogen Processing of Lube Stocks," *J. Inst. Petroleum,* vol. 53, #526, October, pp. 317–327 (1967). With permission.

measuring color before and after oxidation experiments. The curves in Figure 6.18 and Figure 6.19 show that nitrogen removal brings about significant color improvement in oxidation tests. In both cases, nitrogen removal approaching 100% led to small color increases after oxidation. This is supported by the data in Table 6.11,

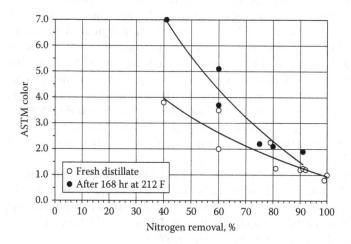

FIGURE 6.19 Effect of nitrogen removal by hydrotreating on the color of fresh and oxidized Tia Juana distillate.
Source: J. B. Gilbert, R. Kartzmark, and L. W. Sproule, "Hydrogen Processing of Lube Stocks," *J. Inst. Petroleum,* vol. 53, #526, October, pp. 317–327 (1967). With permission.

TABLE 6.11

Contributions of Various Chemical Species to the Color and Color Stability of Freshly Distilled SAE 20-Grade Distillate from Tia Juana Crude

| | | Percentage Contribution | | |
| Chemical Species | Mole % in Distillate | Initial Color | Color after Aging at 212°F | |
			48 hr	168 hr
Nitrogen compounds	5	89	98	99
Sulfur and oxygen compounds	19	10	2	0.6
Aromatic hydrocarbons	26	1	0.4	0.4
Paraffins and naphthenes	50	0	0	0

Source: J. B. Gilbert, R. Kartzmark, and L. W. Sproule, "Hydrogen Processing of Lube Stocks," *Journal of the Institute of Petroleum* 53(526):317–327 (1967). With permission.

which shows the effect of removing components from base stocks, nitrogen by fluorosil adsorption, sulfur and oxygen by mild hydrotreatment, and aromatics by column chromatography. The estimate of 99% of color being due to nitrogen compounds is a very strong argument for their removal.

As mentioned in the previous chapter, similar conclusions about the negative effects of nitrogen compounds were found for low amounts of basic nitrogen compounds; specifically, in both solvent refined and group II stocks, their relationship to poor performance in oxidation tests and the formation of varnish in engine tests.[33] Gilbert et al. extended this approach to produce naphthenic base stocks by a single hydrotreatment process step so that an intermediate solvent refining step was no longer necessary. The addition that was necessary was a topping tower, and process conditions were given as 500°F to 800°F catalyst temperatures and pressures up to 400 psig and similar base metal (group VIII and group VIB) catalysts on alumina. Comparative results are provided in Table 6.12 and Table 6.13.

Similar results were reported by BP France[28] using a cobalt/molybdenum catalyst at 20 kg/cm² pressure, temperatures in the range of 200°C to 350°C, and space velocities between 0.5 and 5.0. IFP also developed a hydrofinishing process applicable to naphthenic and solvent refined distillates.[27,34] Bechtel[30] has assumed licensing of Texaco's lube base oil technology.

Chasey and Aczel[35] (ExxonMobil) have studied in detail the compositional changes of the aromatics fraction in a processing scheme (Figure 6.20) that involved a combination of solvent extraction and hydrotreating of a distillate. In this scheme, the distillate is exposed to low and high "severity" extractions and the low severity

TABLE 6.12
Color Properties of Process Oils Made from Tia Juana 102 Distillates

SAE Grade	20			40		
Refining Process	Hydro-treating	Acid and Clay	Solvent Refining and Hydro-finishing	Hydro-treating	Acid and Clay	Solvent Refining and Hydro-finishing
Yield, vol. %	95	90	85	95	84	81
Color, ASTM	2.0	1.3	1.7	3.7	3.0	3.7
Color ASTM after						
16 hr. at 212°F	2.1	2.0	1.9	4.0	4.3	4.2
48 hr. at 212°F	2.3	—	—	4.8	6.0	4.2

Source: J. B. Gilbert, R. Kartzmark, and L. W. Sproule, "Hydrogen Processing of Lube Stocks," *Journal of the Institute of Petroleum* 53(526):317–327 (1967). With permission.

product is hydrotreated at low and high severities, while the high severity extraction product is only given a low severity hydrotreatment step. Dewaxing comes at the end of the process and dewaxed feed is used as the compositional yardstick. Hydrotreatment process conditions were not defined, but from the gross

TABLE 6.13
Inspection Data for Industrial Oils Made from Tia Juana 102 Distillates by Hydrotreating and by Solvent Extraction

SAE Grade	5		20		40	
Process	Hydrotreating	Solvent Extraction and Hydro-finishing	Hydro-treating	Solvent Extraction and Hydro-finishing	Hydro-treating	Solvent Extraction and Hydro-finishing
Yield, vol. %	70	70	70	71	70	68
Nitrogen removal, %	97+	95+	97	94	94	87
Sulfur removal, %	99+	81	99+	78	98	71
Viscosity at 100°F, SUS	100	105	289	297	974	940
VI	64	52	70	67	68	68
Pour point, °F	−35	−30	−20	−10	+10	+20
Color, ASTM	0.1	0.6	1.3	1.6	2.4	3.7
Color, ASTM after 48 hr. at 212°F	0.1	0.9	2.0	1.8	3.8	3.8

Source: J. B. Gilbert, R. Kartzmark, and L. W. Sproule, "Hydrogen Processing of Lube Stocks," *Journal of the Institute of Petroleum* 53(526):317–327 (1967). With permission.

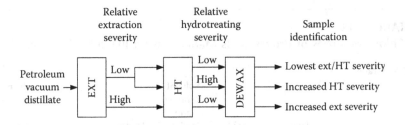

FIGURE 6.20 Processing scheme undertaken to investigate the effect of varying extraction and hydrofinishing severity.
Source: K. L. Chasey and T. Aczel, "Polycyclic Aromatic Structure Distributions by High-Resolution Mass Spectrometry," *Energy & Fuels,* vol. 5, pp. 386–394 (1991). With permission.

compositional changes (Table 6.14) it looks as if these correspond to hydrofinishing (less than 1000 psig) rather than a high pressure case, since mono- and diaromatics are not significantly saturated under these conditions.

It can be seen in Table 6.15 that the total aromatics levels only change significantly in the "severe" extraction case (C), and this is in spite of an increase in the content of the desirable monoaromatic (one ring) structures. In all cases, 1 ring hydrocarbon aromatics increase due to hydrotreatment of benzothiophenes, and monoaromatics increase across the table from left to right. The two-ring aromatics (naphthalenes) increase in the A and B products, but are reduced in the "severely" extracted C case, likely lost in the extract here.

The dewaxed distillate feed contains a range of thiophenes, from benzothiophene and up, but no alkyl thiophenes themselves. The extraction/hydrotreatment process is most effective in removing the two-ring and four-ring or greater thiophenes, and evidently the dibenzothiophenes are the most resistant. The decrease in average Z number is consistent with this. The authors see the decrease in the number of side chain carbons in sample B versus

TABLE 6.14
Clay Gel Aromatics and Heteroatom Contents for Whole Oils

Sample	Extraction Severity	H/T Severity	Clay-Gel Aromatics	Sulfur, wt. %	Nitrogen, ppm
Dewaxed distillate	—	—	44.15	1.76	470
Sample A	Low	Low	42.47	0.97	224
Sample B	Low	High	41.74	0.70	170
Sample C	High	Low	37.40	0.45	86

Source: K. L. Chasey and T. Acel, "Polycyclic Aromatic Structure Distributions by High-Resolution Mass Spectrometry," *Energy & Fuels,* vol. 5, pp. 386–394 (1991). With permission.

TABLE 6.15
Ring Number Distributions for Clay-Gel Aromatic Fractions

Component	Weight Percent			
	Dewaxed Distillate	A	B	C
Hydrocarbon Ring Systems				
1-ring	9.05	13.93	16.94	17.69
2-ring	10.37	11.48	13.25	11.56
3-ring	5.89	5.41	5.10	4.22
4-ring	2.49	1.76	2.51	1.39
5+-ring	0.80	0.31	0.81	0.63
Total	28.60	32.89	38.61	35.49
Thiophene Ring Systems				
1-ring	Traces	Traces	Traces	Traces
2-ring	6.72	2.87	Traces	Traces
3-ring	6.99	5.49	2.80	1.62
4-ring	1.32	0.77	0.33	0.00
5+-ring	0.31	0.06	0.00	0.00
Total	15.34	9.19	3.13	1.62
Furan Ring Systems				
Total	0.21	0.40	Traces	0.30
Overall				
Grand total	44.15	42.47	41.74	37.40
Z number	14.8	13.6	13.3	12.4
Number of carbons in side chains	12.2	13.5	13.3	15.5

Source: K. L. Chasey and T. Aczel, "Polycyclic Aromatic Structure Distributions by High-Resolution Mass Spectrometry," *Energy & Fuels*, vol. 5, pp. 386–394 (1991). With permission.

sample A as being due to hydrocracking, so conditions must have been at the severe end of hydrofinishing. The increased length of the side chains in C fits with an expected increase in VI due to removal of polyaromatics with short side chains. It would have been very instructive to have similar analyses on the extracts, but these were not included in the paper. Overall this paper presents valuable insights not only into the chemistry of the process, but also into the analytical methods employed and the compositional changes occurring during this process.

Limitations on hydrofinishing can be posed by the feed. Those that contain high levels of polyaromatics and nitrogen compounds can cause catalyst deactivation by adsorption on the active sites. If catalyst temperatures have to be increased too much, the low operating pressure of these processes can lead to polyaromatics formation due to reversal of the thermodynamic equilibrium, with loss of product quality.[36]

Catalysts available for hydrofinishing have improved substantially since the appearance of these publications, although the chemistry will be the same or very similar. Both license and catalyst suppliers should be interviewed thoroughly and pilot plant work undertaken where necessary to select the best fit for the specific needs of a particular refiner.

REFERENCES

1. J. Edealanu, "The Refining Process with Liquid Sulphur Dioxide," *Journal of Petroleum Technology* 18:900–920 (1932).
2. A. W. Francis and W. H. King, "Principles of Solvent Extraction," in *The Chemistry of Petroleum Hydrocarbons*, vol. 1, B. T. Brooks, C. E. Boord, S. S. Kurtz, and L. Schmerling, eds. (New York: Reinhold Publishing, 1954).
3. M. R. Cannon and M. R. Fenske, "Composition of Lubricating Oil," *Industrial and Engineering Chemistry* 31:643–648 (1939).
4. B. M. Sankey, D. Bushnell, and D. A. Gudelis, "Exol N: New Lubricants Extraction Process," *Proceedings of the 10th World Petroleum Congress*, Section 4, pp. 407–414 (1980).
5. F. C. Jahnke, "Solvent Refining of Lube Oils. The MP Advantage," presented at the American Institute of Chemical Engineers fall meeting, Miami, Florida, November 2–7, 1986.
6. "Exol N Solvent Extraction Technology," brochure from ExxonMobil Research and Engineering, Fairfax, Virginia.
7. Lubes 'N' Greases, *2005 Guide to Global Base Oil Refining* (Falls Church, VA: LNG Publishing).
8. A. Sequeira, Jr., *Lubricant Base Oil and Wax Processing* (New York: Marcel Dekker, 1994), 105.
9. T. R. Farrell and J. A. Zakarian, "Lube Facility Makes High Quality Lube Oil from Low-Quality Feed," *Oil and Gas Journal* May 19:47–51 (1986).
10. B. M. Sankey, "A New Lubricants Extraction Process," *Canadian Journal of Chemical Engineering* 63:3–7 (1985).
11. G. A. Purdy, *Petroleum: Prehistoric to Petrochemicals*, Vancouver: Copp Clark Publishing, 1958.
12. S. Marple and L. J. Landry, "Modern Dewaxing Technology," in *Advances in Petroleum Chemistry and Refining*, vol. X, J. J. McKetta, Jr., ed., New York: Interscience, 1965.
13. "Propane Dewaxing Technology," brochure from ExxonMobil Research and Engineering, Fairfax, Virginia.
14. R. J. Taylor and A. J. McCormack, "Study of Solvent and Catalytic Lube Oil Dewaxing by Analysis of Feedstocks and Products," *Industrial and Engineering Chemistry Research* 31:1731–1738 (1992).
15. R. J. Taylor and A. J. McCormack, "A Comparison of Solvent and Catalytic Dewaxing of Lube Oils," Symposium on Processing, Characterization, and Application of Lubricant Base Oils, Division of Petroleum Chemistry, meeting of the American Chemical Society, August 23–28, 1992.

16. D. A. Gudelis, J. D. Bushnell and J. F. Eagen, "Improvements in Dewaxing Technology," *API Proceedings* 65:724–737 (1973).
17. D. A. Gudelis, J. F. Eagen and J. B. Bushnell, "New Route to Better Wax," *Hydrocarbon Processing (International Edition)* 52(9):141–146 (1973).
18. V. A. Citerella, E. A. Ruibal, S. Zaczepinski, and B. E. Beasley, "Crystallization Technique to Simplify Dewaxing," *Petroleum Technology Quarterly* Winter:37–43 (1999/2000).
19. N. A. Bhore, R. M. Gould, T. L. Hilbert, M. P. McGuiness, D. McNally, P. H. Smiley, and C. R. Wildemuth, "Membranes Debottleneck Lube and Wax Production," Paper LW-99-128, presented at the Lubricants and Waxes meeting of the National Petrochemical and Refiners Association, Houston, Texas, November 11–12, 1999.
20. N. A. Bhore, R. M. Gould, T. L. Hilbert, B. S. Minhas, S. A. Tabak, A. P. Werner, and C. R. Wildemuth, "Membrane Technology for Wax Deoiling," Paper LW-00-129, presented at the Lubricants and Waxes meeting of the National Petrochemical and Refiners Association, Houston, Texas, November 9–10, 2000.
21. V. L. Kalichevsky and K. A. Kobe, "Petroleum Refining with Chemicals," in *Refining with Adsorption,* (New York: Elsevier, 1956), 244–311.
22. *Hydrocarbon Refining and Static Bed Percolation* (Houston: Porocel Corporation), .
23. C. Go, T. F. Wulfers, M. P. Grosboll, and F. F. McKay, "Treatment of Off-Specification White Mineral Oil Made by Two Stage Hydrogenation," U.S. Patent 5,098,556.
24. W. A. Jones, "Hydrofining Improves Low-Cost-Lube Quality," *Oil and Gas Journal* 53(26):81–84 (1954).
25. J. B. Gilbert, R. Kartzmark, and L. W. Sproule, "Hydrogen Processing of Lube Stocks," *Journal of the Institute of Petroleum* 53(526):317–327 (1967).
26. M. Moret, "Un Example de Production d'Huiles de Base à Partir de Base Hydrocracquées," *Petrole et Techniques* 333:37–42 (1987).
27. R. Dutriau, "Hydrocracking and Hydrorefining of Lubes—I.F.P. Processes," *Chemical Age of India* 17:402–410 (1966).
28. A. Champagnat, J. Demeester, and C. Roit, "Desaromatisation Catalytique des Distillats Legers et Raffinage Hydrogenant des Huiles Lubrifiantes," *Proceedings of the 5th World Petroleum Congress*, Section III, Paper 13 (1959).
29. R. W. Geiser and L. E. Hutchings, "Quality Lubricants from Pennsylvania Grade Crude Oil by the Isomax Process," Preprint 57–73, presented at the 38th midyear meeting of the American Petroleum Institute, Division of Refining, Philadelphia, May 17, 1973.
30. "Bechtel Lube Base Oil Manufacturing Technology," bulletin from Bechtel, Corp., San Francisco.
31. J. B. Gilbert and R. Kartzmark, "Advances in the Hydrogen Treating of Lubricating Oils and Waxes," *Proceedings of the 7th World Petroleum Congress*, Section IV, pp. 193–205 (1967).
32. R. M. Butler, and R. Kartzmark, "Chemical Changes in Lubricating Oil on Hydrofining," *Proceedings of the 5th World Petroleum Congress*, Section III, Paper 11, pp. 151–160 (1959).
33. T. Yoshida, J. Igarashi, H. Watanabe, A. J. Stipanovic, C. Y. Thiel, and G. P. Firmstone, "The Impact of Basic Nitrogen Compounds on the Oxidative and Thermal Stability of Base Oils in Automotive and Industrial Applications," SAE Paper 981405 (Warrendale, PA: Society of Automotive Engineers, 1998).

34. A. Billon, J.-P. Franck, and J.-P. Peries, "Procede d'Hydroraffinage Pour La Production d'Huiles Lubrifiantes," *Proceedings of the 10th World Petroleum Congress*, Section 4, pp. 211–220 (1980).
35. K. L. Chasey and T. Aczel, "Polycyclic Aromatic Structure Distributions by High-Resolution Mass Spectrometry," *Energy & Fuels* 5:386–394 (1991).
36. H. M. J. Bijwaard, W. K. J. Brener, and P. van Doorne, "The Shell Hybrid Process, An Optimized Route for HVI (High Viscosity Index) Luboil Manufacture," presented at the Petroleum Refining Conference of the Japan Petroleum Institute, Tokyo, October 27–28, 1986.

7 Lubes Hydrocracking

7.1 INTRODUCTION

The concept of the widespread production of base stocks by hydrocracking emerged about the same time as the post-World War II development of fuels hydrocracking technology since the catalysts, hardware, and peripherals are all very similar. Standard Oil[1,2] had actually built a plant in the 1920s in Bayway, New Jersey, to make lubricants called "Essolube" by hydrocracking and employing IG Farbenindustrie technology, but this was a short-lived venture. Most of the other pre-World War II work was in connection with coal-to-oil conversions.

In this chapter we will discuss a number of commercial lube plants whose purpose is to make base stocks by hydrocracking and whose design basis has been outlined in papers from either the company that developed the specific technology or by a licensee. In the case of a licensee, the technology would have been adapted (and developed) as needed to meet that company's circumstances, including capital available, feedstocks and their qualities, existing refinery infrastructure, and eventual product marketplace.

This subject area is addressed here most conveniently in a more or less historical manner and begins in the late 1960s. At that time, hydrocracked base stocks with low aromatic levels and almost no sulfur or nitrogen were just coming on the market from the pioneering lube plants using this technology. Refiners were gaining their first manufacturing experience and downstream blenders and formulators were learning how to develop products. There was little consensus as to what these new oils should look like. Should they look like "real" base stocks from solvent refining and be colored and contain substantial aromatics levels or should they be water white and have almost zero aromatics? Could the problems they brought with them—poor additive solubility, the absence of "natural inhibitors," and limited volumes and sources, be overcome?

While the detailed chemistry of hydrocracking is dealt with in the next chapter, a brief outline here will be of use. Hydrocracking is the reaction of a waxy feed, usually a paraffinic distillate or deasphalted oil (DAO), and hydrogen in the presence of a catalyst that promotes molecular reorganization and cracking. The reactions include saturation of aromatics (hydrodearomatization [HDA]) and nearly complete elimination of sulfur (hydrodesulfurization [HDS]) and nitrogen (hydrodenitrification [HDN]) in which some lower molecular weight products are produced by cracking and there is some opening of cycloparaffin rings.

These reactions take place normally in a trickle bed downflow reactor at high hydrogen pressures (1500 to 4000 psi) and temperatures in excess of 600°F. The catalyst is frequently referred to as "dual function" since it promotes both cracking and hydrogenation. The catalyst base material is usually silica-alumina with metal oxides from groups VI and VIII (nickel, cobalt, molybdenum, tungsten). The active forms are the metal sulfides. Fractionation provides the waxy lubes cuts which are subsequently dewaxed and hydrofinished.

The term "hydrocracking" is usually taken to mean that significant molecular weight reduction occurs due to cracking of carbon–carbon bonds, sometimes referred to as "conversion" of feed molecules to lower boiling products. It has been defined as using temperatures greater than 650°F and pressures above 1000 psi with high-activity catalysts.[3] "Hydrotreating" means milder conditions that cause much less cracking, using temperatures of greater than 600°F and pressures above 500 psi.[3] Gulf called their lube process a "hydrotreating" one in reference to the first stage, meaning that waxy lube yields were high due to relatively little cracking, but most people now would likely term it a hydrocracking process. "Hydrofinishing" is usually applied to the final hydroprocessing step, in which only aromatic saturation occurs.

7.2   GROUP II BASE STOCK PRODUCTION

7.2.1   IFP Technology: Empress Nacional Calco Sotelo Refinery in Puertollano, Spain

The first commercial plant[4] to manufacture lubricant base stocks by the hydrocracking process was the Empress Nacional Calco Sotelo refinery in Puertollano, Spain. This plant used Institut Francais du Pétrole's (IFP's) newly developed hydrorefining technology, and was designed to produce base stocks with a viscosity index of 95 or higher. It came online in 1967 using heavy vacuum gas oil (HVGO) and DAO from Aramco Middle East crudes. The second plant producing hydrocracked lubes was Idemitsu Kosan's at their Chiba, Japan, refinery, coming online in 1969, and this in turn was followed by Sun's Yabacoa, Puerto Rico, refinery in 1971. Both of these were Gulf (now part of Chevron) licensees. The schematic of the Spanish plant (Figure 7.1) illustrates the usual unit configuration for lubes hydrocrackers. A high-pressure pump delivers the feedstock to a fired heat exchanger, after which it is mixed with high-pressure recycled hydrogen and delivered to the top of the trickle bed reactor. The schematic indicates five quench ports on the reactor to control exothermic temperature increases in the beds. The operating pressure is given as 2800 psi. The reactor effluent exits the reactor to a high-pressure separator that provides the recycled hydrogen. After removal of the light gases (C_1–C_5, H_2S, and NH_3) in the low-pressure separator, the atmospheric tower fractionates out the naphtha, jet, and diesel products with the waxy bottoms going to a vacuum tower where the waxy lubes are fractionated out. These cuts are sent to tankage, ultimately to be block solvent dewaxed in the

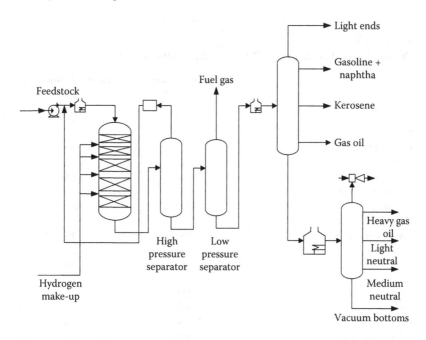

FIGURE 7.1 Process schematic for IFP lube hydrocracking plant in Puertollano, Spain. *Source:* J. Angula, M. Gasca, J. L. Martinez-Cordon, R. Torres, A. Billon, M. Derrien, and G. Parc, "IFP Hydrorefining Makes Better Oils," *Hydrocarbon Processing* 47(6):111–115 (1968). With permission.

final step (not shown). The properties of the two feeds (Table 7.1) show both are quite good quality in VI terms with dewaxed VIs of 53 for the HVGO and 75 for the DAO. (Feed dewaxed VI is one of several measures of feed quality since the hydrocracking process must increase VI.) No base stock stabilization step, by either solvent extraction or hydrofinishing, was included in this groundbreaking plant. The need for a final stabilization step became clearer in the industry with further experience.

The design intention was to make base stocks of a standard VI (95), and by increasing reactor temperature, to produce those with VIs greater than 100 for multigrade automotive products. As it was the intent to make base stocks of high viscosity, the feeds were a HVGO with a viscosity of 14 cSt at 98.9°C and a DAO whose viscosity was 40.3 cSt. As Figure 7.1 shows, the vacuum tower gave three lube cuts with dewaxed viscosities of 5.3, 7.7, and 11 cSt from the HVGO, and from the DAO they were 7.8, 12.8, and 26.5 cSt, all measured at 98.9°C (Table 7.2). Waxy lube yield from the DAO was 87.3% at a medium neutral dewaxed VI of 95. Waxy lube yields decreased with increasing VI, and with DAO as feed, this decrement was about a 1% yield drop per dewaxed VI unit.

After solvent dewaxing, the base stocks had ASTM D1500 colors of less than 2.0, sulfur contents of 0.05 to 0.1 wt. %, and exhibited what is now the familiar

TABLE 7.1
Properties of Feeds to the Puertollano, Spain, Plant

	Aramco Heavy Distillate	Aramco Deasphalted Oil
Gravity, °API	20.0	20.5
Viscosity, cSt at 210°F	14.12	40.31
Dewaxed VI	53	75
Sulfur, wt. %	2.6	2.1
Total nitrogen, wt. %	0.116	0.0619
Ramsbottom carbon, wt. %	0.57	1.58
Asphaltenes, wt. %	0.05	0.05
Distillation, ASTM D1160, °C, vol. %		
Initial boiling point	380	456
5	467	550
10	488	568
30	511	599
50	522	636
70	535	—
90	548	—
Final boiling point	572	—

Source: J. Angula, M. Gasca, J. L. Martinez-Cordon, R. Torres, A. Billon, M. Derrien, and G. Parc, "IFP Hydrorefining Makes Better Oils," *Hydrocarbon Processing* 47(6):111–115 (1968). With permission.

"VI droop" (Table 7.2), in which the distribution of dewaxed VI versus viscosity for the total base stock product shows a sharp drop below about 5 cSt (about 150 SUS) at 100°C. This means that the nominal light base stock has a lower VI than the medium neutral. This is in contrast to solvent refined stocks, whose VIs are generally considered to remain essentially constant.

TABLE 7.2
Typical Properties of Base Stocks from the IFP Process at Puertollano, Spain

	HVGO			DAO		
Feed	Light Oil	Medium Oil	Heavy Oil	Light Oil	Medium Oil	Heavy Oil
Viscosity, cSt at 210°F	5.26	7.72	10.99	7.84	12.78	26.50
VI (ASTM D567)	75	94	98	93	97	98
Pour point, °C	−13	−10	−11	−14	−12	−13

Source: J. Angula, M. Gasca, J. L. Martinez-Cordon, R. Torres, A. Billon, M. Derrien, and G. Parc, "IFP Hydrorefining Makes Better Oils," *Hydrocarbon Processing* 47(6):111–115 (1968). With permission.

TABLE 7.3
IFP Comparison Between Lube Production Yields by Hydrocracking and by Furfural Extraction

	Aramco Heavy Distillate		Aramco Deasphalted Vacuum Residuum	
	Hydrocracking	Furfural	Hydrocracking	Furfural
Yield in total oil, wt. %	83.5	60.0	86.0	61.0
Waxy viscosity at 210°F, cSt	8.7	10.5	16.4	33.6
Dewaxed VI	97	95	98	98
Flash point, °C	225	245	240	280
Product distribution				
>600N	—	—	57.9	61.0
600N	—	—	10.10	—
400N	60.4	60.0	—	—
350N	—	—	9.0	—
200N	18.0	—	—	—
150N	—	—	9.0	—
100N	5.1	—	—	—

Source: J. Angula, M. Gasca, J. L. Martinez-Cordon, R. Torres, A. Billon, M. Derrien, and G. Parc, "IFP Hydrorefining Makes Better Oils," *Hydrocarbon Processing* 47(6):111–115 (1968). With permission.

The hydrocracking process gave dramatically higher total waxy lube yields (83% to 86%) when compared with furfural solvent refining (60% to 61%) at the same total lubes VI (Table 7.3). As well, hydrocracking gave a different base stock yield distribution, since overall viscosity was less due to molecular weight reduction by cracking and by saturation of multiring aromatics to monoaromatics and saturates.

Figure 7.2 shows results published by Chevron[5] that illustrate very well the yield benefit of hydrocracking over solvent extraction as feed quality changes for production, in this case, of waxy 100N (i.e., regardless of feed VI, hydrocracking always gives superior yields of waxy product and the delta increases as feed quality gets worse). It can be seen that for the production of waxy 100 VI base stock, yields for both hydrocracking and solvent extraction decline as feed VI decreases, but the decrease is much greater for the extraction route, so that at a feed VI of 20 to 25, hydrocracking remains a feasible alternative while extraction is no longer possible economically.

The Puertollano plant illustrates a mode of operation for a hydrocracking unit in that each feedstock yields several waxy base stocks after vacuum fractionation. This requires feeds with wide boiling ranges. In such cases, when severity is increased to obtain higher VIs, it is still possible to produce overhead and bottoms base stocks of the same viscosity, but cut point adjustments are required. This will affect base stock yields, particularly of the heaviest base stocks.

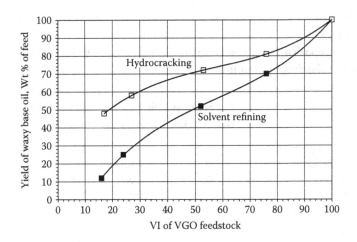

FIGURE 7.2 Comparison of waxy 100N yields by hydrocracking and solvent refining. *Source:* S. J. Miller, M. A. Shippey, and G. M. Masada, "Advances in Lube Base Oil Manufacture by Catalytic Hydroprocessing," Paper FL-92-109, presented at the 1992 National Fuels and Lubricants meeting of the National Petroleum Refiners Association, Houston, Texas, November 5–6, 1992. Figure copyrighted by Chevron Corporation and used with permission.

An alternative mode is for each feedstock to produce a single waxy stock after fractionation out of nonlubes distillates. To operate in this manner, each feedstock must be fractionated in the crude vacuum tower to have the right viscosity and distillation cut points to meet the eventual base stock specifications after hydrocracking/hydrotreating and dewaxing. This reduces VI droop since each base stock can be processed at the correct severity.

Additional process factors influencing VI droop are illustrated in Figure 7.3[6] from results from Gulf's development work on the Gulf hydrotreatment process. This figure shows that VI distributions in the total lube product are dependent on feed source. In this example, a high-quality paraffinic feed and a naphthenic one gave much less VI droop than a highly aromatic one. The VI variation in the higher viscosity ranges are not readily explainable at this time due to insufficient information and may be due to a further inadequacy in the VI method.[8]

Viscosity index droop became very significant when Chevron[7] considered (and used) poor quality Alaskan North Slope oil as feedstock for their new Richmond, California, plant. Table 7.4 shows that the Alaskan North Slope HVGO with a dewaxed VI of only 15 has a droop between the 500N and 100N of 25 units, which is twice that of the same two stocks produced from Arabian light HVGO, with a feed dewaxed VI of 50.

Severity also affects the magnitude of VI droop, with Gulf results[6] that show high severity giving greater VI differences between low and high viscosity stocks

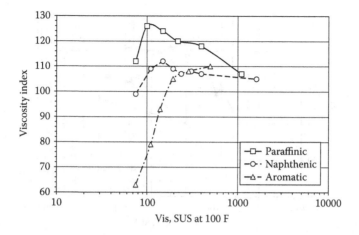

FIGURE 7.3 Effect of feed type on the distribution of VI after bulk hydrocracking. *Source:* M. C. Bryson, W. A. Horne, and H. C. Stauffer, "Gulf's Lubricating Oil Hydrotreating Process," *Proceedings of the Division of Refining of the American Petroleum Institute* 49:439–453 (1969). With permission.

(Figure 7.4), all produced from a common Kuwait DAO. This effect of severity is also seen in data presented by Billon et al.[8] from hydrocracking of Kuwait DAO at mild and high severities (Table 7.5), where it can be seen that high severities produce a wider range in VIs within the product, in contrast to the low severity situation. Within the product distillate cuts, the %C_A increases as boiling point decreases, a not uncommon occurrence, and is essentially zero at the highest

TABLE 7.4
VI Droop When Bulk Hydrocracking Alaskan North Slope and Arabian Light HVGO Feedstocks

	Alaskan North Slope (Feed Dewaxed VI = 15)		Arabian Light (Feed Dewaxed VI = 50)	
	VI	Droop	VI	Droop
500N	100	Base	100	Base
240N	92	8	99	1
100N	75	25	88	12

Source: T. R. Farrell and J. A. Zakarian, "Lube Facility Makes High-Quality Lube Oil from Low-Quality Feed," *Oil and Gas Journal* May 19:47–51 (1986). With permission.

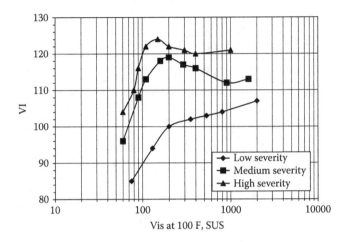

FIGURE 7.4 Effect of severity on dewaxed VI distribution in products from hydrocracking Kuwait DAO.
Source: M. C. Bryson, W. A. Horne, and H. C. Stauffer, "Gulf's Lubricating Oil Hydrotreating Process," *Proceedings of the Division of Refining of the American Petroleum Institute* 49:439–453 (1969). With permission.

severity. In all the cases reported in this paper, the $\%C_P$ increased with increasing viscosity within a product distillate.

Some general guidelines provided by Billon et al. on lubes production via hydrocracking were

- Lube oils with a VI of 120 to 130 can be produced from virtually any crude.
- Hydrocracking broadens the range of crudes that can be used to make base stocks.
- Lube oil yield decreases as feed dewaxed VI decreases.
- Viscosity index droop is a general feature whose extent increases as feed quality (VI) decreases.
- Poor feeds require higher severity and higher hydrogen consumption.

The commercial significance of VI droop lies in the difficulty of producing a slate of base stocks with a constant VI (e.g., the "standard" 95 through the entire viscosity range). Operating the unit to make a 100N with this VI usually will mean that the heavier base stocks can have VIs substantially in excess of 100, which means VI "giveaway." From the foregoing, as well, changing crude feeds can alter the delta between light and heavy, making consistency difficult. Alternatively, having the heavier stocks on target with respect to VI will mean that the lighter stocks may not meet 95 VI, the market minimum for paraffinic stocks, and they will therefore be a commercial liability. Several methods have been taken to circumvent this problem:[6]

TABLE 7.5
Hydrocracking Kuwait DAO at Mild and High Severities

Test	Total Oil at 715°F	Distillation Fractions Mild Severity		
		0–25	25–50	50–100
Viscosity, cSt				
At 100°F	269.25	52.40	239.2	686.4
At 210°F	20.30	6.72	17.96	38.92
VI (ASTM D567)	95.5	86.5	89	100
Pour point, °C	−18	−18	−18	−15
Composition, n-D-M				
C_A, %	7	9.9	6.5	6.0
C_N, %	24.6	29.3	26.3	23.4
C_P, %	68.4	61.8	67.2	70.6

Test	Total Oil at 715°F	High Severity			
		0–25	25–50	50–75	75–100
Viscosity, cSt					
At 100°F	30.36	17.61	23.46	35.88	83.67
At 210°F	5.71	3.81	4.71	6.34	11.64
VI (ASTM D567)	138	133.4	136	134	128
VI_E (ASTM D2270)	144	118	132.3	140	142
Pour point, °C		−18	−18	−18	−15
Composition, n-D-M					
C_A, %	0	0.5	0.7	0	0
C_N, %	15.2	17.1	16	12.3	11.5
C_P, %	84.8	82.4	83.3	87.7	88.5

Source: A. Billon, M. Derrien, and J. C. Lavergne, "Manufacture of New Base Oils by the I.F.P. Hydrofining Process," *Proceedings of the Division of Refining of the American Petroleum Institute* 49:522–548 (1969). With permission.

- Recycle the low VI lighter stock(s) to the reactor to adjust VI upward with reportedly[6] only minor yield loss.
- Recycle the high VI material to extinction.
- Solvent extract the feedstock as per Sun (see below) and others to remove low VI feed components.
- Fractionate the feed into a number of distillation "blocks" such that hydrotreating each block in sequence through the hydrocracker produces one base stock taken as a bottoms from the vacuum tower. Severity for each block is adjusted to yield the target VI.
- Build separate hydrocrackers for light and heavy feeds.

7.2.2 GULF TECHNOLOGY: SUN'S YABACOA, PUERTO RICO, PLANT

Sun Oil, the second largest lubes producer in the United States in 1972, addressed the issue of VI droop for their 12,000 barrels per day (bpd) hydrocracking lube plant (licensed from Gulf Oil) that came online in 1972 in Yabacoa, Puerto Rico[9,10] (this plant was sold in 2001 to Shell Chemicals and converted to other uses). Sun selected hydrocracking as the lube process over solvent extraction for a number of reasons, including (1) reduced dependence on high VI lube crudes, (2) higher base stock yields and more flexibility in the slate of lubes, (3) the ability to make base stocks with VIs greater than 100, and (4) base stocks from hydrocracking had been demonstrated to have superior response to oxidation inhibitors. This was one of the first hydrocracking lube units to be built in North America.

The new plant employed a 6000 bpd solvent extraction unit to remove the low VI feed aromatic components from the HVGO stream. Their removal was thought to bring a number of process and product quality benefits. These components were considered responsible for reducing the VI of lighter lube stocks by being partially cracked into those molecular weight ranges. Figure 7.5 is a schematic of the plant in which the furfural extraction unit (FUR. EXTN.) is used

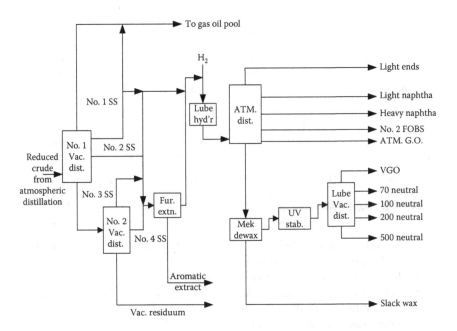

FIGURE 7.5 Plant schematic for Sun's Yabacoa, Puerto Rico, base stock plant.
Source: I. Steinmetz and H. E. Reif, "Process Flexibility of Lube Hydrotreating," *Proceedings of the Division of Refining of the American Petroleum Institute* 53:702–712 (1973). With permission.

TABLE 7.6
Effect of Hydrotreater Feedstock Boiling Range on 100N VI

Nominal Boiling Range of Hydrotreater Feed, °F	VI of 100N Lube Grade Produced
700–825	113
825–875	113
875–925	73
925–1025	63
DAO (1025+ bottoms)	88
Full boiling range feed, 700°F through DAO	86

Source: I. Steinmetz and H. E. Reif, "Process Flexibility of Lube Hydrotreating," *Proceedings of the Division of Refining of the American Petroleum Institute* 53:702–712 (1973). With permission.

to upgrade the heavier feed components—No. 2 SS, No. 3 SS, and No. 4 SS—via a mild extraction prior to their being hydrocracked.

This conclusion, that the "bad actors" were part of the heavy aromatics, came from their pilot plant hydrocracking studies undertaken on narrow boiling range cuts at constant reactor conditions of temperature, pressure, and throughput (Table 7.6). These results were interpreted to mean that some of the low VI aromatic components in the higher boiling range fractions were cracked into the 100N boiling range, where they depressed the VI of the 100N components. An alternative explanation is that the higher boiling range feed cuts contain higher nitrogen levels which depress catalyst activity relative to that experienced by, for example, the 700°F to 825°F cut. The result of course would be less cracking activity due to depressed catalyst acidity and therefore a contribution to lower VI in the product. In addition, we can expect that the light feedstocks will have higher dewaxed oil VIs than the heavier ones.

Whatever the exact explanation, on the process side, the mild solvent extraction step of the heavier feed fractions resulted in no less than doubling the hydrocracker space velocity, a spectacular benefit. Needless to say, feedstock solvent extraction became a popular process step. Lube yields based on feed also improved by 5 vol. % although based on crude, lube yields actually declined (Table 7.7). As IFP had found previously, a decrease in feed dewaxed VI reduced yields at a constant product VI, in Sun's case by about 1.3 volume percent per VI change (Table 7.8).

IFP's paper on their Spanish plant described their hydrocracked stocks as having about the same stability to oxidative storage conditions as those from solvent refining. This has generally not been industry experience elsewhere. In fact, unfinished first-stage hydrocracked base oils, waxy or dewaxed, have been found to darken in color over relatively short times and form haze or insoluble deposits, and these effects are accelerated when samples are exposed to high (about 100°C) temperatures or sunlight in so-called window tests, which were later replaced by more specific exposure to ultraviolet (UV) sunlamps. This was

TABLE 7.7
Effect of Solvent Extraction on Space Velocity and Yields

	Unextracted Distillate Hydrotreater Feed	Partially Extracted Distillate Hydrotreater Feed
5%/95% boiling points of feed	715–940°F	715–940°F
Space velocity, LHSV	X	2X
70+N lube yield		
Vol. % of feed	50	55
Vol. % of crude	10.5	10.1

Source: I. Steinmetz and H. E. Reif, "Process Flexibility of Lube Hydrotreating," *Proceedings of the Division of Refining of the American Petroleum Institute* 53:702–712 (1973). With permission.

a troublesome issue in the early days of hydrocracking and the cause of process oil customer complaints. This instability is now considered to be due to the presence of trace quantities of polyaromatics[11,12] of various types formed in the hydrocracking process either by chemical reactions or due to equilibrium reversal under the operating conditions.

Shell authors[13] have drawn attention to the importance of aromatics "types" in the final base stocks. Daylight instability of hydrocracked stocks has been correlated with UV absorbance in a paper by Pillon.[14] From a very limited suite of group III stocks, those with 260 to 285 nm absorbance of greater than 2 were found to be unstable (less than 5 days to produce floc or precipitate), while low absorbance gave greater stability (e.g., a less than 0.4 absorbance led to failure after 44 days). Accordingly, catalytic hydrofinishing has to be performed at low temperatures, most easily with highly active noble metal catalysts to prevent equilibrium conditions leading to formation of polycyclic aromatics. Some processes have employed solvent extraction to remove the responsible components.

TABLE 7.8
Effect of Feed Dewaxed VI on Lube Yield at Constant 100N VI of 100 at 0°F Pour Point

Crude Source	X	Y	Z
Feed dewaxed VI	75	60	43
Lube yield, vol. % charge	70	55	30

Source: I. Steinmetz and H. E. Reif, "Process Flexibility of Lube Hydrotreating," *Proceedings of the Division of Refining of the American Petroleum Institute* 53:702–712 (1973). With permission.

TABLE 7.9
Thermal Oxidation Stability of Hydrocracked Oil Components

Component	Thermal Oxidation Stability Test (JIS Method)		Content, wt. %
	Judgment	Sludge, mg	
Saturates	Pass	0.1	93.7
Monoaromatics	Pass	0.2	4.7
Diaromatics	No	3.3	0.5
Tri-, tetraaromatics	No	22.7	0.9
Pentaaromatics	Pass	0.2	0.1
Resins	No	6.3	0.1

Source: M. Ushio, K. Kamiya, T. Yoshida, and I. Honjou, "Production of High VI Base Oil by VGO Deep Hydrocracking," Symposium on Processing, Characterization and Application of Lubricant Base Oils, Division of Petroleum Chemistry, American Chemical Society, Washington, DC, August 23–28, 1992. With permission.

Ushio et al.[12] (Nippon Oil) carried out a thorough study of these aspects in their work on hydrocracked lubes. Their JIS thermal stability tests (170°C for 24 hr at 1 atm) on the separated components (Table 7.9) pointed to the tri- and tetra-aromatics as causing instability, these being pyrenes and their dihydro- and tetrahydro derivatives. Addition of phenanthrene, pyrene, and hexahydropyrene to a poly-alpha-olefin (PAO) showed that the hexahydropyrene was the greatest contributor to both color and sludge formation in the thermal stability test. They found that solvent extraction (furfural) was superior to hydrotreating in removing these partially hydrogenated compounds, although this involves a definite yield penalty. Whatever the exact cause, the sure solution is removal of any polyaromatics present.

One of Sun's objectives for the new plant was that the base stocks be at least of the same quality as those from solvent refining, therefore this hurdle had to be overcome. Their solution was to give the total dewaxed hydrocrackate product a light furfural extraction (about 97% raffinate) to remove the tri+ aromatics which appeared to be at the root of the problem.[15,16] Thus their process involved two solvent extraction steps, one before the hydrocracker and one afterwards. This route gave stable base stocks, but with some residual color. An example of the improvement obtained can be seen in Table 7.10, where 100, 200, and 500 SUS base stocks (#1, #2, and #3 in Table 7.10) of approximately 110 VI from hydrocracking a distillate/DAO blend were tested for stability before and after furfural extraction. The results show that the extraction improves color relative to unextracted samples for the immediate hydrocracker products and that their performance in the stability test was improved as well by extraction.

Sun's patent demonstrates the yield and stability improvements that are obtained by feed extraction. It can be seen that this step has quite a significant

TABLE 7.10
Effect of Furfural Extraction on Base Stock Stability

Distillate Fraction	Typical Commercial Product			Product Obtained by Hydrotreating					
				Unextracted			Extracted		
	#1	#2	#3	#1	#2	#3	#1	#2	#3
ASTM D1500 color	0.5	0.75	1.25	0.25	0.75	0.25-1.25	0.25-0.75	0.25-0.75	0.75-1.0
ASTM D1500 color after stability test	2.5	2.5	3.5	5.5	4.5	4.5	3.0	3.0	3.25
Sludge		Light	Haze	None	Heavy	Medium	Haze	Medium	Light	None

Source: S. L. Thompson, "Stabilizing a Hydrocracked Lube Oil by Solvent Extraction," U.S. Patent 3,781,196.

effect on the feed properties (Table 7.11): aromatics are reduced from 50% to 37%, viscosities are reduced due to removal of those (largely polyaromatic) molecules, and waxy and dewaxed VIs increase by 30 to 40 points. Finally, yield on charge to the hydrocracker increases from 39.4 to 45.7 volume percent. The stability test results reported in Table 7.12 show that the extracted feed clearly gives superior quality products in their resistance to oxidation during stability tests as measured by color and sludge formation.

Sun's extraction processes still produced a base stock with some color and color increased with base stock viscosity. Obviously an extraction process can only go so far in removing polyaromatics. Many plants built later put in a second-stage hydrotreater to stabilize products by conversion of the polyaromatics to polycyclic naphthenes rather than by separation. Hydrotreatment also has the capability to give water white base stocks, which many customers prefer. Sun did investigate this finishing route as well but do not appear to have pursued it commercially.[17]

Sun's work on using feedstock extraction prior to hydrocracking was pioneering. This route has also been employed by a number of other companies as fresh generations of lube hydrocracking processes have been developed. Apart from yield and product quality benefits, this adaptation of new and old technologies reduces catalyst temperatures and increases catalyst life, both attractive ends for any refiner. Applications will be seen in some of the following pages.

7.2.3 Shell's Hydroprocessed Lubes

Shell developed its own lubes hydrocracking technology for the Petit-Couronne, France, plant,[13] built in 1972, with a planned output of 1400 bpd of 95+ VI (high

TABLE 7.11
Inspections on Extracted and Unextracted Feed

Inspection Tests	Unextracted	Extracted
LVGO, vol. %	24	29
HVGO, vol. %	29	—
DAO, vol. %	47	—
Extracted HVGO, vol. %	—	24
Extracted DAO, vol. %	—	47
API gravity, 60°F	24	28
IBP	694	706
5%	755	762
10%	784	791
50%	917	940
EP	1030	1030
Receiver	75	67
Conradson carbon, wt. %	1.0	—
Viscosity at 100°F, SUS	217	145
Viscosity at 210°F, SUS	41	35
Waxy VI	79–80	112–113
Dewaxed VI at 0°F pour point	59	About 86

Source: S. L. Thompson, "Stabilizing a Hydrocracked Lube Oil by Solvent Extraction," U.S. Patent 3,781,196.

VI) and 130+ VI (very high VI) base stocks from wax cracking. The success of the plant led to its expansion in 1977 to 6000 bpd. The company's objectives were similar to others as this new technology emerged—namely, to have the ability to make base stocks with VIs exceeding 95, to avoid having to use the restricted suite

TABLE 7.12
Results of Stability Tests on Base Stocks Produced from Extracted and Unextracted Feeds

	Unextracted Feed			Extracted Feed		
	Color, ASTM D1500			Color, ASTM D1500		
	Initial	Final	Sludge	Initial	Final	Sludge
100N	0.75	2.00	Light medium	0.25	2.0	Very light
200N	0.75	1.75	Very light	0.50	2.0	None
500N	1.5	2.25	Haze	1.25	3.25	None
Brightstock	—	—	—	7.0	7.50	None

Source: S. L. Thompson, "Stabilizing a Hydrocracked Lube Oil by Solvent Extraction," U.S. Patent 3,781,196.

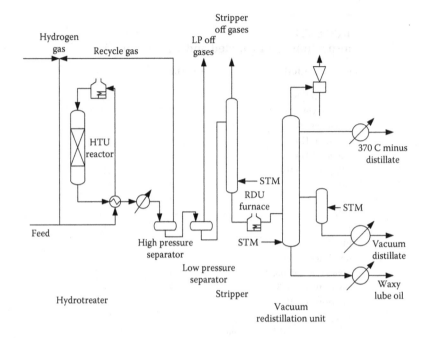

FIGURE 7.6 Schematic for Shell's lube plant at Petit-Couronne, France.
Source: S. Bull and A. Marnin, "Lube Oil Manufacture by Severe Hydrotreatment,"
Proceedings of the 10th World Petroleum Congress 4:221–228 (1980). With permission.

of crudes that solvent extraction demanded, and to produce base stocks that would
be interchangeable with those from existing solvent refined plants (i.e., still having
significant aromatic levels). The plant (see Figure 7.6 for a schematic) used a single
hydrocracking reactor of moderate severity with either vacuum distillates or DAO
(to make brightstock) as feeds. Waxy products were solvent dewaxed. Nonselective
hydroisomerization of wax was used to make the very high VI products. Shell
explored the use of platinum-based catalysts for this purpose.[18]

Subsequently Shell developed the Shell hybrid process[19] to make lubes out of
nonparaffinic feeds and those with a high content of polyaromatics and nitrogen
compounds, both of which reduce catalyst activity by being adsorbed on the catalyst.
Their methodology was to improve the quality of the hydrotreater feed by solvent
extraction, in a manner similar to that used by Sun. In contrast to Sun, Shell blocks
the vacuum distillates and DAO through their plant instead of using a combined feed.
Shell hydrocracks those streams where upgrading is required after solvent extraction.

In their hybrid scheme (Figure 7.7), the light neutral fraction from the
vacuum tower and solvent extraction steps goes directly to the solvent dewaxing
unit, by-passing the hydrotreater, since their experience was that on such good
quality streams, solvent extraction sufficed to make acceptable base stocks. The
other two heavier waxy distillate streams are subjected to quality upgrading

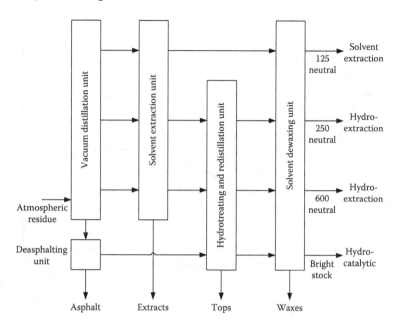

FIGURE 7.7 Schematic of Shell hybrid process for lubes.
Source: H. M. Bijwaard, W. K. J. Bremer, and P. van Doorne, "The Shell Hybrid Process, an Optimized Route for HVI (High Viscosity Index) Luboil Manufacture," presented at the Petroleum Refining Conference of the Japan Petroleum Institute, Tokyo, October 27–28, 1986. With permission.

steps, solvent extraction, and mild hydrocracking, followed by fractionation and solvent dewaxing. This version was termed "hydro-extraction," to include the root names of both technologies. Severity could be adjusted between the solvent extraction and hydrocracking units to afford more flexibility to the process. Their operational strategy was to vary the extraction step severity to provide a constant feed composition, regardless of the crude's origin, to the hydrotreater unit. No finishing step was mentioned.

Figure 7.8 illustrates how Shell hybrid yields from high and low quality feeds compare with those from a stand-alone solvent extraction plant (solid circle and square, respectively) and how they can vary as the preextraction severity is altered. Shell's data indicate that the lube yield (top line) from a high quality crude is marginally increased by extraction, but the yield increase is very sharp for low quality feeds (lower curve) as raffinate yields are decreased. The points for solvent extraction (solid circle and square) reveal the major difference in yields available from good and poor quality feeds. Therefore, using Shell technology, a poor quality feed in the hybrid process can give yields that are superior to solvent extraction of a good quality feed.

If the DAO was of sufficient quality it could be fed directly to the hydrotreater unit without extraction, in what they term a "hydro-catalytic" modification. Where

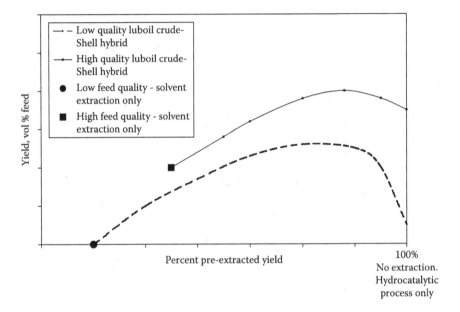

FIGURE 7.8 Shell hybrid process: high VI base oil yield variation with percent preextracted by solvent extraction for low and high quality lube feeds.
Source: H. M. Bijwaard, W. K. J. Bremer, and P. van Doorne, "The Shell Hybrid Process, an Optimized Route for HVI (High Viscosity Index) Luboil Manufacture," presented at the Petroleum Refining Conference of the Japan Petroleum Institute, Tokyo, October 27–28, 1986. With permission.

feedstock quality made it necessary, all feeds could be identically "hydro-extracted" through all the process units (Figure 7.9).

Specific yield results (volume percent of crude) quoted are given in Table 7.13, where relative to solvent extraction, base oil yields are increased by 30% for high quality crude and 66% for a medium to low quality crude. Shell first used this technology at their Geelong plant in Australia in 1986.[20]

7.2.4 GULF TECHNOLOGY: PETRO-CANADA'S MISSISSAUGA REFINERY, CANADA

Petro-Canada licensed the two-stage lube version of the Gulf hydrotreating process and the plant came online in 1979 to make water white base stocks with low levels of aromatics. An excellent description of this plant appears in Cashmore et al.'s SAE paper[21] and a schematic appears in Figure 7.10. The first stage was a lubes hydrocracking unit that produced waxy lubes directly from vacuum gas oil (VGO) or DAO without prior solvent extraction. These, after dewaxing, were hydrofinished in a high-pressure second stage. Feedstocks were light vacuum gas oil (LVGO), HVGO, and DAO from western Canadian crude fed to the first stage as either a 50:50 blend of LVGO/HVGO (light block) or a 60:40 blend of HVGO/DAO (heavy

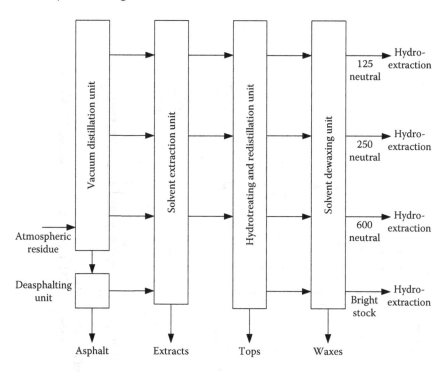

FIGURE 7.9 Shell lubes technology: all hydroextraction mode.
Source: H. M. Bijwaard, W. K. J. Bremer, and P. van Doorne, "The Shell Hybrid Process, an Optimized Route for HVI (High Viscosity Index) Luboil Manufacture," presented at the Petroleum Refining Conference of the Japan Petroleum Institute, Tokyo, October 27–28, 1986. With permission.

TABLE 7.13
Effect of Crude Change on Base Stock Yields by Solvent Extraction and by the Shell Hybrid Process

	Base Oil Yields on Crude, wt. %		
	Solvent Extraction	Shell Hybrid	Percent Change
Light, high quality crude	11.5	14.9	+30
Heavy, medium/low quality crude	8.5	14.1	+66

Source: H. M. Bijwaard, W. K. J. Bremer, and P. van Doorne, "The Shell Hybrid Process, an Optimized Route for HVI (High Viscosity Index) Luboil Manufacture," presented at the Petroleum Refining Conference of the Japan Petroleum Institute, Tokyo, October 27–28, 1986. With permission.

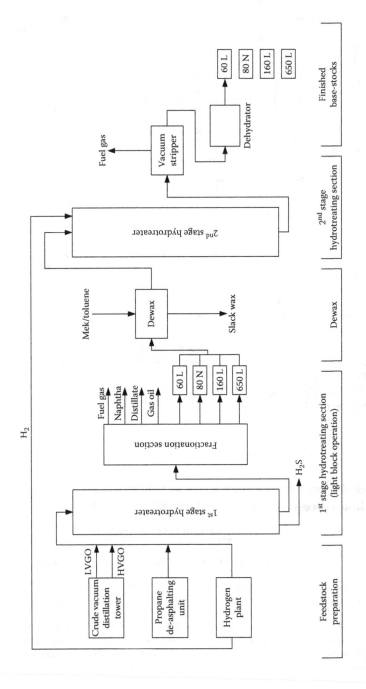

FIGURE 7.10 Schematic of the Petro-Canada two-stage "hydrotreating" lube oil plant.
Source: K. Cashmore, M. Moyle, and P. J. Sullivan, "Hydrotreated Lube Basestocks," SAE Paper 821235 (Warrendale, PA: Society of Automotive Engineers, 1982). With permission.

block). These broad cut feeds each made four waxy products when hydrotreated and vacuum fractionated. Light block made waxy stocks designated as 60L, 80L, 160L, and 650L (after their SUS viscosities), while from the heavy block came waxy 60H, 160H, 650H, and brightstock. First-stage operating conditions are given as 3000 psi total operating pressure at 0.5 liquid hourly space velocity (LHSV), with chemical hydrogen consumption of 1000 to 1500 standard cubic feet per barrel (scf/bbl). The intermediate waxy stocks were sent to tankage and then blocked through a solvent dewax unit to the second-stage hydrotreater, also operating at 3000 psi, a LHSV of 0.5, and a hydrogen consumption of 100 to 200 scf/bbl. Other users of Gulf hydrotreating technology in various forms include Idemitsu Kosan[22] (Chiba, Japan, 1969), Pennzoil[23] (Oil City, Pennsylvania, 1972), Ssangyong (now S-Oil, Onsan, Korea, 1980), and Pennzoil-Atlas (Shreveport, Louisiana, 1983, now owned by Calumet Lubricants). Gulf was taken over by Chevron in 1984.

This plant, in using a hydrotreater as a product stabilizer was essentially a forerunner of many North American lube plant designs to come, manufacturing base stocks with very low aromatics content. Table 7.14 shows that the aromatics contents of 2 to 10 wt. % in dewaxed oil feeds to the second stage are reduced to 0.5 wt. % or less with no change in viscosities or VIs and the remaining nitrogen or sulfur contents are all reduced to les than 2 ppm. One of the advantages cited for all-saturated hydrocarbon base stocks is that for the same viscosity they have higher average boiling points than solvent refined stocks (which contain aromatics). This means that these base stocks have lower volatility than solvent refined stocks, an important feature in an era when engine oil "consumption" numbers were becoming of critical importance.

TABLE 7.14

Physical and Chemical Changes in Second-Stage Hydrotreating at Petro-Canada's Lube Plant

	80L		160L		650L		Brightstock	
Stage	1	2	1	2	1	2	1	2
Aromatics, wt. %								
Mono-	10.43	0.50	10.17	0.22	2.91	0.15	13.12	0.56
Di-	1.96	0.01	1.43	0.01	0.14	0.01	0.99	0.04
Poly-	0.69	<0.01	0.89	<0.01	0.08	<0.01	0.99	<0.01
Total	13.08	0.51	12.49	0.23	3.13	0.16	14.66	0.60
Viscosity, cSt at 40°C	14.66	14.87	34.69	35.09	117.0	116.8	440	439
VI	91	92	110	111	101	101	105	105
Nitrogen, ppm	1.6	<1.0	4.2	1.3	3.0	1.5	10.7	1.8
Sulfur, ppm	<2	<2	<2	<2	7	<2	7	<2

Source: K. Cashmore, M. Moyle, and P. J. Sullivan, "Hydrotreated Lube Basestocks," SAE Paper 821235 (Warrendale, PA: Society of Automotive Engineers, 1982). With permission.

TABLE 7.15
HRMS Group Type Analyses: SR and HT Oils

	80N		160N	
Group Type, wt. %	Solvent Refined	Hydrotreated	Solvent Refined	Hydrotreated
Paraffins, iso- plus n-	24.9	25.5	20.1	26.0
Cycloparaffins				
Monocyclo-	25.0	31.5	23.0	33.4
Condensed	36.6	41.5	39.2	39.4
Total	61.6	73.0	62.2	72.8
Saturates (Total)	86.5	98.5	82.3	98.8
Monoaromatics				
Alkylbenzenes	3.9	1.0	5.5	0.8
Benzocycloparaffins	3.5	0.5	4.6	0.4
Benzodicycloparaffins	3.0	—	3.5	—
Total	10.4	1.5	13.6	1.2
Diaromatics				
Naphthalenes	1.4	—	1.9	—
Naphthocycloparaffins/biphenyls	0.8	—	1.1	—
Naphthadicycloparaffins/fluorenes	0.6	—	0.8	—
Total	2.8	—	3.8	—
Triaromatics	0.1	—	0.1	—

Source: K. Cashmore, M. Moyle, and P. J. Sullivan, "Hydrotreated Lube Basestocks," SAE Paper 821235 (Warrendale, PA: Society of Automotive Engineers, 1982). With permission.

Table 7.15 provides high-resolution mass spectra (HRMS) of solvent refined and hydrotreated 80N and 160N with similar VIs. It can be seen that compositions are not that different if we allow that solvent refined aromatics can be considered as "unsaturated" monocycloparaffins in the comparison.

Research at Gulf[24] represented the chemical steps involved in VI improvement as, first, stepwise saturation of aromatic rings of polyaromatics eventually form low VI polycyclic naphthenes to add to those already present. Some of these, formed or originally present in the feed, can undergo successive ring opening reactions to eventually form single-ring naphthenes with long side chains, probably branched, that would have high VIs and low pour points, as in Figure 7.11. Mass spectra (Table 7.16) of feeds to a lube hydrocracking unit and its product (VIs were not provided) support this view, where condensed compounds (di+-aromatics and condensed cycloalkanes) of poor VI are reduced from about 54% to 27%.

A further feature previously commented on is that regardless of the source, the compositions of hydrotreated base stocks of similar VIs become similar, but

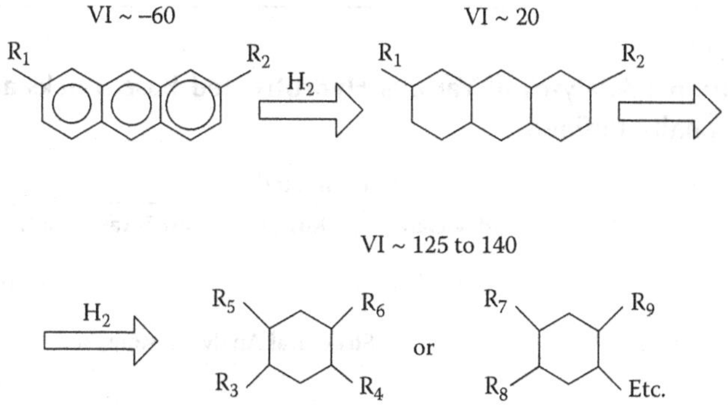

VI ~ −60   VI ~ 20

VI ~ 125 to 140

FIGURE 7.11 Chemical changes proposed to occur during hydrocracking.
Source: H. Beuther, R. E. Donaldson, and A. M. Henke, "Hydrotreating to Produce High Viscosity Index Lubricating Oils," *Industrial and Engineering Chemistry Product Research and Development* 3:174–180 (1964). With permission.

TABLE 7.16
Structural Analysis of Charge Stock and Waxy Lube Oil Product

Type	Compositional Analyses, Mole %	
	Charge	Nondewaxed 725°F Product
Alkanes[a]	1.0	4.1
Noncondensed cycloalkanes	32.6	67.7
Condensed cycloalkanes	19.8	26.5
Monoaromatics	12.6	1.5
Condensed aromatics	29.2	0.2
Benzothiophenes	1.9	0.0
Dibenzothiophenes	1.7	0.0
Naphthobenzothiophenes	1.1	0.0
Total condensed compounds	53.7	26.7

[a] In Tables 7.16–7.18 the n-alkane contents are as reported in the paper, but it should be noted that they are oddly low.

Source: H. Beuther, R. E. Donaldson, and A. M. Henke, "Hydrotreating to Produce High Viscosity Index Lubricating Oils," *Industrial and Engineering Chemistry Product Research and Development* 3:174–180 (1964). With permission.

TABLE 7.17
Compositional Analyses of Various Hydrotreated Base Stocks and of a Conventionally Refined oil

	Hydrotreated			
	Ordovician	Kuwait	West Texas	Solvent Refined
VI	120	125	117	108
	Structural Analysis, Mole %			
Alkanes	1.4	6.9	0.0	3.5
Noncondensed cycloalkanes	69.0	68.2	71.2	58.3
Condensed cycloalkanes				
2-ring	18.0	15.6	18.2	9.2
3-ring	5.9	4.8	5.8	3.9
4-ring	2.3	1.5	1.8	3.6
5-ring	1.1	0.9	0.9	5.7
6-ring	0.5	0.9	0.8	4.7
Total	27.8	23.7	27.5	27.1
Monoaromatics	1.6	0.9	1.0	10.0
Condensed aromatics	0.2	0.3	0.3	1.1

Source: H. Beuther, R. E. Donaldson, and A. M. Henke, "Hydrotreating to Produce High Viscosity Index Lubricating Oils," *Industrial and Engineering Chemistry Product Research and Development* 3:174–180 (1964). With permission.

of course not identical. In the examples cited in Table 7.17, it can be seen that the solvent refined and hydrocracked stocks have about the same levels of naphthenes plus monoaromatics. The differences in distribution of the condensed cycloalkanes is also striking (Table 7.17) in that the number of five- and six-ring components decreases to very low levels after hydrocracking, whereas in solvent refining there are still significant levels of six-ring saturates. These must reflect their levels in the original feed to the extraction unit, since their relative levels may well change little through the extraction process.

These structural features of the hydrocracked products appear to be reasonably consistent through the boiling range as well (Table 7.18) from analyses on 20 vol. % fractions obtained by distillation:

7.2.5 CHEVRON'S HYDROCRACKING TECHNOLOGY FOR THEIR RICHMOND, CALIFORNIA, REFINERY[5,7,25]

Chevron's 1978 decision to build a lube hydrocracking plant at their Richmond, California, refinery was based on the same or similar factors that led others to this

TABLE 7.18
Analysis of Distillation Fractions of a Hydrotreated Oil

Fraction	Total	1	2	3	4
Boiling range, °F	725+	725–811	811–858	858–941	941
Yield, vol. % of base oil	100	28.5	23.6	22.4	25.6
Viscosity, SUV at 210°F	48.7	40.5	45.1	51.5	76.5
VI	120	117	124	121	116
Pour point, °F	0	+5	0	+5	0
Structural analysis, mole %					
Alkanes	1.4	0.0	4.5	3.8	1.1
Noncondensed cycloalkanes	69.0	67.9	68.5	70.6	72.2
Condensed cycloalkanes					
2-ring	18.0	19.1	18.3	16.8	16.7
3-ring	5.9	6.4	4.4	4.4	5.0
4-ring	2.3	2.4	1.6	1.8	1.9
5-ring	1.1	1.4	0.9	1.0	1.3
6-ring	0.5	0.8	0.6	0.6	0.6
Total	27.8	30.1	28.8	24.6	25.5
Monoaromatics	1.6	1.8	1.1	0.9	1.1
Condensed aromatics	0.2	0.2	0.1	0.1	0.1
Average rings/molecule	1.42	1.51	1.34	1.33	1.40

Source: H. Beuther, R. E. Donaldson, and A. M. Henke, "Hydrotreating to Produce High Viscosity Index Lubricating Oils," *Industrial and Engineering Chemistry Product Research and Development* 3:174–180 (1964). With permission.

technology. It undoubtedly was also accentuated by their geographic position in California relative to major refining locations on the Gulf Coast (i.e., there was substantial lubricant demand in California but little base stock production). The technical solution was complicated by their desire to use feed from cheap Alaskan North Slope crude, which was in good supply on the West Coast and whose dewaxed HVGO had a VI of only 15, which sidelined solvent extraction and meant that even hydrocracking would result in diminished yields. This quality also forecast a very steep VI droop. Their solution to obtain the best yields possible, and thereby overcome the droop issue, was a bold one—build two hydrocrackers, one, the light hydrocracker (LC), to make light (100) neutral and some medium (240) neutral from the LC feed, while the second, a heavy hydrocracker (HC), processed the heavy feed to make heavy (500) neutral and the balance of the medium neutral (Figure 7.12). Table 7.19 gives the properties of the feed components for a diet of 100% Alaskan North Slope crude. It can be seen that these are of low wax content (a positive feature for solvent dewaxing), relatively high nitrogen (the HVGO has 2000 ppm), and low de-waxed VI, and are therefore poor quality feeds.

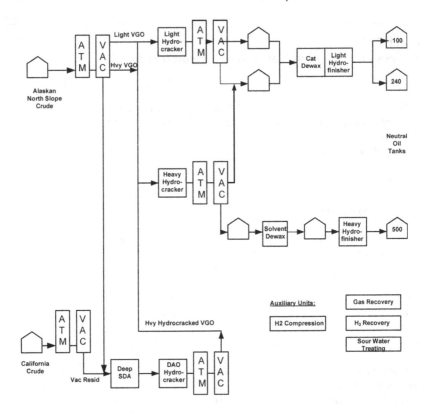

FIGURE 7.12 Plant schematic for Chevron's Richmond, California, lube plant when started. *Source:* T. R. Farrell and J. A. Zakarian, "Lube Facility Makes High-Quality Lube Oil from Low-Quality Feed," *Oil and Gas Journal* May 19:47–51 (1986). With permission.

Table 7.20 gives the properties of some other heavy vacuum gas oils that illustrate the relative quality of Alaskan North Slope HVGO as compared to others available on the West Coast.

The original Richmond lube oil plant (RLOP) configuration is given in Figure 7.12.[5,25] Notable features include

- Stocks are hydroprocessed—hydrocracked, hydrodewaxed, and hydrofinished, except for the 500N, which initially was solvent dewaxed.
- Hot distillates from crude vacuum fractionation were fed directly to the hydrocrackers to conserve energy. There was no intermediate tankage between the crude vacuum towers and the hydrocrackers.
- The two lighter streams were catalytically dewaxed by a ZSM-5-type catalyst, while the heavy stream was solvent dewaxed (in a later modification, all streams were dewaxed by Chevron's Isodewaxing™ process, which isomerizes wax rather than hydrocracking it).
- All streams were hydrofinished to produce stabilized near-water white final base stocks.

TABLE 7.19

Properties of Alaskan North Slope Feeds to Chevron's Richmond, California, Lube Plant

	LVGO from Crude Unit	HVGO from Crude Unit	Blend of 67% LVGO, 33% HVGO (LC Feed)	HVGO from DAO Cracker (33% of HC Feed)
API gravity	21.4	18.6	20.4	22.8
Sulfur, mass %	1.21	1.31	1.26	0.065
Nitrogen, ppm	1220	2030	1700	1015
Viscosity, cSt at 100°C	5.765	14.51	7.662	16.72
Pour point, °F	+85	+105	+95	+110
ASTM D1160, °F				
IBP/5	681/708	737/801	676/738	807/871
10/30	726/755	823/866	751/782	905/939
50	780	896	815	954
70/90	801/840	934/984	852/924	972/1000
95/EP	862/910	1006/1040	960/1013	1013/1029
Percent r Recovery	99	99.5	99	98
Dewaxed oil				
Wax, wt. %	6.2	6.3	6.6	6.3
Pour point, °F	+15	+5	+15	+5
Viscosity, cSt at 100°C	6.208	16.76	8.107	17.88
VI	32	18	38	58

Source: J. A. Zakarian, R. J. Robson, and T. R. Farrell, "All-Hydroprocessing Route for High-Viscosity Index Lubes," *Energy Progress* 7:59–64 (1987). With permission.

Hydrocracking yields are given in Table 7.21 and reflect the poor quality feed employed and the high dewaxed VI targeted in the first stage to allow for the loss of VI in the original catalytic dewaxing step. (Note that Chevron later replaced the ZSM-5-type catalytic dewaxing process with the Chevron Isodewaxing process. This change resulted in the VI loss across the dewax unit becoming a VI gain, which in turn meant a severity reduction on the first stage with a significant accompanying yield increase.)

Properties of the base stocks (Table 7.22) compared with solvent extracted stocks show that the hydrocracked products have better colors, VIs, and flash points. Sulfur and nitrogen levels will be in the low parts per million range.

Chevron attaches considerable practical importance to the results of their in-house "Oxidator BN" oxidation stability test performed on base stocks to which both oxidation catalysts and inhibitors have been added. Chevron's

TABLE 7.20
Properties of Some HVGOs

Crude	Gravity, °API	Sulfur, wt. %	Nitrogen, ppm	Wax, vol. %	Waxy VI	Dewaxed VI
Sumatran light	29.9	0.1	680	54.5	132	67
Arabian light	20.1	2.8	1000	16.7	72	50
Kuwait	19.3	3.1	960	7.4	66	50
Sumatran heavy[a]	21.4	0.2	1750	11.7	57	36
Alaskan North Slope[a]	18.4	1.4	2420	7.0	30	15
California light[b]	17.7	1.0	6200	7.6	33	5
California heavy[c]	11.4	1.2	7400	0.1	−200	−228

[a] These crudes are judged not to be economic sources of high VI lubes using conventional solvent refining.

[b] This is a blend of light California crudes including 77% Elk Hills Stevens.

[c] This crude is not an economic source of high VI lubes even using hydrocracking technology.

Source: T. R. Farrell and J. A. Zakarian, "Lube Facility Makes High-Quality Lube Oil from Low-Quality Feed," *Oil and Gas Journal* May 19:47–51 (1986). With permission.

TABLE 7.21
Typical Yields from RLOP Hydrocrackers

	Light Hydrocracker	Heavy Hydrocracker
H_2 consumption, scf/bbl	1100	1050
Yields, liquid vol. %		
LPG	2.6	2.3
Gasoline	12.7	7.5
Diesel	33.9	20.9
FCC feed	22.3	16.3
Waxy 100N	23.0	—
Waxy 240N	18.0	29.6
Waxy 500N	—	15.6
Total, vol. % feed	112.5	109.8
Controlling product	100N	240N
Dewaxed VI	103	101

Source: T. R. Farrell and J. A. Zakarian, "Lube Facility Makes High-Quality Lube Oil from Low-Quality Feed," *Oil and Gas Journal* May 19:47–51 (1986). With permission.

TABLE 7.22
Chevron RLOP Neutral Oil Quality

	RLOP Specification	RLOP Typical	Solvent Typical
100N			
Viscosity, cSt at 40°C	18.8–20.9	19.31	18.88
Viscosity, cSt at 100°C	—	3.97	3.88
VI	95 minimum	99	95
Color, ASTM D1500	1.0 maximum	<0.5	1.0
Pour point, °F	15 maximum	10	10
Flash point, °F	380 minimum	395	380
240N			
Viscosity, cSt at 40°C	44.0–48.0	46.70	46.50
Viscosity, cSt at 100°C	—	6.80	6.68
VI	95 minimum	99	95
Color, ASTM D1500	1.5 maximum	<0.5	1.5
Pour point, °F	15 maximum	10	10
Flash point, °F	435 minimum	450	415
500N			
Viscosity, cSt at 40°C	—	95.20	95.90
Viscosity, cSt at 100°C	10.7–11.4	11.11	10.75
VI	95 minimum	101	95
Color, ASTM D1500	2.5 maximum	0.5	2.5
Pour point, °F	15 maximum	10	10
Flash point, °F	475 minimum	500	450

Source: T. R. Farrell and J. A. Zakarian, "Lube Facility Makes High-Quality Lube Oil from Low-Quality Feed," *Oil and Gas Journal* May 19:47–51 (1986). With permission.

method measures the time required for 1 L of oxygen uptake by 100 g of base stock at 340°F.[26] This Oxidator BN test sets out to "measure the high temperature oxidation stability of fully formulated lubricants." Some results from this test (Table 7.23) show the performance differences between Chevron's hydrofinished base stocks and solvent refined and naphthenic stocks and the results are in the direction and of the magnitude we'd expect from what we have earlier seen on oxidation stability of base stocks. The results argue that these base stocks, when properly formulated, should produce finished lubes with longer lives or superior performance in demanding environments. When tested using the ASTM turbine oil oxidation stability test (TOST), ASTM D943, the RLOP stocks lasted about 2000 hours longer than solvent refined stocks, a performance gain similar to that seen with other hydrocracked stocks.[10,24]

TABLE 7.23
Chevron Oxidator BN Test Results

Base Stocks	Hours to 1 L O_2 Uptake
Chevron 100N	22+
Chevron 240N	22+
Chevron 500N	22+
Solvent refined 100N	6–8
Solvent refined 600N	6–8
Naphthenic base stocks	2–5

Source: J. A. Zakarian, R. J. Robson, and T. R. Farrell, "All-Hydroprocessing Route for High-Viscosity Index Lubes," *Energy Progress* 7:59–64 (1987). With permission.

7.2.6 ExxonMobil Technologies

For the production of hydroprocessed lube base stocks, ExxonMobil has developed two technologies:[27]

- Lubes hydrocracking (LHDC) with Mobil selective dewaxing (MSDW), which converts gas oils to lubes by an all-hydrogen catalytic route, and
- Raffinate hydroconversion (RHC), which can be viewed as a drop-in process upgrade for an existing solvent extraction plant.

Both of these produce group II base stocks and the LHDC route can also lead to group III as required.

The LHDC technology is exemplified in ExxonMobil's new (1997) 8000 bpd lube unit at their Jurong, Singapore, refinery, which produces light and heavy group II base stocks via hydrocracking followed by wax isomerization and hydrofinishing to produce base stocks with more than 98% saturates.[27,28] Figure 7.13 is a schematic of the lube production train. It is reported that the plant has operated successfully on some 30 different crudes. Some typical base stock properties are given in Table 7.24. Since coming online, the plant has added production of group III light neutrals to their slate. It is also reported to be considering production of medicinal-grade white oils,[29] not a surprise given the all-hydroprocessing nature of the plant.

The aging curve for the hydrocracking catalyst, based on EHC-110 production, can be seen (Figure 7.14) to have over a three-year period the usual initial de-edging period followed by a much slower rate whose slope declines with time.

The raffinate hydroconversion process, developed by Exxon Research and Engineering,[30] was developed to upgrade the solvent refining process at Exxon's

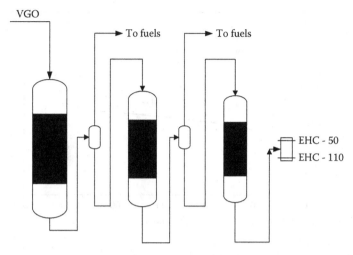

Process:	Lubes hydrocracking	Hydrodewaxing	Hydrofinishing
Catalyst:	HDC	MSDW™	HDF
Objective:	VI Nitrogen	Pour point	Saturates color

FIGURE 7.13 Schematic of ExxonMobil Jurong, Singapore, lubes unit.
Source: W. B. Genetti, A. B. Gorshteyn, A. Ravella, T. L. Hilbert, J. E. Gallagher, C. L. Baker, S. A. Tabak, and I. A. Cody, "Process Options for High Quality Base Stocks," presented at the 3rd Russian Refining Technical Conference, Moscow, Russia, September 25–26, 2003. With permission.

TABLE 7.24
Typical Properties of Jurong, Singapore, Base Stock

Base Stock	EHC-50	EHC-110
Viscosity, cSt at 100°C	5.6	10.5
VI	115	95
Noack volatility, wt. %	15 (maximum)	3
Pour point, °C	−18	−15
Saturates, wt. %	98+	98+

Source: W. B. Genetti, A. B. Gorshteyn, A. Ravella, T. L. Hilbert, J. E. Gallagher, C. L. Baker, S. A. Tabak, and I. A. Cody, "Process Options for High Quality Base Stocks," presented at the 3rd Russian Refining Technical Conference, Moscow, Russia, September 25–26, 2003. With permission.

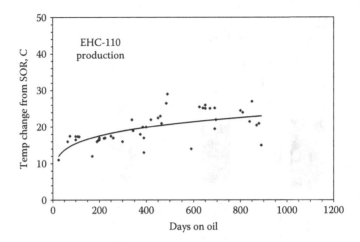

FIGURE 7.14 Commercial LHDC™ aging data for Jurong, Singapore, refinery based on EHC-110 production.
Source: W. B. Genetti, A. B. Gorshteyn, A. Ravella, T. L. Hilbert, J. E. Gallagher, C. L. Baker, S. A. Tabak, and I. A. Cody, "Process Options for High Quality Base Stocks," presented at the 3rd Russian Refining Technical Conference, Moscow, Russia, September 25–26, 2003. With permission.

Baytown, Texas, refinery. Figure 7.15 and Figure 7.16 provide schematic process diagrams. The process employs mild extraction using the existing solvent extraction unit to prepare the feed for the subsequent hydrotreating[31] unit, whose output is then hydrofinished directly. As expected, there is a higher raffinate yield than for solvent refined lubes. After hydrofinishing, the product then proceeds to a vacuum tower to remove distillates and establish the volatility required. The waxy base stock is subsequently solvent dewaxed—in this scenario, the light lubes stream meets current quality demands without using the RHC capability, as does the heavy stream. Undoubtedly an attractive feature of this process must have been the potential opportunity to license this technology to other solvent extraction-based refiners worldwide, who still account for more than 75% of world lubes production and who inevitably will face the need to produce base stocks with higher standards.

The advantages cited[31] include

- Low investment by integration with existing solvent lube plant.
- Preserves wax volume and quality.
- Upgrades only the stocks selected.
- Achieves low viscosity and volatility using efficiently fractionated solvent neutrals from the lubes vacuum pipestill via one base stock per feed.
- Affords cycle length flexibility by appropriate tailoring of solvent extracted feed.

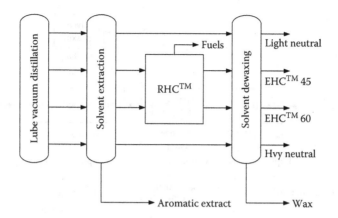

FIGURE 7.15 Schematic of RHC™ lube plant.
Source: J. E. Gallagher, Jr., I. A. Cody, and A. A. Claxton, "Raffinate Hydroconversion. Development and Commercialization of Raffinate Hydroconversion—A New Technology to Manufacture High Performance Basestocks for Crankcase and Other Applications," Paper LW-99-121, presented at the National Petroleum Refiners Association meeting, , 1999. With permission.

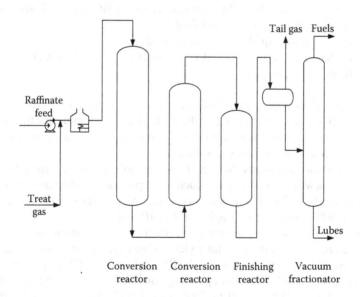

FIGURE 7.16 Details of RHC™ unit.
Source: J. E. Gallagher, Jr., I. A. Cody, and A. A. Claxton, "Raffinate Hydroconversion. Development and Commercialization of Raffinate Hydroconversion—A New Technology to Manufacture High Performance Basestocks for Crankcase and Other Applications," Paper LW-99-121, presented at the National Petroleum Refiners Association meeting, 1999. With permission.

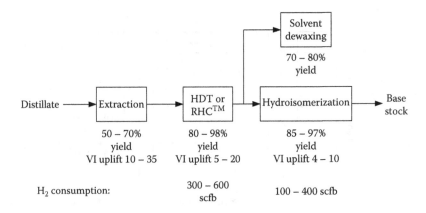

FIGURE 7.17 RHC™ process flow and yield and VI changes.
Source: D. S. McCaffrey, J. P. Andre, and S. A. Tabak, "Process Options for Producing Higher Quality Base Stocks," Proceedings of the Third International Symposium on Fuels and Lubricants, ISFL-2002, New Delhi, India, October 7–9, 2002. With permission.

- Gives saturates flexibility through unique RHC reactor design and tailoring of solvent extracted feed.
- Permits poorer quality feeds to be used.
- Underextraction debottlenecks the extraction tower, allowing more distillate to be processed.

Typical process flows and yields are shown in Figure 7.17.[32] Note the relatively low hydrogen consumption in the RHC step, which indicates that there is relatively little cracking, and the VI increase must depend to a significant extent on the conversion of aromatics to saturates.

This scheme maximizes the use of existing solvent refining equipment and the solvent dewax unit if wax production is profitable. Alternatively, solvent dewaxing can be replaced by wax isomerization, which improves dewaxing yields to 85% to 97% and gives a VI uplift of 4 to 10 units.

The first catalyst in the reactor sequence (Figure 7.16) is one of low acidity to reduce cracking and improve lube yields. One expects the primary objectives in this first unit are to perform as much HDA, HDS, and HDN as possible while preserving lube yield. Preferred operating conditions are given[33,34] as 340°C to 400°C (644°F to 752°F), 800 to 2000 psig, LHSV 0.3 to 3.0, and a hydrogen:feed ratio of 2000 to 4000 scf/bbl. Conditions for the hydrofinisher are given as 290°C to 350°C (554°F to 662°F), 800 to 2000 psig, LHSV of 0.7 to 3.0 and a hydrogen:feed ratio of 2000 to 4000 scf/bbl. Both sulfide and noble metal catalysts are indicated for use in the hydrofinisher and selection will obviously depend on hydrogen sulfide levels carried over from the first system and sulfur in the liquid feed.

7.3 GROUP III BASE STOCKS

7.3.1 BACKGROUND

Group III base stocks are defined by the American Petroleum Institute's (API's) interchangeability guidelines as having VIs of 120 or greater, 90 wt. % or greater saturates, and less than 0.03 wt. % sulfur. Group III+ has emerged as a non-API sanctioned marketing subgroup with a VI requirement of 130 to 150. A confusing cacophony of names has been given to these at various times and by various companies:

- VHVI: very high viscosity index; VI between 120 and 130.[5]
- UCBO: unconventional base oil; VI between 120 and 140.[5,35]
- VHVI: applied by Shell[13] to base oils with VIs greater than 145 at one time.
- UHVI: ultrahigh viscosity index; VI greater than 130.[5]
- XHVI: extra high viscosity index, another Shell term; VI of 140 or greater.[36]

Fortunately the API group system has brought some badly needed uniformity and simplicity to the nomenclature, and at least the new terms II+ and III+ fit in with the system and what they mean is easily recognized or inferred.

There are three routes to manufacture these base stocks, which are basically routes to meet the VI targets since the saturates and sulfur contents fall into place "automatically" due to the processes employed. These are

- Severe hydrocracking of distillates and DAO. Higher severity results in higher VI, but the penalties are viscosity and yield. These transformations are usually associated with the bottoms streams from fuel hydrocrackers.
- Isomerization of petroleum waxes.
- Isomerization of Fischer-Tropsch waxes.

The compositional goal is to make base stocks whose structures are dominated by isoparaffins and monocycloparaffins with long hydrocarbon chains attached. These molecules do not exist in natural distillates in sufficient concentration to obtain them by solvent extraction (and no solvent with the requisite selectivity appears to have been developed), therefore catalytic methods must be used. Group III+ base stocks will be largely isoparaffins in composition and therefore will use wax, preferably of Fischer-Tropsch origin, as feed to an isomerization process unit. In this section we will consider only the hydrocracking option; the other options will be discussed elsewhere.

Hydrocracking was used in Europe to produce the first commercial quantities of group III stocks[4] (VIs greater than 120 by ASTM D2270). Europe has been a early group III producer since these base stocks can be blended with the predominant group I stocks to achieve volatility targets. They first appeared in the form of a 125 VI base stock (measured by ASTM D567) of 8.8 cSt at 100°C from severe hydrocracking of DAO from the IFP-licensed plant in Puertollano, Spain.

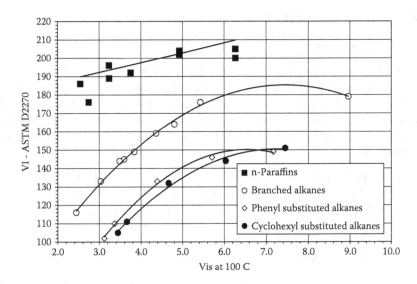

FIGURE 7.18 VI versus viscosity at 100°C for model compounds of several different compositional types.

Source: A. Billon, M. Derrien, and J. C. Lavergne, "Manufacture of New Base Oils by the I.F.P. Hydrofining Process," *Proceedings of the Division of Refining of the American Petroleum Institute* 49:522–548 (1969). With permission.

The starting high viscosity feed means that, in the parallel processes of molecular weight reduction and VI increase through ring opening and isomerization, enough steps in molecular reorganization can take place to achieve the high VI without complete loss of viscosity or yield.

IFP[8] attempted to predict the maximum VI attainable using model compound data—their results are shown in Figure 7.18, where it can be seen that the most favorable structures for high VI are the branched alkanes, which form a curve of their own below that of the (unfortunately) unusable n-alkanes. Below that are two other curves, close together, belonging to alkyl benzenes and alkylcyclohexanes. As viscosities increase beyond 4 cSt at 210°F, group III+ base stocks (VI greater than 130) become easier to obtain, perhaps because high molecular weight simply increases chain length and therefore VI. These curves plus the other VI versus compositional data we have seen predict that group III and III+ base stocks will have structures dominated by isoparaffins and monocyclonaphthenes with only a minor presence of low VI components having polycyclonaphthene or polycyclic aromatic frameworks. This figure also predicts that high VI base oils are possible that contain largely monoaromatics with long chains. In practice, these are not produced because the process conditions employed to reach the VI are severe enough to saturate most monoaromatics.

TABLE 7.25

Variation in VI and Composition with Severity for Hydrocracking DAOs

	Mild Treatment of Kuwait DAO			
	Total Oil, 715°F	Distillation Fractions		
		0–25	25–50	50–100
Viscosity, cSt at 210°F	20.3	6.72	17.96	38.92
VI, ASTM D2270	95	87	89	100
Pour point, °C	−18	−18	−18	−15
C_A, %	7.0	9.9	6.5	6.0
C_N, %	24.6	29.3	26.3	23.4
C_P, %	68.4	61.8	67.2	70.6

	Medium Treatment of Pennsylvania DAO			
	Total Oil, 715°F	Distillation Fractions		
		0–25	25–50	50–100
Viscosity, cSt at 210°F	12.83	5.42	10.17	23.10
VI, ASTM D2270	116	113	113	114
Pour point, °C	−18	−20	−18	−15
C_A, %	2.0	2.6	1.6	1.1
C_N, %	24.0	25.5	23.7	23.9
C_P, %	74.0	71.9	74.7	75

	Severe Treatment of Kuwait DAO				
	Total Oil, 715°F	Distillation Fractions			
		0–25	25–50	50–75	75–100
Viscosity, cSt at 210°F	5.71	3.81	4.71	6.34	11.64
VI, ASTM D2270	144	118	133	140	141
Pour point, °C	−18	−18	−18	−18	−15
C_A, %	0.0	0.5	0.7	0.0	0.0
C_N, %	15.2	17.1	16.0	12.3	11.5
C_P, %	84.8	82.4	83.3	87.7	88.5

Source: A. Billon, M. Derrien, and J. C. Lavergne, "Manufacture of New Base Oils by the I.F.P. Hydrofining Process," *Proceedings of the Division of Refining of the American Petroleum Institute* 49:522–548 (1969). With permission.

These compositional trends are seen through a different analytical lens in the n-d-M data published by IFP, where the percent C_P (paraffinic carbon) increases with increasing VI (Table 7.25) and C_N and C_A both decline. The effect of increased severity on viscosity can be seen here very clearly in the Kuwait case, where low severity (95 VI on total oil) gives a product with a viscosity at 210°F of 20.3 cSt, whereas the high severity case produced a total oil with a VI of 144 and the viscosity at 210°F decreased to 5.71 cSt.

7.3.2 SHELL

Shell has produced group III base stocks with VIs greater than 145 at their Petit Couronne, France, refinery since the 1970s. The feed used was slack wax which was catalytically isomerized and then dewaxed, and presumably finished if that last step was part of the process—no details appear to have been published.

7.3.3 BRITISH PETROLEUM

More recently, British Petroleum (BP)[37–39] produced two group III stocks, a 4 cSt BP HC-4 and a heavier HC-6 from the fractionator bottoms of their fuels hydrocracker at BP's Lavera, France, refinery. The total bottoms are solvent extracted to stabilize the final products, fractionated, and finally solvent dewaxed. The hydrocracker is operated in a severe mode (relative to lube hydrocracking) at a once-through conversion of 90% (in comparison, lube hydrocrackers may operate at conversions of only about 20%, and perhaps less with a very good quality feed). Like other hydrocracker-sourced group III products, their compositions are virtually independent of feed source due to the extent of the molecular reorganization that occurs and the molecular structures that are required for those VIs. The properties of the HC-4 are compared in Table 7.26 with those of some competitive group I, II, and IV base stocks. It can be seen that the HC-4 closely resembles the more

TABLE 7.26
Physical Properties of Base Oils: 80N Severely Hydrotreated, 100N Solvent Refined, HC-4 and PAO 4 Typical Properties

	80N	100N	HC-4	PAO-4
Viscosity, cSt at 100°C	3.5	3.81	3.95	3.89
VI	99	92	130	124
Pour point, °C	−21	−18	−27	−54
Cold cranking simulator viscosity, −20°C	520	670	459	413
Noack	40	32	15.3	13
Sulfur, %	0.008	0.46	0.11	0.0
Total nitrogen, ppm	5	17	12	0
Aromatics, %	0	18.2	7.7	0
Aliphatics, %	100	81.8	92.3	100

Source: G. R. Dobson, N. P. Wilkinson, and N. C. Yates, "Hydrocracked Base Oils—An Important Class of Synthetics," *Proceedings of the 13th World Petroleum Congress*:163–169 (1991). With permission.

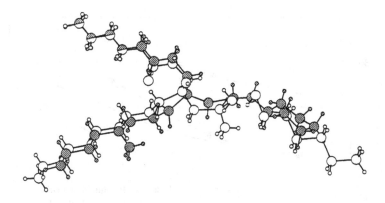

FIGURE 7.19 Comparison of average carbon molecular structures of BP HC-4 molecule (shaded circles) and decene trimer (open circles).
Source: G. R. Dobson, N. P. Wilkinson, and N. C. Yates, "Hydrocracked Base Oils—An Important Class of Synthetics," *Proceedings of the 13th World Petroleum Congress*:163–169 (1991). With permission.

expensive PAO-4 in physical properties, particularly in those key ones of VI, low-temperature viscosity, and volatility. The average chemical structure, deduced from field ionization mass spectroscopy and nuclear magnetic resonance (NMR), concluded that the overall average molecular shape is quite similar to that of a decene trimer in a PAO molecule (i.e., the average HC-4 molecule can be considered as having the same "star shape" as a PAO (see Figure 7.19).[37] From mass spectroscopic analyses, it was deduced that the components were predominantly alkyl-substituted monocycloparaffins plus some dicycloparaffins and isoparaffins.

7.3.4 Nippon Oil

Ushio et al. of Nippon Oil (Japan) have described some of their research work undertaken to produce group III stocks via hydrocracking distillates.[12] Their process was similar to BP's, employing a fuels hydrocracker to produce the waxy lubes stream(s) by downstream fractionation of the hydrocracker bottoms. Base oil products were stabilized by furfural extraction.

Interestingly, mass spectroscopic analyses on the bottoms product over a range of conversions in their pilot plant showed that low space velocities, and therefore low reactor temperatures, consistently favored the formation of two of the high VI components, the isoparaffins (VI approximately 155) and the mono-cycloparaffins (VI approximately 142) (Figure 7.20).

They concluded that there must be two competing temperature-sensitive chemical routes (Figure 7.21), with the low temperature one leading to greater selectivity for complete saturation of polyaromatics to polynaphthenes and subsequent ring opening to monocyclic naphthenes and isoparaffins. The high temperature route was seen as a cracking one, leading to the formation of lower

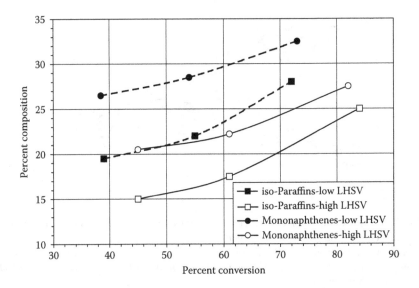

FIGURE 7.20 Hydrocracking distillates for lubes: bottoms composition—isoparaffins and mononaphthene levels versus conversion and space velocity.
Source: M. Ushio, K. Kamiya, T. Yoshida, and I. Honjou, "Production of High VI Base Oil by VGO Deep Hydrocracking," Symposium on Processing, Characterization and Application of Lubricant Base Oils, Division of Petroleum Chemistry, American Chemical Society, Washington, DC, August 23–28, 1992. With permission.

molecular weight products which would fall in the fuels molecular weight range. Compositions of the 100N produced by this process and solvent refining are compared in Table 7.27—the differences in isoparaffins and mononaphthene levels are readily apparent.

7.3.5 MITSUBISHI

Mitsubishi made low (4.0 cSt at 100°C) and medium (6.83 cSt at 100°C) viscosity group III base oils, also from hydrocracker bottoms.[11] The waxy lube cuts were subsequently both solvent extracted and hydrofinished to stabilize them and finally solvent dewaxed according to their schematic in Figure 7.22.

The properties of these base stocks are compared to some corresponding PAOs in Table 7.28 and are very similar or better in terms of VI, volatility, and UV and oxidation stability, but of course neither is as competitive with respect to pour point.

Mitsubishi employed two feeds, a 4.25 cSt heavy gas oil (HGO) and a 6.26 cSt vacuum gas oil (VGO) to make the group III stocks, with the lighter HGO giving a higher yield of the greater than 125 VI 4 cSt product—the heavier feed exhibited more severe VI droop in the 4 cSt region, which consequently excluded this feed from producing the light product with the desired VI. Figure 7.23 illustrates the basis

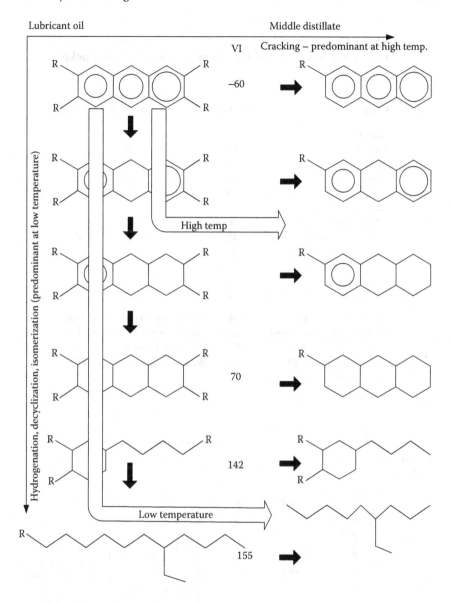

FIGURE 7.21 Possible routes for VGO hydrocracking reaction.
Source: M. Ushio, K. Kamiya, T. Yoshida, and I. Honjou, "Production of High VI Base Oil by VGO Deep Hydrocracking," Symposium on Processing, Characterization and Application of Lubricant Base Oils, Division of Petroleum Chemistry, American Chemical Society, Washington, DC, August 23–28, 1992. With permission.

TABLE 7.27
Composition of SR and Hydrocracked 100N Base Stocks

Components	SR 100N	Nippon High VI HC 100N
Saturates		
Isoparaffins	15	40
Mononaphthenes	25	35
Dinaphthenes+	35	20
Total	75	95
Monoaromatics		
Alkylbenzenes	8	2
Naphthenoaromatics	15	3
Total	23	5
Polyaromatics	2	0.0
Sulfur, mass %	0.13	0.00

Source: M. Ushio, K. Kamiya, T. Yoshida, and I. Honjou, "Production of High VI Base Oil by VGO Deep Hydrocracking," Symposium on Processing, Characterization and Application of Lubricant Base Oils, Division of Petroleum Chemistry, American Chemical Society, Washington, DC, August 23–28, 1992. With permission.

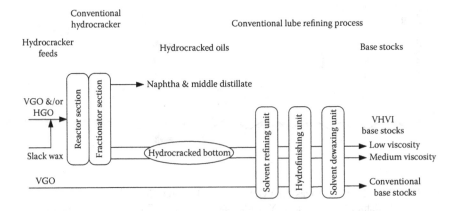

FIGURE 7.22 Mitsubishi group III base stock production scheme.
Source: M. Takizawa, T. Takito, M. Noda, K. Inaba, Y. Yoshizumi, and T. Sasaki, "Commercial Production of Two Viscosity Grades VHVH Basestocks," paper presented at the 1993 National Fuels and Lubricants meeting of the National Petroleum Refiners Association, Houston, Texas, November 4–5, 1993. With permission.

TABLE 7.28

Typical Properties of Mitsubishi Group III Base Stocks of Two Viscosity Grades Compared to Those of PAOs

	Group III Base Stocks		PAOs	
	Low Viscosity	Medium Viscosity	Low Viscosity	Medium Viscosity
Viscosity, cSt at 100°C	4.00	6.83	3.90	5.90
VI	133	140	123	135
Pour point, °C	−15	−15	−73	−68
Volatility, Noack, wt. %	13.8	4.5	14.9	5.8
UV stability	Excellent	Excellent	Excellent	Excellent
Oxidation stability (RBOT and TOST)	Excellent	Excellent	Excellent	Excellent

Source: M. Takizawa, T. Takito, M. Noda, K. Inaba, Y. Yoshizumi, and T. Sasaki, "Commercial Production of Two Viscosity Grades VHVI Basestocks," paper presented at the 1993 National Fuels and Lubricants meeting of the National Petroleum Refiners Association, Houston, Texas, November 4–5, 1993. With permission.

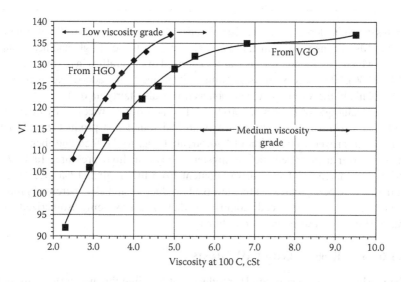

FIGURE 7.23 Hydrocracker bottoms from HGO and VGO: VI distribution versus viscosity at 100°C.
Source: M. Takizawa, T. Takito, M. Noda, K. Inaba, Y. Yoshizumi, and T. Sasaki, "Commercial Production of Two Viscosity Grades VHVH Basestocks," paper presented at the 1993 National Fuels and Lubricants meeting of the National Petroleum Refiners Association, Houston, Texas, November 4–5, 1993. With permission.

TABLE 7.29
Typical Properties of Hydrocracker Feeds

	Gas Oils		Slack Wax	
	HGO	VGO	Low Viscosity	Medium Viscosity
Density, g/cm³ at 15°C	0.90	0.91	0.83	0.86
Viscosity, cSt at 100°C	4.25	6.26	3.93	7.69
Distillation, ASTM D2887, °C				
IBP	244	254	313	367
10%	344	352	394	450
50%	414	444	430	498
90%	468	527	457	546
Final boiling point	525	588	476	606

Source: M. Takizawa, T. Takito, M. Noda, K. Inaba, Y. Yoshizumi, and T. Sasaki, "Commercial Production of Two Viscosity Grades VHVI Basestocks," paper presented at the 1993 National Fuels and Lubricants meeting of the National Petroleum Refiners Association, Houston, Texas, November 4–5, 1993. With permission.

for this decision, depicting the VI distributions for the two feeds from 10% distillation cuts. Feedstock inspections in Table 7.29 also include those of light and heavy slack waxes, which they found to be partially isomerized to isoparaffins in the hydrocracker (Table 7.30). The increase in isoparaffin content due to the added wax was small, and it is likely the VI improvement obtained was also fairly modest.

The technique of improving product VI by addition of wax to a hydrocracker feed was also investigated by Gulf,[40] who had found that product dewaxed VI could be improved by addition of hydrotreated wax to the feed and then hydrocracking over a fluorided nickel/tungsten catalyst at high pressure. Table 7.31 shows that increasing the feed wax content from the original 12% to 27%, and particularly to 52%, increased the dewaxed product VI and also increased product yield based on original feed, confirming that wax was being isomerized (and cracked) under these process conditions.

7.3.6 THE KOREAN GROUP III GIANTS

The world's two largest group III producers at the time this was written are both located in South Korea, SK Corp (14,000 bpd) at Ulsan and S-Oil, formerly Ssangyong (9,000 bpd), at Onsan.[41] Both operate very large refineries (more than 1,000,000 and 525,000 bpd, respectively) and both employ fuels hydrocrackers to generate waxy high VI intermediates which are subsequently fractionated into waxy lube streams, dewaxed, and hydrofinished.

TABLE 7.30
Effect of Slack Wax Injection on Isoparaffin Product Levels (Low Viscosity Group III Production)

	Slack Wax Injection		Effect of Slack Wax Injection,
	No	Yes (10% vol. %)	Increase in Isoparaffins
Hydrocracker feed			
Isoparaffin, wt. %	18.7	19.3	+0.6
Hydrocracker bottoms			
Isoparaffin, wt. %	57.2	61.4	+4.2

Source: M. Takizawa, T. Takito, M. Noda, K. Inaba, Y. Yoshizumi, and T. Sasaki, "Commercial Production of Two Viscosity Grades VHVI Basestocks," paper presented at the 1993 National Fuels and Lubricants meeting of the National Petroleum Refiners Association, Houston, Texas, November 4–5, 1993. With permission.

TABLE 7.31
Influence of Added Hydrotreated Wax to Hydrocrackate Properties

Charge Operating Conditions	100% Ordovician Residue			80% Ordovician Residue + 15% Wax		60% Ordovician Residue + 40% Wax	
Run Number	1	2	3	4	5	6	7
Temperature, °F	730	741	750	750	730	730	750
Pressure, psig	3000	3000	3000	3000	3000	3000	3000
LHSV	0.5	0.5	0.5	0.5	0.5	0.5	0.5
H_2 rate, scf/bbl	5000	5000	5000	5000	5000	5000	5000
Yields							
Vol. % of total charge	—	—	—	51.9	52.8	49.2	44.6
Vol. % of original charge	69.8	59.5	57.2	61.7	62.8	84.3	76.4
725°F DWO inspections							
VI	110	116	120	125	116	121	132
Wax yield, vol. % charge	10.9	9.0	7.9	11.0	24.6	27.3	15.4

Source: A. M. Henke and R. E. Peterson, "Process for Preparing an Improved Lubricating Oil," U.S. Patent 3,046,218.

7.3.6.1 SK Corporation (Formerly Yukon Limited)

In two papers,[42,43] SK Corporation outlined the development steps of their process, which first came online in 1995. The process takes unconverted oil (UCO, otherwise known as hydrocracker bottoms) from a two-stage fuels hydrocracker, fractionates it, and then blocks the waxy 70N, 100N, and 150N streams through a catalytic dewaxing unit, a hydrofinisher, and the final product stripper (Figure 7.24). The products were termed Yubase VHVI base stocks, with VHVI being defined as having a VI greater than 120.

Initially the catalytic dewaxing technology they employed was ZSM-5 based. When wax isomerization technologies became available, a catalyst change to the Chevron Isodewaxing™ catalyst technology was made, resulting in an immediate VI increase (+11 for the 150N) and yield improvements[43], greatest also for the 150N (approximately 20%) and decreasing through the 100N and 70N (Table 7.32).

Comparison (Table 7.33) of compositional analyses on Yubase 6 with group I, II, and IV base stocks of similar viscosities shows the expected compositional differences—the Yubase 6 has a much higher isoparaffin content than either groups I or II and less 2+ ring cycloparaffins.[44] The other benefits expected for group III stocks, namely higher flash point and decreased volatility as measured by the Noack test, also materialized.

This change was also accompanied by a change in hydrofinishing catalyst, the previously employed sulfided nickel/tungsten catalyst being replaced by a noble metal one. This made production of food-grade white oils possible because of greater catalyst activity which permitted lower temperatures. More details are provided in a paper by Sung et al.[43] Properties of the white oils are shown in Table 7.34.

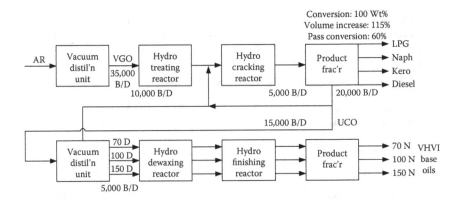

FIGURE 7.24 Process schematic for SK Oil lubes process.
Source: H. Y. Sung, S. H. Kwon, and J. P. Andre, "VHVI Base Oils and White Oils from Fuels Hydrocracker Bottoms," presented at the Asia Fuels and Lubes Conference, January 25–28, 2000. With permission.

TABLE 7.32
First Yubase Upgrading Project: Results for 150N Case

	Specification	Isodewaxing	CDW	Change
Dewaxing chemistry	—	Isomerization	Cracking	—
Pour point °C	Maximum −15	−15	−15	
VI	Minimum 125	132	121	+11
Flash point, °C	Minimum 220	240	232	+8
Noack	Maximum 10	6.2	7.5	1.3
Yields				
Naphtha		10.5 %	33.8%	−23.3
Distillate		4.5%	6.0%	+1.5
Lubes		85.5%	64.5%	+21.0
Make-up H$_2$	-	1900 nm³/hr	4500 nm³/hr	−2600

Source: H. Y. Sung, S. H. Kwon, and J. P. Andre, "VHVI Base Oils and White Oils from Fuels Hydrocracker Bottoms," presented at the Asia Fuels and Lubes Conference, Singapore, January 25–28, 2000. With permission.

TABLE 7.33
Comparison of Yubase 6 Properties with Those of Group I and II 150N

	Solvent Refined, Group I	Hydrocracked, Group II	Yubase 6, Group III	PAO, Group IV
Viscosity, cSt at 100°C	5.1	5.1	6.0	5.9
VI	95	99	133	135
Flash point, °C	216	222	234	240
Pour point, °C	−12	−12	−15	−60
Noack, wt. %	17.0	16.5	7.8	7.0
Sulfur, ppm	5800	300	<10	<10
Nitrogen, ppm	12	4	<1	<1
Composition by mass spectra				
Paraffins	27.6	33.4	55.5	100
Cycloparaffins				
1-ring	20.8	30.2	20.4	
2-ring	25.9	17.2	12.1	
3-ring	2.9	9.3	9.1	
4-ring	0.3	5.1	2.1	
5-ring	0.0	1.1	0.0	
Aromatics	22.5	3.5	0.8	

Source: W. S. Moon, Y. R. Cho, C. B. Yoon, and Y. M. Park, "VHVI Base Oils from Fuels Hydrocracker Bottoms," presented by Y.-R, Cho at the China Oil and Gas Producers' Conference, June 1998. With permission.

TABLE 7.34
Test Results on UCO Process Products as White Oils

	Phazol-3	Phazol-4	Phazol-6
Viscosity, cSt at 40°C	12.3	19.1	32.5
Viscosity, cSt at 100°C	3.1	4.2	6.0
VI	115	126	133
Sulfur, ppm	<1	<1	<1
UV absorbance, DMSO at 260–350 nm	0.018	0.020	0.021
Readily carbonizable substances	Pass	Pass	Pass

Source: W.-S. Moon, Y.-R Cho, and J.-S. Chun, "Application of High Quality (Group II,III) Base Oils to Specialty Lubricants," presented at the 6th annual Fuels and Lubes Asia Conference, Singapore, January 28, 2000. With permission.

TABLE 7.35
Properties of Base Stocks Produced from S-Oil's Fuels Hydrocracker

Product	ASTM	Ultra-S2	Ultra-S3	Ultra-S4	Ultra-S6
Color, Saybolt	D156	+30	+30	+30	+30
Viscosity, cSt at 100°C	D445	2.2–2.3	3.0–3.2	4.1–4.3	5.4–5.7
VI	D2270	109	117	123	130
Flash point, °C (COC)	D92	158	204	228	232
Pour point, °C	D97	−37.5	−25	−20	−17.5
Sulfur, ppm	D5453	<1	<1	<1	<1
Composition, n-D-M					
C_A, wt. %	D3238	0.2	0.2	0.3	
C_N, wt. %		21.9	17.3	17.7	17.2
C_P, wt. %		78.0	82.5	82.1	82.5
Noack, wt. %	D5800	—	30–34	14.5	8.5
Saturates, wt. %	D2007	>99	>99	>99	>99
CCS viscosity	D5293	—	—	1500–1550 cP at −30°C	2930–3170 cP at −30°C, 1670–1800 cP at −25°C
UV absorbance, 260–350 nm	D2269	<0.1 (typical 0.043)	<0.1 (typical 0.045)	<0.1 (typical 0.048)	<0.1 (typical 0.052)
Hot acid test	D565	Pass	Pass	Pass	Pass
Thermal stability at 24 hr	JIS K2540	<0.5 BC	<0.5 BC	<0.5 BC	<0.5 BC

Source: S-Oil and ExxonMobil, "Successful Conversion of a Fuels Hydrocracker to Group III Lube Production," presented at the ARTC 7th Annual Meeting and Reliability Conference, Singapore, April 2004, p. 15. With permission.

7.3.6.2 S-Oil (Formerly Ssangyong)

S-Oil and ExxonMobil have described their joint work on modifying a fuels hydrocracker to make group III lubes, in which the bottoms stream was fractionated, dewaxed using MSDW2 hydroisomerization technology followed by aromatic saturation with ExxonMobil's MAXSAT catalyst.[44] The properties of the base stocks are shown in Table 7.35, and like SK Corporation's products, these also meet U.S. food-grade white oil specs.

REFERENCES

1. G. G. Pritzger, "Production of Synthetic Lubricating Oils by Hydrogenation Reactions," *Petroleum Processing* 2:205–208 (1947).
2. R. T. Haslam and W. C. Bauer, "Production of Gasoline and Lubricants by Hydrogenation," *SAE Journal* XXVIII:307–314 (1931).
3. D. C. Kramer, B. K. Lok, R. R. Krug, and J. M. Rosenbaum, "The Advent of Modern Hydroprocessing—The Evolution of Base Oil Technology—Part 2," *Machinery Lubrication*, May 2003.
4. J. Angula, M. Gasca, J. L. Martinez-Cordon, R. Torres, A. Billon, M. Derrien, and G. Parc, "IFP Hydrorefining Makes Better Oils," *Hydrocarbon Processing* 47(6):111–115 (1968).
5. S. J. Miller, M. A. Shippey, and G. M. Masada, "Advances in Lube Base Oil Manufacture by Catalytic Hydroprocessing," Paper FL-92-109, presented at the 1992 National Fuels and Lubricants meeting of the National Petroleum Refiners Association, Houston, Texas, November 5–6, 1992.
6. M. C. Bryson, W. A. Horne, and H. C. Stauffer, "Gulf's Lubricating Oil Hydrotreating Process," *Proceedings of the Division of Refining of the American Petroleum Institute* 49:439–453 (1969).
7. T. R. Farrell and J. A. Zakarian, "Lube Facility Makes High-Quality Lube Oil from Low-Quality Feed," *Oil and Gas Journal* May 19:47–51 (1986).
8. A. Billon, M. Derrien, and J. C. Lavergne, "Manufacture of New Base Oils by the I.F.P. Hydrofining Process," *Proceedings of the Division of Refining of the American Petroleum Institute* 49:522–548 (1969).
9. I. Steinmetz and H. E. Reif, "Process Flexibility of Lube Hydrotreating," *Proceedings of the Division of Refining of the American Petroleum Institute* 53:702–712 (1973).
10. R. P. Bryer, H. E. Reif, I. Steinmetz, and J. R. Thomas, "Sun Oil Company's New Lube Refinery," Paper F&L-72-43, presented at the National Fuels and Lubricants Meeting of the National Petroleum Refiners Association, New York, September 14–15, 1972.
11. M. Takizawa, T. Takito, M. Noda, K. Inaba, Y. Yoshizumi, and T. Sasaki, "Commercial Production of Two Viscosity Grades VHVI Basestocks," paper presented at the 1993 National Fuels and Lubricants meeting of the National Petroleum Refiners Association, Houston, Texas, November 4–5, 1993.
12. M. Ushio, K. Kamiya, T. Yoshida, and I. Honjou, "Production of High VI Base Oil by VGO Deep Hydrocracking," Symposium on Processing, Characterization and Application of Lubricant Base Oils, Division of Petroleum Chemistry, American Chemical Society, Washington, DC, August 23–28, 1992.

13. S. Bull and A. Marnin, "Lube Oil Manufacture by Severe Hydrotreatment," *Proceedings of the 10th World Petroleum Congress* 4:221–228 (1980).

14. L. Z. Pillon, "Use of UV Spectroscopy to Predict the Daylight Stability of Hydrocracked Base Stocks," *Petroleum Science and Technology* 19(9–10):1263–1271 (2001).

15. S. L. Thompson, "Stabilizing a Hydrocracked Lube Oil by Solvent Extraction," U.S. Patent 3,781,196.

16. A. Sequiera, Jr., *Lubricant Base Oil and Wax Processing* (New York: Marcel Dekker, 1996), 134.

17. S. L. Thompson, R. F. Kvess, A. T. Olenzak I. Stein mets "Oil Stabilizing Sequential Hydrocracking and Hydrozenation Treatment" Canadian Patent 908,590.

18. J. Koome and G. M. Good, "Production of Oils from Waxes," U.S. Patent 2,817,693, R. J. Moore and B. S. Greensfelder, "Hydrocarbon Conversion," U.S. Patent 2,475,358, and G. M. Good, J. W. Gibson, and B. S. Greensfelder, "Isomerization of Paraffin Wax," U. S. Patent 2,668,790.

19. H. M. Bijwaard, W. K. J. Bremer, and P. van Doorne, "The Shell Hybrid Process, an Optimized Route for HVI (High Viscosity Index) Luboil Manufacture," presented at the Petroleum Refining Conference of the Japan Petroleum Institute, Tokyo, October 27–28, 1986.

20. J. M. L. M. Vlemmings, "Supply and Demand of Lube Oils. A Global Perspective," Paper AM-88-19, presented at the meeting of National Petroleum Refiners Association, San Antonio, Texas, March 22, 1988.

21. K. Cashmore, M. Moyle, and P. J. Sullivan, "Hydrotreated Lube Basestocks," SAE Paper 821235 (Warrendale, PA: Society of Automotive Engineers, 1982).

22. "First Lubricant-Oil Cracker has Trouble-Free Record," *Oil and Gas Journal* June 12:94–97 (1972).

23. W. D. Thomas, "Pennzoil Hydrotreater for Lube Production put on Stream," *Oil and Gas Journal* February 12:82–84 (1973).

24. H. Beuther, R. E. Donaldson, and A. M. Henke, "Hydrotreating to Produce High Viscosity Index Lubricating Oils," *Industrial and Engineering Chemistry Product Research and Development* 3:174–180 (1964).

25. J. A. Zakarian, R. J. Robson, and T. R. Farrell, "All-Hydroprocessing Route for High-Viscosity Index Lubes," *Energy Progress* 7:59–64 (1987).

26. R. J. Robson, "Base Oil Composition and Oxidation Stability," *Proceedings of the American Chemical Society, Division of Petroleum Chemistry* 29:1094–1100 (1984).

27. W. B. Genetti, A. B. Gorshteyn, A. Ravella, T. L. Hilbert, J. E. Gallagher, C. L. Baker, S. A. Tabak, and I. A. Cody, "Process Options for High Quality Base Stocks," presented at the 3rd Russian Refining Technical Conference, Moscow, Russia, September 25–26, 2003.

28. R. G. Wuest, R. J. Anthes, R. T. Hanlon, S. M. Jacob, L. Loke, and C. T. Tan, "Singapore All-Catalytic Lube Plant Performs Well," *Oil and Gas Journal* July 19:70–73 (1999).

29. J. E. Gallagher, Jr., I. A. Cody, S. A. Tabak, R. G. Wuest, A. A. Claxton, L. Loke, and C. T. Tan, "New ExxonMobil Process Technology for Producing Lube Basestocks," presented at the Asia Pacific Refining Technology Conference, Kuala Lumpur, Malaysia, March 9, 2000.

30. P. S. Adam, "Three Refineries, One Strategy. Exxon Invests to Hold North American Base Oil Lead," *Lubes 'N' Greases* July 20:20–24 (1997).

31. J. E. Gallagher, Jr., I. A. Cody, and A. A. Claxton, "Raffinate Hydroconversion. Development and Commercialization of Raffinate Hydroconversion—A New Technology to Manufacture High Performance Basestocks for Crankcase and Other Applications," Paper LW-99-121, presented at the National Petroleum Refiners Association meeting, November 11–12, 1999, Houston, TX.

32. D. S. McCaffrey, J. P. Andre, and S. A. Tabak, "Process Options for Producing Higher Quality Base Stocks," Proceedings of the Third International Symposium on Fuels and Lubricants, ISFL-2002, New Delhi, India, October 7–9, 2002.

33. I. A. Cody, D. R. Boate, S. J. Linek, W. J. Murphy, J. E. Gallagher, and G. L. Harting, "Raffinate Hydroconversion Process," U.S. Patent 6,325,918.

34. I. A. Cody, D. R. Boate, S. J. Linek, W. J. Murphy, J. E. Gallagher, A. Ravella, and R. A. Demmin, "Hydroconversion Process for Making Lubricating Oil Basestocks," U.S. Patent 6,322,692.

35. D. C. Kramer, B. K. Lok, and R. R. Krug, "The Evolution of Base Oil Technology," in *Turbine Lubrication in the 21st Century*, ASTM STP #1407, W. R. Herguth and T. M. Warne, eds. (West Conshohocken, PA: American Society for Testing and Materials, 2001).

36. M. Moret, "Un exemple de production d'huiles de base à partir de bases hydro-craquées," *Petroles et Techniques* 333, 37, 1987.

37. G. R. Dobson, N. P. Wilkinson, and N. C. Yates, "Hydrocracked Base Oils—An Important Class of Synthetics," *Proceedings of the 13th World Petroleum Congress*: 163–169 (1991).

38. T. E. Kiovsky, N. C. Yates, and J. R. Bales, "Fuel Efficient Lubricants and the Effect of Special Base Oils," *Lubrication Engineering* 50:307–312 (1994).

39. N. C. Yates, T. E. Kiovsky, and J. R. Bales, "The Formulation of Improved Fuel Efficient 5W30 Automotive Crankcase Lubricants Using Hydrocracked (HC) Base Oils," presented at the Division of Petroleum Chemistry, meeting of the American Chemical Society, 1992.

40. A. M. Henke and R. E. Peterson, "Process for Preparing an Improved Lubricating Oil," U.S. Patent 3,046,218.

41. Lubes 'N' Greases, *2005 Guide to Global Base Oil Refining—Supplement* (Falls Church, VA: LNG Publishing, 2005).

42. J.-P. Andre, S.-K. Hahn, S.-H. Kwon, and W. Min, "An Economical Route to High Quality Lubricants," Paper AM-96-38, presented at the National Petroleum Refiners Association annual meeting, San Antonio, Texas, March 17–19, 1996.

43. H. Y. Sung, S. H. Kwon, and J. P. Andre, "VHVI Base Oils and White Oils from Fuels Hydrocracker Bottoms," presented at the Asia Fuels and Lubes Conference, Singapore, January 25–28, 2000.

44. S-Oil and ExxonMobil, "Successful Conversion of a Fuels Hydrocracker to Group III Lube Production," presented at the ARTC 7th Annual Meeting and Reliability Conference, Singapore, April 2004, p. 15.

8 Chemistry of Hydroprocessing

8.1 INTRODUCTION

The chemistry that occurs within a hydrocracking reactor or severe hydrotreater is complex due to the variety of chemical structures that make up the feed and the number of different types of catalytic reactions that occur. As in many instances, simplifications have, of necessity, been introduced to develop an understanding of the processes and these in turn have been studied using model compounds to develop or confirm hypotheses.

The main reactions involved are

- Aromatic saturation—hydrodearomatization (HDA)
- Sulfur reduction or removal—hydrodesulfurization (HDS)
- Nitrogen reduction or removal—hydrodenitrification (HDN)
- Hydrocracking/isomerization (HCR)

In the sections which follow, the basics of each of these steps that are relevant to lubes will be discussed.

8.2 HYDRODEAROMATIZATION (HDA)

Aromatic levels in lubricant feedstocks and base stocks are obvious parameters of interest to both the processor and the user. Table 8.1 shows how the compositions of fractions can vary depending on their distillation positions in a crude oil, in this case heavy crude.[1] The analysis here should not be taken as representative of any crude used for lubes, but the general trends are typical for most crudes, that is, what are usually labeled as the "impurities"—nitrogen, sulfur, aromatics, and polars—all increase as boiling point increases. These components are essentially those which must be reduced in whatever lubes manufacturing process is employed. In hydroprocessing, these steps all consume hydrogen.

As we have seen, reduction of aromatics levels in lubricating oil base stocks relative to the feed has always been a significant part of their overall processing. The extraction step in the traditional solvent refining technology removes some aromatics, particularly the low VI polycyclic aromatics that contribute to oxidation instability and deposit formation. Hydrofinishing further reduces the levels of polynuclear aromatics, since conditions are generally too mild to reduce mono and

TABLE 8.1
Compositional Changes with Boiling Point for a Heavy 22° API Crude Oil

Fraction	Boiling Point Range, °F	Yield, wt. % Crude	Hydrogen: Carbon Ratio	Sulfur, wt. %	Oxygen, wt. %	Nitrogen, ppm	Compound Groups, wt. %		
							Saturates	Aromatics	Polars
Light naphtha	Initial boiling point–300	8	2.05	0.1	0.01	<1	88	12	—
Heavy naphtha	300–400	8	1.95	0.4	0.03	2	80	20	—
Kerosene	400–500	5	1.85	1.0	0.10	15	75	25	—
AGO	500–600	12	1.80	2.0	0.12	300	65	34	1
LVGO	650–800	15	1.7	2.8	0.18	1500	45	50	5
HVGO	800–1000	10	1.6	3.3	0.24	2000	32	58	10
SHVGO	1000–1300	15	1.5	4.2	0.28	3500	15	70	15
Residual	>1300	27	1.3	5.9	1.00	8700	2	8	90

Source: K. H. Altgelt and M. M. Boduszynski, *Composition and Analysis of Heavy Petroleum Fractions* (New York: Marcel Dekker, 1993), chap. 10. With permission.

diaromatics. However, solvent refined lubes may still contain 25% or more total aromatics[2] together with significant amounts of the original sulfur and nitrogen.

High pressure lubes hydrocracking brings together the catalysts and process conditions that can essentially eliminate nitrogen- and sulfur-containing molecules and also can effectively reduce the levels of mono- and diaromatic compounds, the most difficult types, to low levels. The extent of aromatics reduction in a lubes hydrocracker is generally limited by residence time in the reactor (i.e., space velocity), the usually limited HDA activity of the sulfided metal oxide catalyst, and by thermodynamic considerations—beyond specific temperatures and pressures, aromatics saturation may be reversed. Where necessary, high pressure hydrofinishing, using noble metal catalysts with feeds whose nitrogen and sulfur contents have been reduced to near zero, have largely removed these constraints.

Examples of both solvent refined and hydrocracked-hydrofinished finished base stock compositions are shown in Table 8.2,[3] where it can be seen that hydrocracked oils have exceptionally low levels of sulfur and aromatics compared to those obtained by solvent refining.

With the current growth of lubes hydroprocessing, knowledge of the chemistry behind hydroprocessing is important to understanding, operating, and troubleshooting the process. Much of the groundbreaking work on chemistry has been in relation to distillate hydrotreating, whose operating conditions are thermodynamically more marginal, and coal upgrading, but the concepts are readily transferable to lubes processing.

TABLE 8.2
Composition of Some Commercial 95 to 105 VI Neutral Oils

| Base Oil | Processing | Weight % | | | |
		Sulfur	Aromatics	Paraffins	Naphthenes
A	HC-SD-HR	0.002	4.50	25.60	69.90
B	HC-CD-HR	0.002	5.60	23.50	70.90
C	HC-SD-SR	0.010	7.60	20.70	71.70
D	SR-SD	0.050	9.30	18.80	71.90
E	SR-SD	0.740	28.90	25.00	46.10
F	SR-HF-SD	0.550	29.80	25.20	45.00
G	SR-HF-SD	0.366	28.10	19.50	52.40
H	SR-HF-SD	0.256	23.60	19.60	56.80
I	SR-CD-HF	0.590	27.20	24.30	48.50
J	SR-CD-HF	0.240	27.00	20.50	52.50

CD, catalytic dewaxing; HC, hydrocracking; HF, hydrogen finish; HR, hydrogen refined; SD, solvent dewaxing; SR, solvent refining.

Source: A. Sequiera, Jr., *Lubricant Base Oil and Wax Processing* (New York: Marcel Dekker, 1994), chap. 2, Table 2-19. With permission.

8.3 HDA: KINETIC ASPECTS

Hydrodearomatization is a thermodynamically reversible process:

$$k_1$$

$$ArH + nH_2 \leftrightarrow Naphthene,$$

$$k_2$$

which means that naphthenes can lose hydrogen to form aromatics if conditions are appropriate.

The equilibrium constant (K) for the reaction is given by

$$K_{eq} = (Naphthene)/(ArH)*(H_2)^n = k_1/k_2,$$

where the terms in parentheses are the concentrations (or more accurately the activities) at equilibrium. High hydrogen partial pressure drives the reaction to the right and, since aromatics saturation is exothermic, high temperatures increase the aromatics level at equilibrium. Equilibrium is not going to be established in a commercial reactor (reaction time is insufficient), but the effect of equilibrium aromatics levels can sometimes be seen in the product.

Figure 8.1 illustrates the curve that is frequently used to illustrate the behavior of product total aromatics from a hydrotreater as catalyst temperature is increased and hydrogen partial pressure is maintained constant. As the figure shows, increased reactor temperature initially causes product aromatics to decrease as

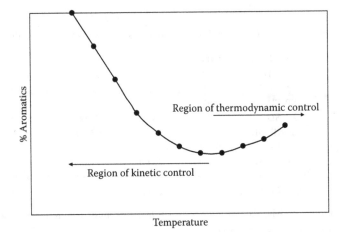

FIGURE 8.1 HDA: change from kinetic control to thermodynamic control as reactor temperature is increased.

we would expect from an increase in severity, but eventually the product aromatics level goes through a minimum and aromatics begin to increase again as temperature is further increased.

For HDA reactions, if raising the reactor temperature decreases product aromatics levels, the reaction is said to be under "kinetic control," (i.e., the rate of the forward hydrogenation reaction, k_1, has been increased and is high relative to the reverse). Conversely, if raising the reaction temperature causes an increase in aromatics, the rate k_2 of the reverse reaction has increased and $k_2 > k_1$, and the reaction is under "thermodynamic control."

Undesirable (in most cases) thermodynamic control for HDA kicks in if the hydrogen partial pressure is too low or the temperature is too high. The position of the minimum varies with aromatic type. It occurs at lower pressure and temperature for polyaromatics than for monoaromatics. For situations where product specifications demand very low aromatics (including polyaromatics), high hydrogen pressures and low catalyst temperatures are necessary to stay in the kinetic control zone. Accordingly, high activity noble metal or massive nickel catalysts are used for this type of application (e.g., in the production of many water white base stocks and particularly in white oil manufacturing). These catalysts require very low feed sulfur levels, since sulfur is a poison to these catalysts. Two-stage catalysts with separate gas systems are therefore usually needed for white oil manufacturing, the first to remove nitrogen and sulfur, and to perform some HDA, while the second low temperature system removes remaining aromatics.

To pursue this subject more generally and with more detailed background of what is involved in HDA we need to have information on a number of aspects of the reactions:

- What are the chemical steps in aromatic saturation?
- How do the rates of these reactions depend on the aromatics in the feedstock?
- Can we calculate or estimate the equilibrium positions at specific temperature and pressure conditions?
- How do these equilibria vary with chemical structure?
- How much useful thermodynamic data is available?
- Are there other factors that influence the process involved?

The overall objective is to develop ways to model the reactions, eventually leading to predictive technologies. Several excellent reviews of aromatics saturation are available.[4-6]

It is accepted[5,7,8] that at least for simple unsubstituted aromatics (these are the usual "model" compounds that are employed; they are recognized as an oversimplification but a necessary first step in this kind of work), they are hydrotreated to the corresponding naphthenes, with stepwise saturation of the aromatic rings in the case of polyaromatics. Figure 8.2 shows the reactions that can occur with (a) benzene to cyclohexane, (b) naphthalene to tetralin to decalin, and (c) phenanthrene to the dihydro-, tetrahydro-, octahydro-, and perhydrophenanthrene forms.

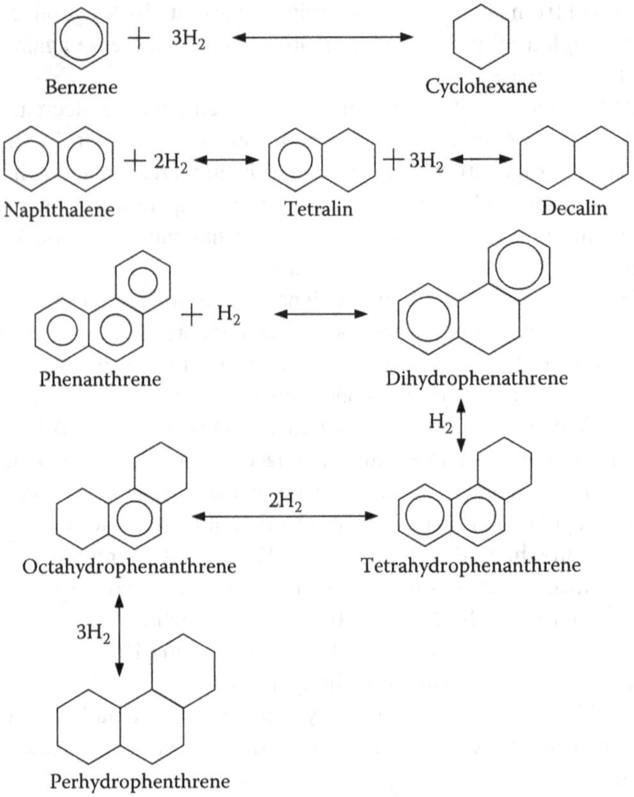

FIGURE 8.2 Discrete steps in the hydrotreatment of typical aromatic structures.

Furthermore, under conditions of kinetic control, the rate of saturation of fused polyaromatic rings has been found to be faster than for monoaromatics over sulfided molybdenum and tungsten catalysts, and it appears that relative rates can be generalized as

Triaromatics > diaromatics >> monoaromatics,

which means in practice that triaromatics are converted to diaromatics faster than the latter are converted to monoaromatics.

For the three model aromatic compounds—benzene, naphthalene, and anthracene[7]—naphthalene underwent hydrotreatment over sulfided nickel/tungsten catalyst at 340°C/70 bar H_2 to the corresponding monoaromatic (tetralin) an order of magnitude faster than benzene saturates to cyclohexane (Table 8.3).

The rate of disappearance of anthracene (to a product containing two or fewer aromatic rings) was two to three times faster than that of naphthalene. Directionally

TABLE 8.3
Relative Rate Constants for Hydrogenation of Benzene, Naphthalene, and Anthracene over MoS_2 and Sulfided Nickel-Tungsten Catalyst at 340°C and 70 bar H_2

Compound	Relative Rate over MoS_2	Relative Rate over NiW
Benzene	1	1
Naphthalene	23	18
Anthracene	62	40

Source: C. Moreau, C. Aubert, R. Durand, N. Zmimita, and P. Geneste, "Structure-Activity Relationships in Hydroprocessing of Aromatic and Heteroaromatic Compounds Over Sulfided $NiO-MoO_3\gamma-Al_2O_3$ and $NiO-WO_3/-Al_2O_3$ Catalysts: Chemical Evidence for the Existence of Two Types of Catalyst Sites," *Catalysis Today* 4:117–131 (1988). With permission.

similar results were reported by Kokayeff[9] for naphthalene, biphenyl, and benzene during an investigation of diesel hydrotreating over a sulfided nickel/molybdenum catalyst at 43 bar/190°C to 305°C with the pure aromatic components dissolved in a saturated polyolefin. As would be expected from its structure, in this and other work, biphenyl hydrogenates at rates more similar to benzene than naphthalene.

The effects of these relative rates of hydrogenation are seen in hydrocracked and hydrofinished lube samples by the product aromatics content being predominantly monoaromatics with small amounts of diaromatics and trace polyaromatics. Much of the published work has been reported however for nonlube applications. For example, Fafet and Magné-Drisch (IFP),[10] as part of their development of kinetic models for HDA, HDS, and HDN, found for a C11 to C25 light cycle oil distillate, that poly- and triaromatics were almost completely removed on hydrotreatment, and diaromatics were much decreased. The monoaromatics increased dramatically since the higher aromatics were all reduced to the monostage. Conversions calculated are given in Table 8.4 and results from actual analyses by mass spectroscopy are in Table 8.5.

Similar relative reactivity of polycyclic aromatics is found with noble metal catalysts.[11,12] The development of noble metal catalysts more resistant to sulfur and nitrogen compounds has led to their more prevalent use, particularly by those refiners who wish to have low aromatic base stocks. Nickel catalysts can also be used for this purpose, but do not appear to have achieved wide acceptance except historically in white oil applications.

Sapre and Gates[8] (University of Delaware) studied the hydrotreatment of benzene, biphenyl, naphthalene, and 2-phenyl naphthalene over $CoO-MoO_3/\gamma-Al_2O_3$

TABLE 8.4

Conversions of Aromatic Types during Hydrotreatment

Aromatic Types	Percent Conversion
Diaromatics	74
Tri+ aromatics	90
Sulfur-containing aromatics	99

Source: A. Fafet and J. Magne-Drisch, "Analyse Quantitative Détaillée des Distillats Moyens par Couplage CG/MS— Application à l'Étude des Schémas Réactionnels du Procédé d'Hydrotraitemant," *Revue de L'Institut Francais due Pétrole* 50:391–404 (1995). With permission.

sulfided catalysts (Figure 8.3), where the pseudo first-order rate constants they determined are given for the forward and reverse reactions at 325°C and 75 atm. It can be seen here that the sole benzene ring in benzene itself and the first one in biphenyl undergo saturation at about the same rate. The phenyl group as a benzene substituent does not have a marked effect. The second ring in biphenyl reacts even more slowly under these conditions. In the case of naphthalene, where the second ring is fused, the first ring reacts at about 20 times the benzene rate, and the second, now simply substituted benzene, reacts at a rate about the same as benzene. It should be noted as well that under these conditions the rates of the aromatics saturation reactions are much higher than those of the reverse. Higher temperatures or lower hydrogen partial pressures would be expected to alter these relative rates.

In the case of 2-phenyl naphthalene (Figure 8.3), the fourth structure investigated, only the hydrotreatment of the naphthalene part proceeds at a significant rate. Of the two naphthalene rings involved, the ring attached to the phenyl group reacts more quickly, with the 2- and 6-phenyl tetralins being the primary products formed from the two competitive routes. Both initial hydrogenation steps were slower than the corresponding one for naphthalene. Hydrotreatment of the phenyl substituent is not a pathway that could be verified in this case. Rearrangement and isomerization during or subsequent to hydrotreatment can also occur, which enlarges the product slate. From the simple compound biphenyl, three different methyl-substituted cyclopentyl benzenes are believed to be among the products (Figure 8.4) formed by isomerization of the saturated cyclohexyl ring.

Gates also investigated hydrotreatment of the three-ring compound fluoranthene, which incorporates a five-membered carbon ring plus naphthalene and benzene systems. The catalyst employed was a commercially available sulfided nickel/tungsten one operated at 2250 psi hydrogen and temperatures in the range 320°C to 380°C. For this compound, the first ring of the naphthalene group underwent saturation about an order of magnitude faster than the second one of

TABLE 8.5
Mass Spectroscopic Analyses of Hydrotreater Feed and Product

Hydrocarbon type	Charge	Product
Paraffins	23.5	27.3
Naphthenes		
1-ring	4.2	5.1
2-ring	3.0	7.2
3-ring	2.0	6.2
4-ring	1.5	2.5
5-ring	0.3	0.1
6-ring	0.0	0.0
Total naphthenes	11.0	21.1
Total saturates	34.5	48.4
Monoaromatics		
CnH_2n-6	5.6	9.2
CnH_2n-8	3.7	19.6
CnH_2n-10	2.3	12.3
Total monoaromatics	11.6	41.1
Diaromatics		
CnH_2n-12	21.7	3.5
CnH_2n-14	8.0	3.5
CnH_2n-16	6.4	2.3
Total diaromatics	36.1	9.3
Triaromatics	8.7	0.6
Polyaromatics	1.8	0.5
Sulfur compounds		
CnH_2n-10S	3.3	0.0
CnH_2n-12S	4.0	0.1
Total aromatics	65.5	51.6

Source: A. Fafet and J. Magne-Drisch, "Analyse Quantitative Détaillée des Distillats Moyens par Couplage CG/MS—Application à l'Étude des Schémas Réactionnels du Procédé d'Hydrotraitemant," *Revue de L'Institut Francais due Pétrole* 50:391–404 (1995). With permission.

the naphthalene moiety or of the final ring. The reaction pathways and rate constants measured at 380°C are summarized in Figure 8.5.

It should be noted that these chemical changes all are in the direction of viscosity index (VI) increases—examples of these are given in Figure 8.6 for a number of polycyclic aromatics and their perhydrogenated naphthenes[13] where

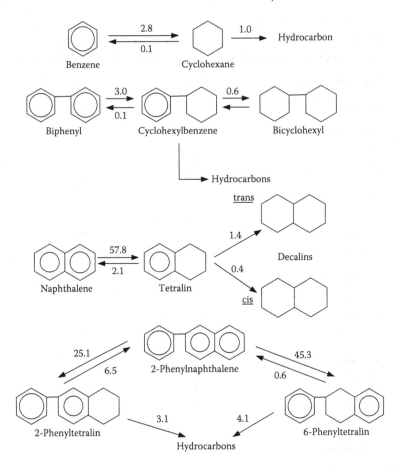

FIGURE 8.3 Pseudo first-order rate constants for the hydrotreatment of benzene, biphenyl, naphthalene, and 2-phenylnaphthalene over sulfided cobalt/molybdenum catalyst at 325°C and 75 atm.

Source: A. V. Sapre and B. C. Gates, "Hydrogenation of Aromatic Hydrocarbons Catalyzed by Sulfided CoO-MoO₃ γ-Al₂O₃. Reactivities and Reaction Networks," *Industrial and Engineering Chemistry Process Design and Development* 20:68–73 (1981). With permission.

FIGURE 8.4 Methyl-substituted isomeric phenylcyclopentanes possibly originating from biphenyl as by-products of hydrotreatment.

Source: A. V. Sapre and B. C. Gates, "Hydrogenation of Aromatic Hydrocarbons Catalyzed by Sulfided CoO-MoO₃ γ-Al₂O₃. Reactivities and Reaction Networks," *Industrial and Engineering Chemistry Process Design and Development* 20:68–73 (1981). With permission.

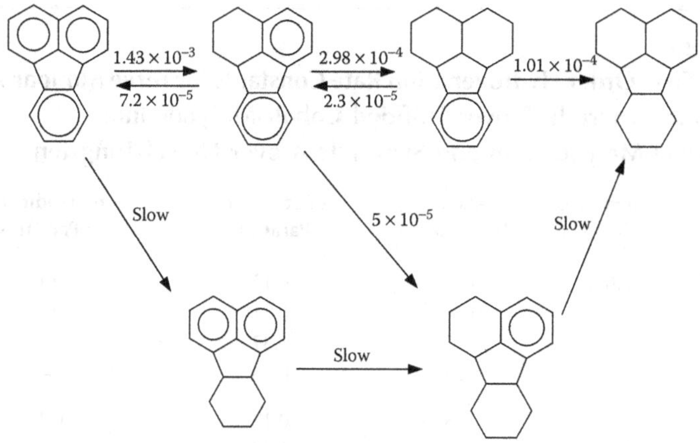

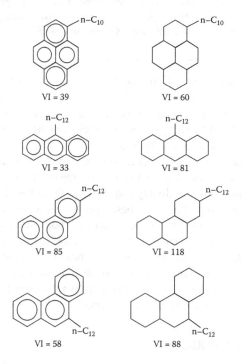

FIGURE 8.5 Pathways and rate constants for the individual steps in the hydrotreatment of fluoranthene over a nickel/tungsten catalyst at 320°C to 380°C and 2250 psi hydrogen. *Source:* M. J. Girgis and B. C. Gates, "Reactivities, Reaction Networks, and Kinetics in High-Pressure Catalytic Hydroprocessing," *Industrial and Engineering Chemistry Research* 30:2021–2058 (1991). With permission.

FIGURE 8.6 Viscosity index changes possible on hydrotreatment.
Source: Properties of Hydrocarbons of High Molecular Weight, Research Project 42, 1940–1966, America Petroleum Institute, New York. With permission.

TABLE 8.6
Pseudo First-Order Hydrogenation Rate Constants for Three Monoaromatic Groups in Syncrude B over Sulfided Cobalt-Molybdenum and Nickel-Molybdenum and Syncrude A over Nickel-Tungsten

Catalyst	Temperature, °C	Alkyl Benzenes	Benzocyclo-Paraffins	Benzodicyclo-Paraffins
Co-Mo	340	0.09	0.12	0.18
	360	0.14	0.21	0.20
	380	0.20	0.37	0.25
	400	0.26	0.38	—
Ni–Mo	340	0.08	0.13	0.14
	360	0.18	0.34	0.34
	380	0.47	0.70	0.70
Ni-W	340	0.33	0.31	0.24
	360	0.54	0.45	0.30
	380	0.66	0.53	0.38

Source: I. P. Fisher and M. F. Wilson, "Kinetic and Thermodynamics of Hydrotreating Synthetic Middle Distillates," presented at the Symposium on Advances in Hydrotreating, Division of Petroleum Chemistry, American Chemical Society Meeting, April 8–10, 1987, pp. 310–314. With permission.

no isomerization has occurred. It can be seen that the VI increases can be substantial.

Worth mentioning as well, as an illustration of what can be achieved using sophisticated analytical technology (but a stretch at being model compounds for lube range materials), Fisher and Wilson,[14] using mass spectroscopy, studied the rates of disappearance of alkylbenzenes, benzocycloparaffins, and benzodicycloparaffins, all monoaromatics, but the latter two being naphtho- and dinaphthobenzenes (mass spectroscopy can distinguish between these based on their different Z numbers). The hydrotreatment of these species, native in a syncrude diesel, was studied over cobalt-molybdenum, nickel-molybdenum, and nickel-tungsten catalysts at between 340°C and 400°C and at 2500 psi hydrogen. The results (Table 8.6) show that rate differences between these monoaromatic types are not very great, but quite consistently with the cobalt-molybdenum and nickel-molybdenum catalysts the alkylbenzenes react slower than the other two types, but with nickel-tungsten catalysts the results are reversed.

8.4 HDA: EQUILIBRIA

Equilibrium is related to the free energy change in the reaction

$$\Delta G^\circ = -RT\ln(K_{eq}).$$

Accurately determined free energies of petroleum aromatics are difficult to obtain by calculation methods alone—there are methods for estimation,[15] but they are not considered sufficiently accurate since small errors in G^o can cause significant errors in K_{eq} values. Most of the frequently quoted values of equilibrium constants have therefore come from experimental results on model compounds.

Since

$$\Delta G^o = \Delta H^o - T\Delta S^o,$$

where H^o and S^o are the standard enthalpy and entropy changes for the reaction, then

$$-RT\ln(K_{eq}) = H^o - T\Delta S^o,$$

the variation in the equilibrium constant with temperature is expressed by the equation

$$\ln K_{eq} = -(\Delta H^o/R)1/T + \Delta S^o/R.$$

Consequently knowledge of ΔH^o (which varies little with temperature), permits calculation of the variation of $\ln K_{eq}$ and therefore K_{eq}, if known, with temperature.

Much of the basic information available on thermochemical aspects of HDA came initially from academic studies[16-20] on pure compounds, undertaken to establish some of the basic chemistry of hydrogenations in general. Kistiakowsky et al.[16] calorimetrically established that saturation of aromatic rings was exothermic and that the enthalpies of hydrogenation ($H_{355°K}$) of a number of monoaromatics decreased with increasing alkyl substitution—benzene (49.8 kcal/mole), ethylbenzene (48.9 kcal/mole), o-xylene (47.25 kcal/mole), and 1,3,5-trimethylbenzene (47.62 kcal/mole).

Frye[21,22] (American Oil Co) experimentally studied the gas phase equilibria involving hydrogen and the polycyclic aromatics: naphthalene, phenanthrene, diphenyl, indene, acenaphthenes, and fluorene. This work established that aromatics formation from naphthenes was favored by high temperatures and low hydrogen pressures. Equilibrium concentrations could be estimated from the data. The catalyst employed in this work was platinum on -alumina and products were analyzed by gas chromatography. Figure 8.7 illustrates how the equilibrium concentrations of naphthalene, its tetrahydro derivative, tetralin, and the perhydronaphthalenes, cis- plus trans-decalin, vary with temperature at 2500 psi hydrogen partial pressure (these values are calculated from Frye's experimental data). At low temperatures (less than 700°F, less than 371°C), the fully saturated perhydro product decalin constitutes 99.5 mole % of the hydrocarbons present. Between 700°F and 800°F, equilibrium begins to favor both tetralin and naphthalene, with tetralin reaching a maximum at about 980°F, where the composition is 40% tetralin, 40% naphthalene, and 20% decalin. At still higher temperatures, dehydrogenation of tetralin to naphthalene is favored, and beyond 1150°F there

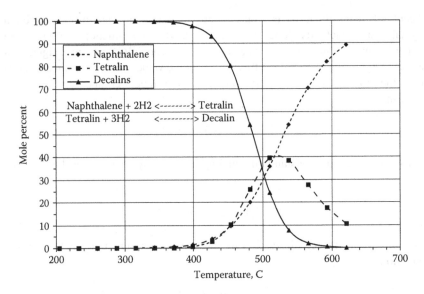

FIGURE 8.7 Variation in naphthalene, tetralin, and decalin concentrations as functions of temperature at 2500 psi hydrogen partial pressure.

is no decalin, only 10% tetralin remains, and the equilibria favors the hydrocarbon with the least hydrogen. Lower hydrogen partial pressures would shift these curves toward lower temperatures.

Some values of K_{eq} and $H°$ for aromatic saturation reactions of interest are given in Table 8.7. High values of K correspond to the hydrogenated component being favored under the conditions specified. The purpose in developing the thermodynamic terms is to provide a generalized framework within which experimental data can be compared and predictions made for similar reactions involving other polycyclic hydrocarbons. In particular, accurate thermodynamic data on a reaction allows equilibrium concentrations to be calculated.

Figure 8.8 shows the equilibrium concentrations of the individual aromatics from phenanthrene to tetralin calculated from hydrogenation data measurements by Frye and standardized to 2500 psi hydrogen partial pressure. It can be seen that higher temperatures favor reversible formation of aromatics, and polyaromatics in particular. This is what is normally observed in practice. As catalyst temperatures increase from low levels, polyaromatics are initially stepwise hydrogenated, an aromatic ring at a time, to the corresponding polycyclic naphthenes and then they reappear as catalyst temperatures are further increased, with the polycyclic aromatics being the first to appear. This of course is only true when the naphthenes with the correct number of rings are present or have been formed in the initial hydrogenation step. This thermodynamic feature establishes a maximum catalyst temperature for aromatics conversion for specific hydrogen pressures and leads to the minimum in the curve in Figure 8.1. A product can therefore fail a PNA-dependent specification for one of two reasons: (1) the reaction is being kinetically controlled

TABLE 8.7
Calculated Values of the Equilibrium Constant, K (atm^{-n}) for Saturation of Aromatics

Conversion	K at 600°F	at 800°F	H (kJ/mole)
Naphthalene to tetralin	7.37E-02	7.0E-04	142.0
Tetralin to decalin	6.6E-03	5.67E-06	217.8
Naphthalene to decalin	4.85E-04	4.19E-09	359.9
Diphenyl to cyclohexylbenzene	1.56E-02	1.42E-05	216.1
Cyclohexylbenzene to cyclohexyl cyclohexane	3.46E-03	3.21E-06	215.6
Phenanthrene to dihydrophenathrene	2.05E-02	3.98E-03	50.7
Phenanthrene to tetrahydrophenanthrene	7.7E-03	1.42E-04	132.1
Phenanthrene to perhydrophenanthrene	9.20E-8	3.88E-14	453.1

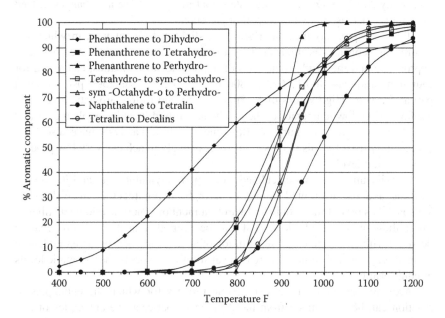

FIGURE 8.8 Hydrogenation of various aromatics at 2500 psi hydrogen: variation of aromatics content at equilibrium with temperature.

TABLE 8.8
Naphthene:Aromatic Ratio at Equilibrium at 2500 psi Hydrogen

Temperature, °F	Naphthene:Aromatic Ratio = $K_{eq} \cdot (H_2)^n$			
	400	600	800	1000
Naphthalene to tetralin + $2H_2$	2.02E+06	2.38E+03	2.39E+01	8.45E-01
Tetralin to decalin + $3H_2$	2.38E+15	9.08E+07	7.85E+02	1.66E-01
Phenanthrene +H_2 to dihydrophenanthrene	3.86E+01	3.49E+00	6.77E-01	2.06E-01
Phenanthrene + $2H_2$ to tetrahydrophenanthrene	8.52E+04	2.47E+02	4.59E+00	2.54E-01
Phenanthrene + $7H_2$ to perhydrophenanthrene	1.22E+18	5.56E+08	2.35E+02	5.53E-03
Tetrahydrophenanthrene + $2H_2$ to octahydrophenanthrene	1.20E+05	2.51E+02	3.74E+00	1.76E-01
Octahydrophenanthrene + $3H_2$ to perhydrophenanthrene	1.44E+08	1.26E+04	2.13E+01	2.07E-01

but the catalyst temperature is not high enough, or (2) the reaction is under thermodynamic control and temperature should be reduced to return it to kinetic control (if activity suffices). Other remedial steps that may be of use are changes in feed rate and an increase in hydrogen partial pressure.

Table 8.8 presents these numbers in a slightly different way, in which the equilibrium constant is multiplied by the pressure term to whatever power is appropriate and the result is the ratio of naphthenes to aromatics, calculated for four temperatures—400°F, 600°F, 800°F, and 1000°F—at 2500 psi hydrogen partial pressure. High values of $K_{eq}*pH_2$ correspond to low aromatic levels and are favored by low temperatures, hence the use of highly active noble metals and nickel catalysts for this purpose. Base metal sulfide catalysts have also been successfully used for this purpose, but are less likely to achieve very low product polyaromatic levels.

Equilibrium constants calculated from thermodynamic data at 170 atm for benzene and its substituted homologs have been put together by Fisher and Wilson[14,23] as part of their study on hydrotreatment of syncrude middle distillates. From these numbers (Table 8.9) it can be seen that heats of hydrogenation decrease with either increased substitution of an aromatic ring or increased length of an alkyl group substituent. Increased substitution favors higher aromatic levels at equilibrium.

Aromatic saturation is an exothermic process for which the enthalpies of reaction can be determined from the differences between the enthalpies of formation of the product and starting material:

$$\Delta H = \sum \Delta H_f(\text{products}) - \sum \Delta H_f(\text{reactants}).$$

TABLE 8.9
Equilibrium Constants, K_p, Heats of Hydrogenation, and Molar Ratios of Cycloparaffins (C_p) to Monoaromatics (Ar)

Aromatic	Cycloparaffin	K_p, atm³		ΔH, kcal/mole	C_p/Ar at 170 atm	
		620°F	800°F		620°F	800°F
Benzene	Cyclohexane	2.97E+02	5.55E-05	52.4	1.46E+05	2.73E+02
Toluene	Methyl-cyclohexane	1.27E-02	2.56E-05	51.8	6.25E+04	1.25E+02
n-Hexylbenzene	n-Hexyl-cyclohexane	2.58E-03	6.52E-06	50.4	1.27E+04	3.21E+01
n-Hexadecylbenzene	n-Hexadecyl-cyclohexane	2.67E-03	6.56E-06	50.1	1.31E+04	3.23E+01
m-Xylene	1,3-Dimethyl-cyclohexane	1.30E-03	3.46E-06	49.8	6.40E+03	1.67E+01
Mesitylene	1,3,5-Trimethyl-cyclohexane	5.38E-04	1.38E-06	49.8	2.65E+03	6.80E+00
Middle distillate study, nickel-tungsten catalyst			1.34E-06 at 824°F			6.6E+00 at 824°F
Middle distillate study, cobalt-molybdenum catalyst			1.45E-06 at 824°F			7.1E+00 at 824°F

Source: I. P. Fisher and M. F. Wilson, "Kinetic and Thermodynamics of Hydrotreating Synthetic Middle Distillates," presented at the Symposium on Advances in Hydrotreating, Division of Petroleum Chemistry, American Chemical Society Meeting, April 8–10, 1987, pp. 310–314. With permission.

Examples of these[24] are given in Table 8.10, where it can be seen that enthalpies of reaction of monoaromatics decrease with increasing substitution and increase with the number of aromatic rings being saturated.

Magnabosco[25] (EniChem International) modeled catalytic hydrogenation of aromatics types in jet fuel and light gas oil using methods and parameters that perhaps can also be employed in lubricants, if required. Figure 8.9 and Figure 8.10 are examples of the equilibrium networks that were employed. For phenanthrenes and lower aromatics, there exist sufficient experimental data to determine values of K_p and H^o (Figure 8.9). For four- or more ring polyaromatics, estimates of K_p and H^o must be used (Figure 8.10). The K_p calculated may not be accurate in those cases.

Miller and Zakarian[26] have studied the kinetics of the HDA (hydrofinishing) of hydrocracked and dewaxed 500Ns over noble, base metal, and nickel-tin catalysts, all on a silica-alumina base, at 15.3 MPa total pressure and at temperatures between 220°C and 260°C. The feedstocks employed contained trace quantities of sulfur (up to 19 ppm) and nitrogen (3 to 4 ppm). Since a correlation

TABLE 8.10
Enthalpies of Formation for Reactants and Products in Aromatic Saturation Reactions (All in Liquid Phase)

Reactant	ΔH_f(reactant), kJ/Mole	Product	H_f(product), kJ/Mole	H, kJ/Mole
Benzene	49.0	Cyclohexane	−156.4	−205.4
Toluene	12.4	Methylcyclohexane	−190.1	−202.5
Ethylbenzene	−12.3	Ethylcyclohexane	−211.9	−199.6
Cyclohexylbenzene	−76.6	Cyclohexylcyclohexane	−273.7	−197.1
o-Xylene	−24.4	cis-1,2-Dimethylcyclohexane	−211.8	−187.4
o-Xylene	−24.4	trans-1,2-Dimethylcyclohexane	−218.2	−193.8
Tetrahydronaphthalene	−29.2	cis-Decahydronaphthalene	−219.4	−190.2
Tetrahydronaphthalene	−29.2	trans-Decahydronaphthalene	−230.6	−201.4
Naphthalene	−77.9	trans-Decahydronaphthalene	−230.6	−308.5

Source: David R. Lide, ed., *CRC Handbook of Chemistry and Physics*, 73rd ed. (Boca Raton, FL: CRC Press, 1992). With permission.

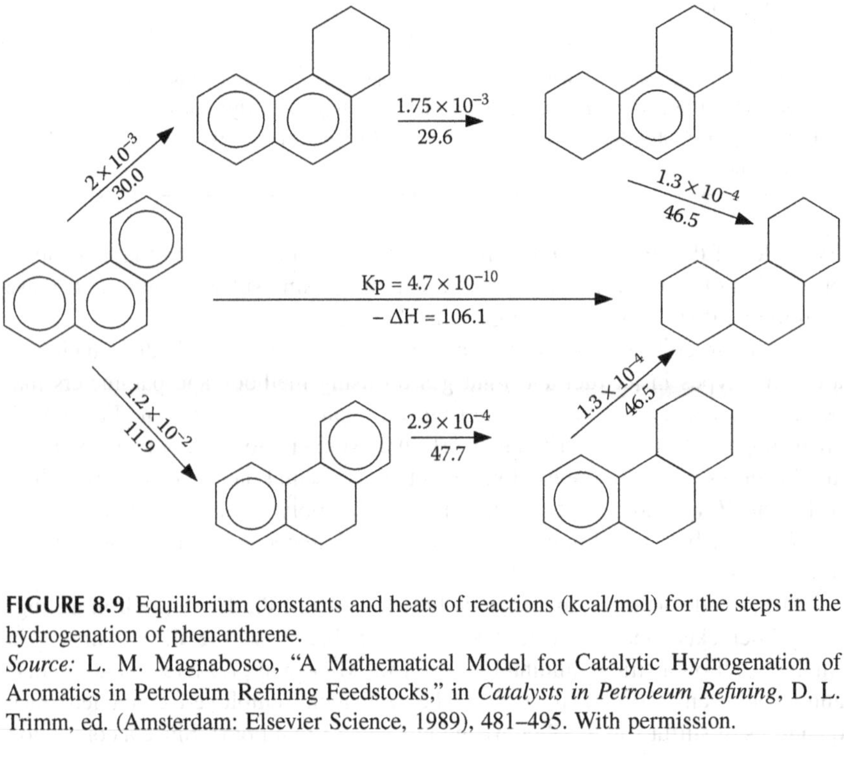

FIGURE 8.9 Equilibrium constants and heats of reactions (kcal/mol) for the steps in the hydrogenation of phenanthrene.
Source: L. M. Magnabosco, "A Mathematical Model for Catalytic Hydrogenation of Aromatics in Petroleum Refining Feedstocks," in *Catalysts in Petroleum Refining*, D. L. Trimm, ed. (Amsterdam: Elsevier Science, 1989), 481–495. With permission.

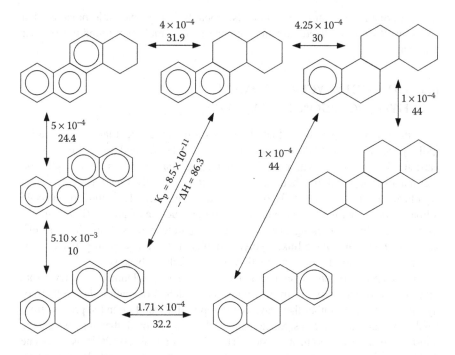

FIGURE 8.10 Equilibrium constants and heats of reactions for hydrogenation of polyaromatics (Chrysene).
Source: L. M. Magnabosco, "A Mathematical Model for Catalytic Hydrogenation of Aromatics in Petroleum Refining Feedstocks," in *Catalysts in Petroleum Refining*, D. L. Trimm, ed. (Amsterdam: Elsevier Science, 1989), 481–495. With permission.

exists between ultraviolet (UV) absorbance at 226 nm ($A_{(226)}$) of the finished base stock and Chevron's oxidation stability test, they measured the kinetics using the absorbance on the feed and products. This study found that first-order irreversible kinetics which included an expression for deactivation (Equation 8.1) best fitted their results:

$$k = \{LHSV/S_0 exp^{[-kd*t]}\}ln(\{A_{226}\}_{feed}/\{A_{226}\}_{product}), \qquad (8.1)$$

where

    k = rate constant for change in absorbance at 226 nm,
    k_d = deactivation rate constant,
    S_0 = initial number of active sites, and
    LHSV = space velocity.

They were able to use this method to track commercial catalyst activity decline. Their work concluded that not only were nitrogen and sulfur compounds catalyst poisons (the sulfur compounds probably being reversible poisons), but also as well were oxidized base stock components, which they found in the polar fraction.

8.5   HDA: POLYCYCLIC AROMATIC HYDROCARBON FORMATION

It is important to be aware that there is strong evidence that polyaromatics (polycyclic aromatic hydrocarbons [PAHs]) can be formed in hydrocracking, particularly in severe hydrocracking, and these may play a role in light and thermal instability of finished base stocks. It is unlikely that this feature is a significant factor in a normally operating lubes hydrocracking unit to make 95 VI base stocks, but it could be a factor during production of group III base stocks using severe hydrocracking in a fuels unit. Separate papers from Nippon Oil[27] and Mitsubishi Oil[28] have linked polyaromatics or "partially hydrogenated" aromatics to this feature in the production of very high VI base stocks.

These new PAHs are not considered to be native to petroleum. It is believed that they are not formed directly from naphthenes by dehydrogenation due to thermodynamic factors, but rather they have been proposed to build up in the process from smaller polyaromatic species. Since many of these are colored, they may have to be removed in a final stabilization step. They were first observed[29,30] as coronene derivatives (Figure 8.11) in the recycle streams of two-stage fuels hydrocrackers during gasoline production and are seen as an outcome of the high temperatures involved. Conditions leading to their production were said to be temperatures greater than 700°F and pressures of less than 2000 psi. The PAHs thus formed interfere with the process by plating out in heat exchangers (due to their very low solubility) and by reducing catalyst life through coke formation. Since they are red in color, their effect has sometimes been termed the "Red Death" for fuels hydrocrackers.

The most extensive work published has been from Chevron, which, through the development of analytical methodologies, has identified a number of these polycyclic aromatics[31–35] and tracked them through the sequence of hydroprocessing steps.[36,37] Specifically they examined the polycyclic aromatic levels in start-of-run (SOR) and end-of-run (EOR) vacuum gas oil (VGO) product samples from a residuum hydrocracker (RDS unit) using a battery of methods including high-pressure liquid chromatography (HPLC), mass spectroscopy, and UV spectroscopy. They found that the EOR products from the RDS unit had substantially higher PAH levels than those from SOR conditions (final column in Table 8.11).

The PAHs included pyrenes and phenanthrenes and their hydrogenated analogs, with the EOR products being shifted to lower carbon numbers—typically in the range C_{15} to C_{40} in SOR products to predominantly C_{15} to C_{30}, attributed to cracking off short hydrocarbon substituents.

These VGOs were subsequently fed to a pilot unit simulating a two-stage hydrocracker in a jet fuel operating mode with recycle to extinction (Figure 8.12). Figure 8.13 shows the distribution of aromatics in RDS-VGO, second-stage feed, and recycle oils, and it can be seen that six- and seven-ring PAHs are more

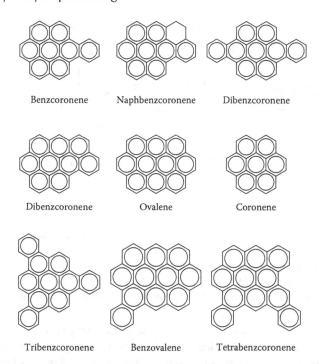

Benzcoronene   Naphbenzcoronene   Dibenzcoronene

Dibenzcoronene   Ovalene   Coronene

Tribenzcoronene   Benzovalene   Tetrabenzcoronene

FIGURE 8.11 Polyaromatic coronene derivatives found or suspected to be in fuels hydrocracker bottoms streams.

Source: F. C. Wood, C. P. Reeg, A. E. Kelley, and G. D. Cheadle, "Recycle Hydrocracking Process for Converting Heavy Oils to Middle Distillates," U.S. Patent 3,554,898.

TABLE 8.11
HPLC Results for the SOR and EOR RDS-VGOs

HPLC Fraction	Concentration, wt. %, in VGO		EOR:SOR Ratio
	SOR RDS-VGO	EOR RDS-VGO	
Saturates	57.8	53.0	0.92
Monoaromatics	27.50	28.46	1.03
Diaromatics	13.04	14.35	1.10
Triaromatics	1.198	2.862	2.39
Tetraaromatics	0.187	0.729	3.90
Penta+-aromatics	0.134	0.380	2.84

Source: R. F. Sullivan, M. M. Boduszynski, and J. C. Fetzer, "Molecular Transformations in Hydrotreating and Hydrocracking," Energy & Fuels, Vol 3, pp 603–612 (1989). With permission.

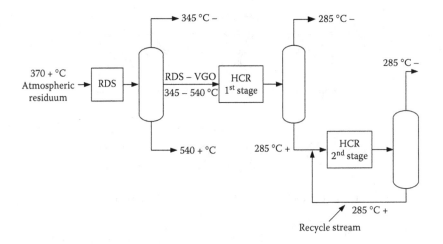

FIGURE 8.12 Schematic of the RDS-VGO hydrocracking process.
Source: R. F. Sullivan, M. M. Boduszynski, and J. C. Fetzer, "Molecular Transformations in Hydrotreating and Hydrocracking," *Energy and Fuels* 3:603–612 (1989). With permission.

significant in the recycle stream where they have effectively become concentrated. A scheme, dubbed the "naphthalene zigzag" reaction pathway, whereby these PAHs and those with higher numbers of aromatic rings arise, is shown in Figure 8.14. The "peri" ring closure path is considered as the preferred route.

Ring formations arise from the alkyl substituents (not shown), with an alkyl chain having at least two carbons being responsible for the "peri" ring closures,

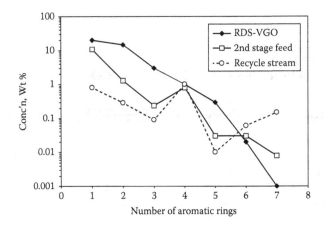

FIGURE 8.13 Aromatics Distribution by number of aromatic rings in EOR RDS-VGO product streams from Chevron studies.
Source: R. F. Sullivan, M. M. Boduszynski, and J. C. Fetzer, "Molecular Transformations in Hydrotreating and Hydrocracking," *Energy and Fuels* 3:603–612 (1989). With permission.

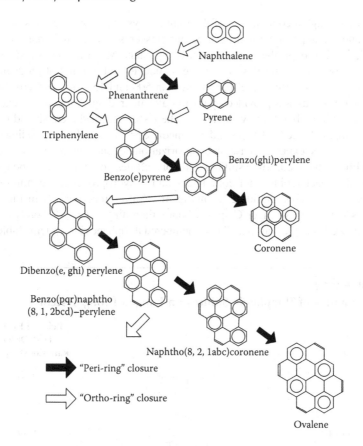

FIGURE 8.14 "Naphthalene zigzag" reaction pathway.
Source: R. F. Sullivan, M. M. Boduszynski, and J. C. Fetzer, "Molecular Transformations in Hydrotreating and Hydrocracking," *Energy and Fuels* 3:603–612 (1989). With permission.

and one with four carbons or more taking part in the "ortho" ring closures. These reactions have previously been established in the hydrocracking chemistry of simpler model compounds.

8.6 HYDRODESULFURIZATION

We have already discussed some aspects of desulfurization under "hydrofinishing." In that section (Chapter 6), mild hydrotreatment of solvent refined stocks was found by Imperial Oil and others to lead to the relatively easy desulfurization of benzothiophenes to alkyl benzenes and dibenzothiophenes to biphenyls and therefore the enhancement of the monoaromatic content of the finished base stock.

In the production of base stocks by hydrocracking and feed preparation for dewaxing by current hydroisomerization technology, very low sulfur levels are achieved through a combination of the right catalyst, the correct catalyst temperatures, and the appropriate hydrogen partial pressures. As in other areas, a lot has been learned through the use of model compounds. Much of this investigational work was originally undertaken with the particular goal in mind of production of low and ultralow sulfur diesel fuels (less than 500 and 15 ppm, respectively) and additional impetus was provided by coal liquids upgrading. A number of excellent reviews on this subject are available.[38–40] The lessons learned there should in principle be applicable to feedstocks and intermediate products in the lube boiling range.

As might be expected, ease of desulfurization depends on the type of carbon-sulfur bonds involved. Alkyl sulfides and polysulfides possess weak carbon-sulfur bonds and react rapidly and completely under hydroprocessing conditions, as their uses in catalyst sulfiding agents testify. Sulfur incorporated in thiophene systems have both stronger $C(sp^2)$–S bonds than alkyl $C(sp^3)$–S bonds and the sulfur as well is in the "aromatic" five-membered thiophene ring system. Table 8.12

TABLE 8.12
Reactivities of Thiophenic Sulfur in a Range of Ring Systems

Reactant	Structure	Relative Pseudo First-Order Rate Constant, L/g
Thiophene		1.00
Benzothiophene		0.59
Dibenzothiophene		0.04
Benzo[b]naphtha[2,3-d] thiophene		0.12
7,8,9,10-Tetrahydrobenzo[b] naphtha [2,3-d]thiophene		0.06

Source: N. K. Nag, A. V. Sapre, D. H. Broderick, and B. C. Gates, "Hydrodesulfurization of Polycyclic Aromatics Catalyzed by Sulfided CoO-MoO₃ γ-Al₂O₃: The Relative Reactivities," *Journal of Catalysis* 57:509–512 (1979). With permission.

TABLE 8.13

Relative Rates of Disappearance of Nonthiophenic Sulfur-Containing Aromatic Compounds over Nickel-Molybdenum Catalyst at 340°C and 70 Atm

Reactant	Structure	Relative Pseudo First-Order Rate Constant
Thianthrene		12.5
Phenothiazine		11.7
Phenoxathlin		10.8
Thioxanthene		9.2
Dibenzothiophene		1

Source: C. Aubert, R. Durand, P. Geneste, and C. Moreau, "Hydroprocessing of Dibenzothiophene, Phenothiazine, Phenoxanthlin, Thianthrene, and Thioxanthene on a Sulfided NiO-MoO$_3$/γ-Al$_2$O$_3$," *Journal of Catalysis* 97:169–176 (1986). With permission.

shows that rates of disappearance of thiophenes with increasing numbers of aromatic rings decrease over a cobalt-molybdenum catalyst at 71 atm and 300°C.[41]

Nonthiophenic aromatic sulfur compounds react more quickly than dibenzothiophenes (Table 8.13).[42] In these cases, the sulfur lone pair of electrons is not delocalized to the extent it is in thiophenes.

Particular attention has been focused on certain alkyl substituted dibenzothiophenes because of their slowness to react, thus their persistence in

TABLE 8.14
Relative Reactivities of Methyl-Substituted Dibenzothiophenes Relative to the Parent Compound over Cobalt-Molybdenum Catalyst at 300°C and 102 Atm

Reactant	Structure	Relative Pseudo First-Order Rate Constant
Dibenzothiophene		1.00
2,8-Dimethyldibenzothiophene		0.91
3,7-Dimethyldibenzothiophene		0.48
4,6-Dimethyldibenzothiophene		0.07
4-Methylbenzothiophene		0.09

Source: M. Houalla, D. H. Broderick, A. V. Sapre, N. K. Nag, V. H. J. de Beer, B. C. Gates, and H. Kwart, "Hydrodesulfurization of Methyl-Substituted Dibenzothiophenes Catalyzed by Sulfided CoO-MoO$_3$/γ-Al$_2$O$_3$," *Journal of Catalysis* 61:523–527 (1980). With permission.

hydrotreated products and consequent difficulties in meeting very low sulfur specifications. Table 8.14 demonstrates that dibenzothiophenes with one or more alkyl substituents adjacent to the sulfur reacts about an order of magnitude slower than those with the alkyl group further removed.[43] Product analyses have shown that the sulfur-containing compounds remaining in diesel fuels are almost entirely these 4-alkyl and 4-, 6-dialkyl dibenzothiophenes.[44,45]

It has been found that dibenzothiophenes can react via two pathways (Figure 8.15). One is a hydrogenolysis in which both carbon-sulfur bonds in the five-membered ring are replaced by carbon-hydrogen bonds, leading to ring opening and the direct formation of biphenyl; in the second, hydrogenation of one of the two aromatic rings occurs initially to either the tetrahydro- or hexahydro- product, considered to be in equilibrium, and this intermediate product subsequently undergoes desulfurization to produce cyclohexylbenzene. The direct hydrogenolysis route requires two moles of hydrogen, whereas the route leading to cyclohexylbenzene needs five moles. These two routes have also been confirmed in the HDS of benzo[b]naphthothiophenes.[46] Cobalt-molybdenum catalysts have been found to favor the direct hydrogenolysis route, whereas nickel-molybdenum catalysts are better for the intermediate saturation of one of the aromatic rings. Catalyst approaches to this issue of producing low-sulfur fuels have included using both types. The Nebula catalysts developed jointly by Akzo Nobel, ExxonMobil, and Nippon Ketjen are reported to completely eliminate such sterically hindered dibenzothiophenes in a single reactor stage under appropriate conditions.[47] This catalyst is of interest in the present context because it is reported to be in use in a proprietary lubricants operation.[48]

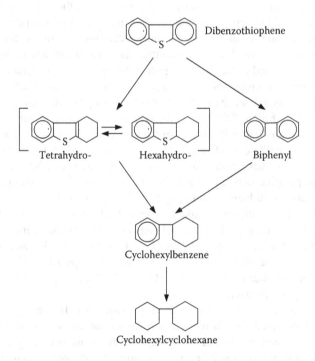

FIGURE 8.15 Competing routes in HDS of hindered dibenzothiophenes.
Source: M. J. Girgis and B. C. Gates, "Reactivities, Reaction Networks, and Kinetics in High-Pressure Catalytic Hydroprocessing," *Industrial and Engineering Chemistry Research* 30:2021–2058 (1991). With permission.

8.7　HYDRODENITRIFICATION

Nitrogen levels in base stocks generally do not get the attention that sulfur and aromatics levels do. Nitrogen measurements do not usually appear as part of the inspections data for base stocks, and unlike sulfur and aromatics, nitrogen is not one of the criteria in the API Base Oil Interchangeability Guidelines. Nevertheless, nitrogen compounds remaining in base stocks do appear to play a role.

In solvent refined base stocks, nitrogen levels may amount to several hundred parts per million and vary with the feedstock and the severity of the solvent extraction step, and they are obviously reduced by the finishing unit, whether a clay treater or a hydrofinisher. Hydrocracking generally reduces nitrogen levels to 10 ppm or less. We have already discussed the evidence in Chapter 5 that nitrogen compounds can affect color development and lacquer formation during lubricant use and that the finishing steps (clay treatment and hydrofinishing) to remove these impurities are reported to improve quality in this area. In hybrid processes where a solvent extraction step precedes a hydrotreating or hydro-cracking one, the reduction in nitrogen levels in the feed to the catalyst can have a marked effect on catalyst activity due to reduced adsorption of nitrogen com-pounds on the acid sites of the catalyst. In cases such as these, feed nitrogen can cause loss of cracking activity, which requires higher catalyst temperatures/lower space velocities and usually result in shorter catalyst life. In the Sun Oil lube plant in Yabacoa, Puerto Rico, for example, the solvent extraction step on their heavy feeds permitted a doubling of space velocity in the hydrocracking unit.

In feedstocks to either solvent refined plants or lube hydrocrackers, nitrogen levels will usually fall in the range of 300 to 2000 ppm, depending on boiling range and crude. Figure 8.16 shows the type of nitrogen compounds present in feedstocks. The amounts will vary from crude to crude. Table 8.15 provides some numbers from distillate fractions from a California crude oil[49,50] and it can be seen that most of these compound types are present. It can be seen that both pyrrole and pyridine derivatives are present and these are the compounds that will be removed in solvent refining, hydrofinishing, or hydrocracking.

Hydrotreating studies undertaken on model nitrogen compounds have been directed at determining the kinetics and pathways undertaken during reactions of both pyridine (basic) and pyrrole (acidic) derivatives. HDN rates over both nickel/tungsten[51] and nickel/molybdenum[52] catalysts appear to change relatively little with the number of fused rings (Figure 8.17 and Figure 8.18), with the exception of pyridine itself.

The current proposals on the chemical pathways for HDN are that saturation of the five-membered ring in pyrrole derivatives and the six-membered ring in pyridine and its derivatives and the contiguous rings are necessary for nitrogen removal. That is, the predominant route requires that the bond being broken is a C(sp3)–N bond. There appears to be no "direct" or hydrogenolysis route for removal of nitrogen as there is for sulfur. The initial steps are considered to be equilibria in which commercial pressures are sufficient to drive it to the right, but sufficiently high temperatures can slow down HDN rates. Favored pathways

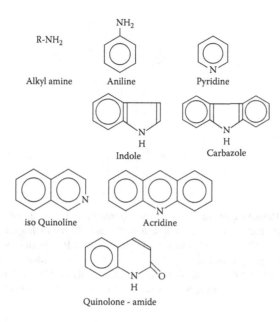

FIGURE 8.16 Representative nitrogen-containing compounds found in petroleum. *Source:* K. H. Altgelt and M. M. Boduszynski, *Composition and Analysis of Heavy Petroleum Fractions* (New York: Marcel Dekker, 1993), chap. 10. With permission.

TABLE 8.15
Distribution of Nitrogen Compounds Found in Three Heavy Distillates from a California Crude Oil

	Wt. % in Distillate		
Boiling Range, °C	205–370	370–455	455–540
Boiling Range, °F	400–700	700–850	850–1000
Indoles	0.07	0.59	0.75
Carbazoles	0.28	3.40	4.08
Benzcarbazoles	0.00	0.50	1.28
Pyrolle derivatives	0.35	4.49	6.11
Pyridines	0.35	0.66	1.30
Quinolines	0.21	1.74	2.00
Benzquinolines	0.03	0.26	1.60
Pyridine derivatives	0.59	2.66	4.9
Pyridones, quinolones	0.20	1.20	2.00
Azaindoles	0.00	0.10	0.40

Source: K. H. Altgelt and M. M. Boduszynski "Composition and Analysis of Heavy Petroleum Fractions", Marcel Dekker, New York, 1993. With permission.

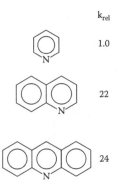

FIGURE 8.17 Relative rate constants for hydrogenation of pyridine, quinoline, and acridine over sulfided nickel/tungsten catalyst at 340°C and 70 bar hydrogen.
Source: C. Moreau, C. Aubert, R. Durand, N. Zmimita, and P. Geneste, "Structure-Activity Relationships in Hydroprocessing of Aromatic and Heteroatomic Model Compounds Over Sulphided NiO-MoO3 γ-Al₂O₃ and NiO-WO₃ γ-Al₂O₃ Catalysts: Chemical Evidence for the Existence of Two Types of Catalytic Sites," *Catalysis Today* 4:117–131 (1988). With permission.

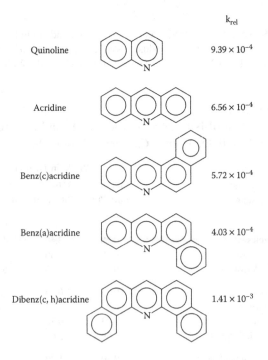

FIGURE 8.18 Reactivities for the HDN of selected basic nitrogen heterocyclic compounds at 376°C and 136 atm over sulfided nickel-molybdenum catalyst.
Source: M. J. Girgis and B. C. Gates, "Reactivities, Reaction Networks, and Kinetics in High Pressure Catalytic Hydroprocessing," *Industrial and Engineering Chemistry Research* 30:2021–2058 (1991). With permission.

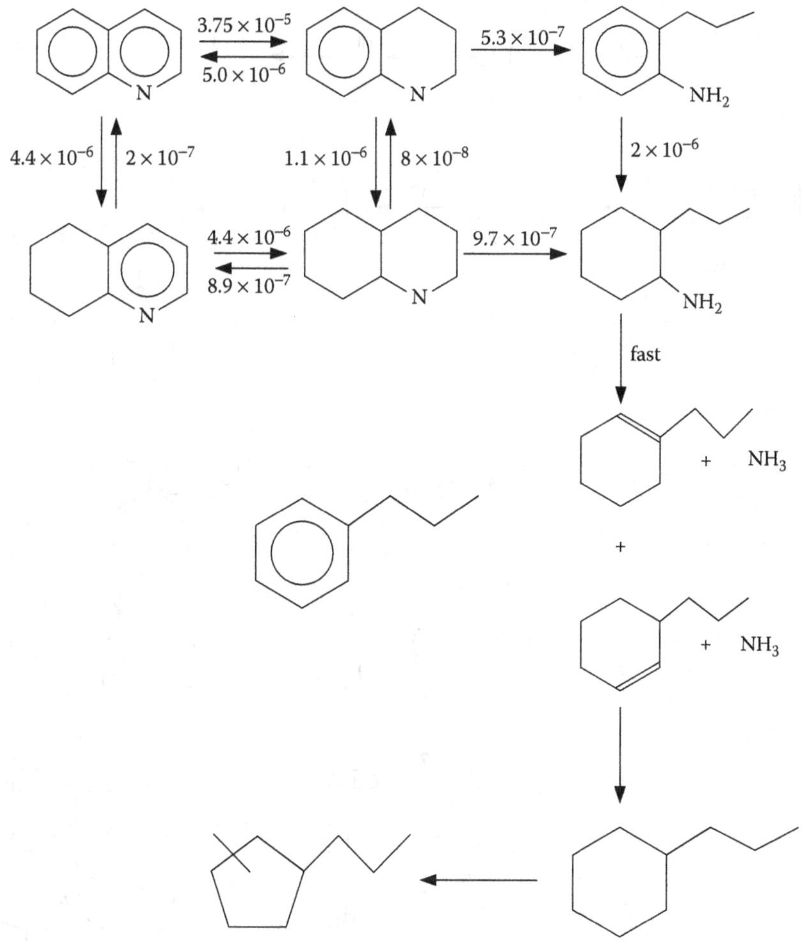

FIGURE 8.19 Reaction network and rate constants at 375°C for quinoline HDN. *Source:* C. N. Satterfield and S. H. Yang, "Catalytic Hydrodenitrogenation of Quinoline in a Trickle-Bed Reactor. Comparison with Vapor Phase Reaction," *Industrial and Engineering Chemistry Product Research and Development* 23:11–19 (1984). With permission.

illustrating these are outlined in Figure 8.19 for pyridine-base components and in Figure 8.20 for carbazoles.

8.8 HYDROCRACKING

Hydrocracking is the catalytic process under hydrogen pressure and high temperatures where part or all of the petroleum molecules are reduced in molecular weight (e.g., the production of gasoline and diesel from VGOs). Lube hydrocracking is generally regarded as a mild form of this since, at least to make group

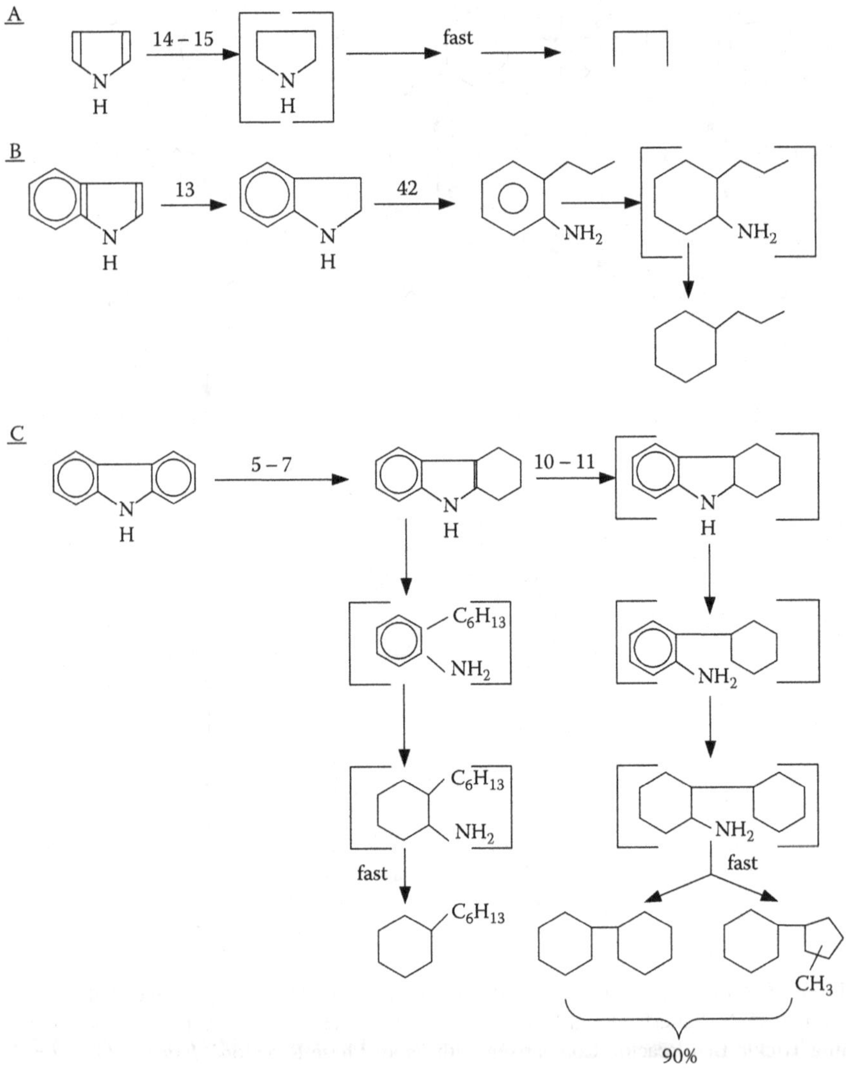

FIGURE 8.20 Reaction networks and rate constants ($\times\ 10^3$/min/g catalyst) for hydroprocessing of pyrrole, indole, and carbazole over sulfided nickel-molybdenum HR 346 (A,B,C) or nickel-tungsten 6 (C) and nickel-tungsten 8 (C) catalysts at 340°C and 70 bar hydrogen. *Source:* C. Moreau, C. Aubert, R. Durand, N. Zmimita, and P. Geneste, "Structure-Activity Relationships in Hydroprocessing of Aromatic and Heteroaromatic Compounds Over Sulfided NiO-MoO$_3$/γ-Al$_2$O$_3$ and NiO-WO$_3$/γ-Al$_2$O$_3$ Catalysts: Chemical Evidence for the Existence of Two Types of Catalyst Sites," *Catalysis Today* 4:117–131 (1988). With permission.

II products with high quality feeds, conversions of feed molecules to products with boiling points less than 650°F are low. Hydroisomerization together with HDA, HDS, and HDN are very much part of this process. Hydrocracking obviously gives yields of less than 100%, as chemical transformations to produce higher VI molecules (e.g., conversion of polycyclic naphthenes to mono- and dicyclic naphthenes) inevitably lead to loss of less than C_{20} moieties. With less favorable feed chemical structures, more extensive chemical changes are required to meet basic product specifications, and yields are reduced. Obviously any comparison of different catalysts means that one criterion is comparison of base stock yields at some target product quality, usually the VI of one of the base stocks. On the other hand, hydrocracking also shifts the molecular weight distribution of feed molecules downward such that base stock distribution is shifted, relative to solvent refining, to that matching current demand for lighter base stocks. Loss of a C_6 to C_{10} fragment is sufficient to reduce a 500N to a 100N to 240N. The overall benefits, and more specifically the improvement in light base stock yields, can be seen in Table 8.16 from a Bechtel[53] estimation of a comparison of an upgraded solvent refined plant with that of one using the same feed but with the installation of a severe hydrocracking unit. In this instance it can be seen that whereas solvent refining produced bright stock, hydrocracking the same feed is projected to give increased overall lube yield, no bright stock, and increased yields of 150N and particularly 240N, all at VIs 20 units higher.

TABLE 8.16
Comparison of Base Stock Yields and Their Distributions from 20,000 bpd Reduced Crude Feed to a Solvent Refined Lube Plant and One Using a Lubes Hydrocracker

Base Stock	Solvent Refined Plant		Lubes Hydrocracker	
	Base Stock Yield	VI	Base Stock Yield	VI
150N	9.88	95	10.75	115+
260N	4.11	95	12.11	115+
600N	7.38	95	8.65	115+
750N			5.11	115+
Bright stock	8.02	95	—	—
Total	29.39		36.62	

All yields at a pour point of 10°F.

Source: V. E. Arnold, "Lubricants Processing Strategies," Paper FL-97-111, presented at the Fuels and Lubes Meeting of the National Petroleum Refiners Association, Houston, Texas, November 6–7, 1997.

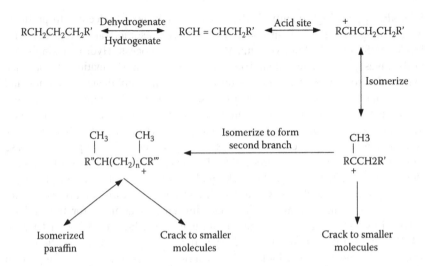

FIGURE 8.21 Illustration of acid-catalyzed routes for isomerization and hydrocracking of an n-paraffin.

A number of books and articles have reviewed the chemistry of hydrocracking,[54–62] a truly complex arena in which only the largest commercial players can afford to participate as process developers and licensors. The catalytic process, really a multiplicity of chemical reactions occurring simultaneously, is generally understood to proceed via carbenium as (very) short-lived intermediates. The catalysts used are generally described as being "dual-function," that is, they have two types of catalytic sites: acidic sites, which promote ion formation, and hydrotreating sites, which can effect hydrogenation-dehydrogenation reactions. The relative strength of these sites can be varied within catalysts.

As in other cases, details of the reactions have been mostly studied with model compounds, simple paraffins, cycloalkanes, and aromatics. Using an n-paraffin as an example, the sequence of reactions can be illustrated (Figure 8.21) in which the reaction steps are dehydrogenation to an olefin, protonation to form a carbenium ion, isomerization to produce a single methyl branch, and subsequently further branches, cracking to produce smaller molecules. This type of reaction sequence will also apply, in general terms, to cycloparaffins and naphthenoaromatics. In practice, in lube hydrocracking, neither isomerization nor hydrocracking of n-paraffins is a desirable outcome because the former will result in a softer wax and the latter will reduce overall waxy lube yield and lower wax yield as well.

Polycyclic naphthenes undergo ring opening[63] to form naphthenes with fewer rings, a sequence that is clearly seen in severe hydrocracking to produce high VI base stocks, which contain significant levels of monocycloparaffins but few

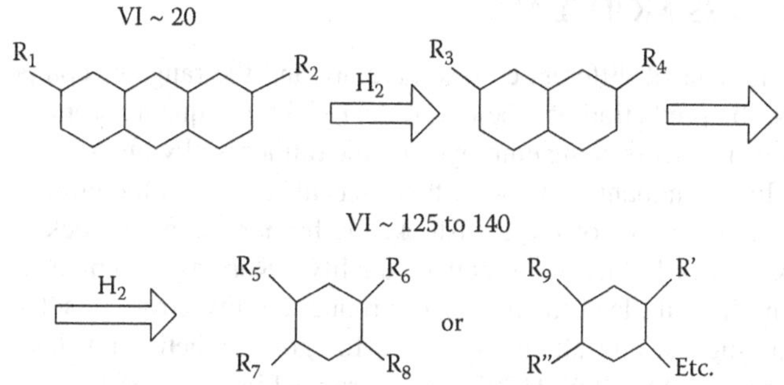

VI ~ 20

VI ~ 125 to 140

FIGURE 8.22 Stepwise ring opening of polycycloparaffins to high VI monocycloparaffins. *Source:* H. Beuther, R. E. Donaldson, and A. M. Henke, "Hydrotreating to Produce High Viscosity Index Lubricating Oils," *Industrial and Engineering Chemistry Product Research and Development* 3(3):174–180 (1964). With permission.

polycycloparaffins (Figure 8.22). As has been discussed previously, polycyclo-paraffins have low VIs and their presence degrades oxidative stability.

Naphthenes with six-membered rings and sufficient alkyl substituents are also known to undergo a more specific "paring" reaction in which ring contraction occurs to form a cyclopentane with elimination of an isobutane molecule (Figure 8.23).[64–66]

Polyaromatics react under hydrocracking conditions to undergo partial or complete saturation of the aromatic rings, together with isomerization and cracking of intermediate and perhydro products. Phenanthrene, for example, can lead to tetralin and methyl cyclohexane as the principal products. With a platinum/silica-alumina catalyst,[67] the major products are isomerized perhydrophenanthrenes, which includes some adamantanes and cracked products, the total number of products exceeding 100.

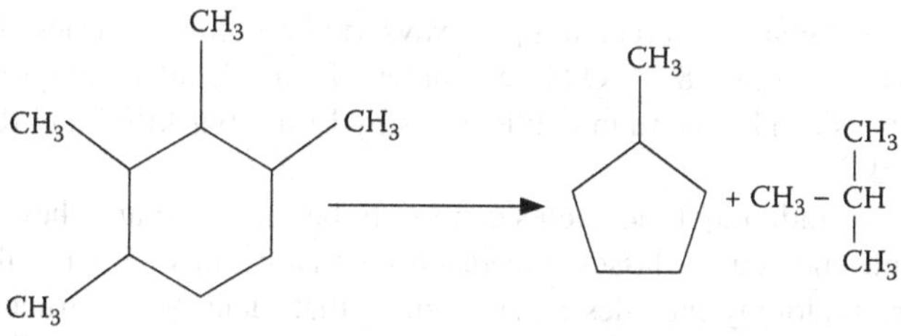

FIGURE 8.23 Illustration of the "paring" reaction.
Source: C. J. Egan, G. E. Langlois, and R. J. Watts, "Selective Hydrocracking of C9- to C12-Alkylcyclohexanes on Acidic Catalysts. Evidence for the Paring Reaction," *Journal of the American Chemical Society* 84:1204–1212 (1962). With permission.

8.9   PROCESS MODELING

Given the number of different crudes there are and the range of composition that this entails, the prediction of lube oil product yield and quality from any feed and from whatever process being employed by the refiner is obviously a very desirable objective. In a nonquantitative way, there has always been the guideline that the higher the dewaxed VI of the feedstock, the higher the base stock yield, and a base stock with a higher VI is better "quality." Roberts[68] pointed out that this reliance on VI could lead to significant product quality errors in solvent refining but did not suggest any alternatives. The foregoing kinetic and thermodynamic information on HDA, HDS, HDN, and hydrocracking lays the basis for a quantitative description of the chemistry of lube processes.

One example of a sophisticated approach is that taken by Jacob et al.[69] at ExxonMobil, where they developed a proprietary molecular-based approach that they claim leads to successful modeling of the process steps involved in refining and, in particular, in lubes production. These steps include solvent extraction and dewaxing, catalytic dewaxing, hydrocracking, and hydrofinishing and the model is also said to predict base stock quality. The prerequisite is high-detail hydrocarbon analyses (HDHAs) on the crude. An example cited of the economic benefits of this modeling technique is an additional 50,000 to 70,000 bpd crude processed for lubricant production at a Mobil refinery.

Conventional modeling beyond the gasoline boiling range traditionally employs compositional "lumps," such as "paraffins," "naphthenes," and "aromatics," since individual components are no longer identifiable. These "lumps" are simple (and useful[70]), but they ignore the distinctions within paraffins (n-paraffins, mono- and multiply substituted branched paraffins), naphthenes (mono-, di-, etc., cycloparaffins), and aromatics (alkyl- and naphthenobenzenes, naphthalenes, triaromatics and higher). The approach employed by ExxonMobil, termed "structure oriented lumping" (SOL), incorporates molecular detail obtained by a combination of analytical techniques, including liquid chromatography, field ionization mass spectroscopy, and gas chromatography/mass spectroscopy.[71,72] Each molecule can be described by group contributions organized into a 22-increment vector per molecule. These vectors lend themselves to assignment of both molecular properties and behavior in reaction pathways via homologous series. The groups are illustrated in Figure 8.24. Other examples of very detailed analyses of petroleum distillates can be found in publications by Boduszynski[73,74] and Rønningsen and Skjevrak.[75]

In practice individual molecules can rarely be identified and they are instead collected into molecular classes. For reactions that occur in a particular process step, the methodology includes reaction rules that identify the structures which will undergo reaction and converts the reactant vector into that of the corresponding product(s). In addition, the appropriate kinetic parameters are selected for each reactant vector for that reactant. The physical properties of each product vector can be calculated so that product properties can be arrived at and the "quality" estimated as well as yield.

A6	A4	A2	N6	N5	N4	N3	N2	N1	R	br	me	H	A_A	S	RS	AN	NN	RN	O	RO	O=

	Increment stoichiometry																				
C: 6	4	2	6	5	4	3	2	1	1	0	0	0	0	-1	0	-1	-1	0	-1	0	0
H: 6	2	0	12	10	6	4	2	0	2	0	0	2	-2	-2	0	-1	-1	1	-2	0	-2
S: 0	0	0	0	0	0	0	0	0	0	0	0	0	0	1	1	0	0	0	0	0	0
N: 0	0	0	0	0	0	0	0	0	0	0	0	0	0	0	0	1	1	1	0	0	0
O: 0	0	0	0	0	0	0	0	0	0	0	0	0	0	0	0	0	0	0	1	1	1

Structural assembly

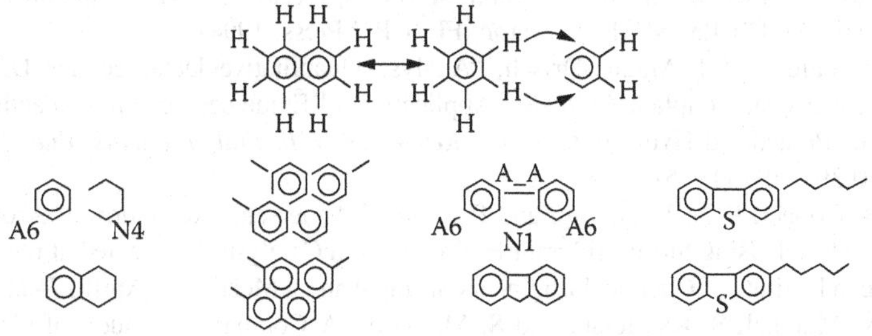

FIGURE 8.24 Molecular groups identified for use in structure oriented lumping.
Source: R. J. Quann and S. B. Jaffe, "Building Useful Models of Complex Reaction Systems in Petroleum Refining," *Chemical Engineering Science* 51:1615–1635 (1996). With permission.

ExxonMobil has successfully applied these concepts to both separation and conversion steps in lube processing and uses this technology to rapidly assess crudes for the most economic applications.[69] Doubtless other companies have developed and use their own methodologies for the different types of lube processes.

REFERENCES

1. K. H. Altgelt and M. M. Boduszynski, *Composition and Analysis of Heavy Petroleum Fractions* (New York: Marcel Dekker, 1993), chap. 10.
2. See, for example, A. Sequiera, Jr., *Lubricant Base Oil and Wax Processing* (New York: Marcel Dekker, 1994), chap. 2, Tables 2-15 through 2-19.
3. A. Sequiera, Jr., *Lubricant Base Oil and Wax Processing* (New York: Marcel Dekker, 1994), chap. 2, Table 2-19.
4. M. J. Girgis and B. C. Gates, "Reactivities, Reaction Networks, and Kinetics in High-Pressure Catalytic Hydroprocessing," *Industrial and Engineering Chemistry Research* 30:2021–2058 (1991).
5. A. Stanislaus and B. H. Cooper, "Aromatics Hydrogenation Catalysis: A Review," *Catalysis Reviews—Science and Engineering* 36:75–123 (1994).

6. B. H. Cooper and B. B. L. Donnis, "Aromatic Saturation of Distillates: An Overview," *Applied Catalysis A: General* 137:203–223 (1996).

7. C. Moreau, C. Aubert, R. Durand, N. Zmimita, and P. Geneste, "Structure-Activity Relationships in Hydroprocessing of Aromatic and Heteroaromatic Compounds Over Sulfided NiO-MoO₃/γ-Al₂O₃ and NiO-WO₃/γ-Al₂O₃ Catalysts: Chemical Evidence for the Existence of Two Types of Catalyst Sites," *Catalysis Today* 4:117–131 (1988).

8. A. V. Sapre and B. C. Gates, "Hydrogenation of Aromatic Hydrocarbons Catalyzed by Sulfided CoO-MoO₃/γ-Al₂O₃. Reactivities and Reaction Networks," *Industrial and Engineering Chemistry Process Design and Development* 20:68–73 (1981).

9. P. Kokayeff, "Aromatics Saturation Over Hydrotreating Catalysts: Reactivity and Susceptibility to Poisons," in *Catalytic Hydroprocessing of Petroleum and Distillates*, M. Oballa, ed. (Boca Raton, FL: CRC Press, 1994).

10. A. Fafet and J. Magne-Drisch, "Analyse Quantitative Détaillée des Distillats Moyens par Couplage CG/MS—Application à l'Étude des Schémas Réactionnels du Procédé d'Hydrotraitemant," *Revue de L'Institut Francais due Pétrole* 50:391–404 (1995).

11. B. Cooper, P. N. Hannerup, and P. Sogaard-Andersen, "Reduction of Aromatics in Diesel Using Sulfur-Tolerant Hydrogenation Catalysts," presented at the American Institute of Chemical Engineers Spring Annual Meeting, April 17–21, 1994.

12. N. Marchal, S. Kasztelan, and S. Mignard, "A Comparative Study of Catalysts for the Deep Aromatic Reduction in Hydrotreated Gas Oil," in *Catalytic Hydroprocessing of Petroleum and Distillates*, M. Oballa, ed. (Boca Raton, FL: CRC Press, 1994), 315–327.

13. "Properties of Hydrocarbons of High Molecular Weight," Research Project 42, 1940–1966, American Petroleum Institute, New York.

14. I. P. Fisher and M. F. Wilson, "Kinetic and Thermodynamics of Hydrotreating Synthetic Middle Distillates," presented at the Symposium on Advances in Hydrotreating, Division of Petroleum Chemistry, American Chemical Society Meeting, April 8–10, 1987, pp. 310–314.

15. D. R. Stull, E. F. Westrum, and G. R. Sinke, *The Chemical Thermodynamics of Organic Compounds* (New York: John Wiley & Sons, 1969).

16. G. B. Kistiakowsky, H. Romeyn, Jr., J. R. Ruhoff, H. A. Smith, and W. E. Vaughan, "Heats of Organic Reactions. I. The Apparatus and the Heat of Hydrogenation of Ethylene," *Journal of the American Chemical Society* 57:65–75 (1935).

17. G. B. Kistiakowsky, J. R. Ruhoff, H. A. Smith, and W. E. Vaughan, "Heats of Organic Reactions. II. Hydrogenation of Some Simpler Olefinic Hydrocarbons," *Journal of the American Chemical Society* 57:876–883 (1935).

18. G. B. Kistiakowsky, J. R. Ruhoff, H. A. Smith, and W. E. Vaughan, "Heats of Organic Reactions. III. Hydrogenation of Some Higher Olefins," *Journal of the American Chemical Society* 58:137–145 (1936).

19. G. B. Kistiakowsky, J. R. Ruhoff, H. A. Smith, and W. E. Vaughan, "Heats of Organic Reactions. IV. Hydrogenation of Some Dienes and Benzene," *Journal of the American Chemical Society* 58:146–153 (1936).

20. M. A. Dolliver, T. L. Gresham, G. B. Kistiakowsky, and W. E. Vaughan, "Heats of Organic Reactions. V. Heats of Hydrogenation of Various Hydrocarbons," *Journal of the American Chemical Society* 59:831–841 (1937).

21. C. G. Frye, "Equilibria in the Hydrogenation of Polycyclic Aromatics," *Journal of Chemical and Engineering Data* 7:592–595 (1962).

22. C. G. Frye and A. W. Weitkamp, "Equilibrium Hydrogenations of Multi-Ring Aromatics," *Journal of Chemical and Engineering Data* 14:372–376 (1969).
23. I. P. Fisher and M. F. Wilson, "Kinetics and Thermodynamics of Aromatics Hydrogenation in Distillates from Athabasca Syncrudes," *Energy and Fuels* 2:848–555 (1988).
24. David R. Lide, ed., *CRC Handbook of Chemistry and Physics*, 73rd ed. (Boca Raton, FL: CRC Press, 1992).
25. L. M. Magnabosco, "A Mathematical Model for Catalytic Hydrogenation of Aromatics in Petroleum Refining Feedstocks," in *Catalysts in Petroleum Refining*, D. L. Trimm, ed. (Amsterdam: Elsevier Science, 1989), 481–495.
26. S. J. Miller and J. A. Zakarian, "Determination of Lube Hydrofinishing Catalyst Kinetics Using Ultraviolet Absorbance," *Industrial and Engineering Chemistry Research* 30:2507–2513 (1991).
27. M. Ushio, K. Kamiya, T. Yoshida, and I. Honjou, "Production of High VI Base Oil by VGO Deep Hydrocracking," presented at the Symposium on Processing, Characterization and Application of Lubricant Base Oils, Division of Petroleum Chemistry, American Chemical Society Meeting, August 23–25, 1992.
28. M. Takizawa, T. Takito, M. Nada, K. Inaba, Y. Yoshizumi, and T. Sasaki, "Commercial Production of Two Viscosity Grades VHVI Basestocks," Paper FL-93-118, presented at the National Fuels and Lubricants Meeting of the National Petroleum Refiners Association, Houston, Texas, November 1993.
29. F. C. Wood, C. P. Reeg, A. E. Kelley, and G. D. Cheadle, "Recycle Hydrocracking Process for Converting Heavy Oils to Middle Distillates," U.S. Patent 3,554,898.
30. G. W. Hendricks, E. C. Attane, and J. W. Wilson, "Hydrocracking Process with Benzcoronenes Bleedstream," U.S. Patent 3,619,407.
31. J. C. Fetzer and W. R. Biggs, "A Review of the Large Polycyclic Aromatic Hydrocarbons," *Polycyclic Aromatic Compounds* 4:3–17 (1994).
32. J. C. Fetzer and W. R. Biggs, "Identification of a New Eight-Ring Condensed Polycyclic Aromatic Hydrocarbon," *Polycyclic Aromatic Compounds* 5:193–199 (1994).
33. J. R. Kershaw and J. C. Fetzer, "The Room Temperature Fluorescence Analysis of Polycyclic Aromatic Compounds in Petroleum and Related Compounds," *Polycyclic Aromatic Compounds* 7:253–268 (1995).
34. J. C. Fetzer, "The Use of Synchronous-Scanning Fluorescence Spectrometry for Detection of Dicoronylene in Hydrocracker Streams," *Polycyclic Aromatic Compounds* 7:269–274 (1995).
35. J. C. Fetzer, "The Production of Large Polycyclic Aromatic Hydrocarbons During Catalytic Hydrocracking," in *Catalysts in Petroleum Refining and Petrochemical Industries*, M. Absi-Halabi, J. Beshara, H. Qabazard, and A. Stanislaus, eds. (Amsterdam: Elsevier Science, 1996).
36. R. F. Sullivan, M. M. Boduszynski, and J. C. Fetzer, "Molecular Transformations in Hydrotreating and Hydrocracking," presented at the Symposium on Recent Developments and Challenges in Hydrotreating, Division of Industrial and Engineering Chemistry, 3rd Chemical Congress of North America and 195th National Meeting of the American Chemical Society, Toronto, Canada, June 5–10, 1988.
37. R. F. Sullivan, M. M. Boduszynski, and J. C. Fetzer, "Molecular Transformations in Hydrotreating and Hydrocracking," *Energy and Fuels* 3:603–612 (1989).
38. M. J. Girgis and B. C. Gates, "Reactivities, Reaction Networks, and Kinetics in High Pressure Catalytic Hydroprocessing," *Industrial and Engineering Chemistry Research* 30:2021–2058 (1991).

39. D. D. Whitehurst, T. Isoda, and I. Mochida, "Present State of the Art and Future Challenges in the Hydrodesulfurization of Polyaromatic Sulfur Compounds," *Advances in Catalysis* 42:345–471 (1998).

40. P. T. Vasudevan and J. L. G. Fierro, "A Review of Deep Hydrodesulfurization Catalysis," *Catalysis Reviews—Science and Engineering* 38:161–188 (1996).

41. N. K. Nag, A. V. Sapre, D. H. Broderick, and B. C. Gates, "Hydrodesulfurization of Polycyclic Aromatics Catalyzed by Sulfided CoO-MoO$_3$/γ-Al$_2$O$_3$: The Relative Reactivities," *Journal of Catalysis* 57:509–512 (1979).

42. C. Aubert, R. Durand, P. Geneste, and C. Moreau, "Hydroprocessing of Dibenzothiophene, Phenothiazine, Phenoxanthlin, Thianthrene, and Thioxanthene on a Sulfided NiO-MoO$_3$/γ-Al$_2$O$_3$," *Journal of Catalysis* 97:169–176 (1986).

43. M. Houalla, D. H. Broderick, A. V. Sapre, N. K. Nag, V. H. J. de Beer, B. C. Gates, and H. Kwart, "Hydrodesulfurization of Methyl-Substituted Dibenzothiophenes Catalyzed by Sulfided CoO-MoO$_3$/γ-Al$_2$O$_3$," *Journal of Catalysis* 61:523–527 (1980).

44. M. C. Hu, Z. Ring, J. Briker, and M. Te, "Rigorous Hydrotreater Simulation," *Petroleum Technology Quarterly* Spring:85–91, 2002.

45. A. P. Lamourelle, J. McKnight, and D. E. Nelson, "Clean Fuels: Route to Low Sulfur Low Aromatic Diesel," Paper AM-01-28, presented at the annual meeting of the National Petroleum Refiners Association, March 18–20, 2001.

46. M. L. Vrinat, "The Kinetics of the Hydrodesulfurization Process; A Review," *Applied Catalysis* 6:137–158 (1983).

47. D. A. Pappal, F. L. Plantagena, R. A. Bradway, G. Chitnis, W. J. Tracy, and W. E. Lewis, "Stellar Improvements in Hydroprocessing Catalyst Activity," Paper AM-03-59, National Petroleum Refiners Association, Houston, Texas, March 23–25, 2003; available at www.prod.exxonmobil/refiningtechnologies/pdf/AM-03-59NPRApaperforweb.pdf.

48. Albemarle Catalysts, "Nebula 20—The Next Step into Deep Space," brochure, available at www.albemarle.com/TDS/HPC/NEBULA20_MIB_8Nov05.pdf.

49. K. H. Altgelt and M. M. Boduszynski, *Composition and Analysis of Heavy Petroleum Fractions* (New York: Marcel Dekker, 1993), chap. 10.

50. L. R. Snyder, "Nitrogen and Oxygen Compound Types in Petroleum," *Analytical Chemistry* 41:315–323 (1969).

51. M. J. Girgis and B. C. Gates, "Reactivities, Reaction Networks, and Kinetics in High Pressure Catalytic Hydroprocessing," *Industrial and Engineering Chemistry Research* 30:2021–2058 (1991).

52. C. Moreau, C. Aubert, R. Durand, N. Zmimita, and P. Geneste, "Structure-Activity Relationships in Hydroprocessing of Aromatic and Heteroatomic Model Compounds Over Sulphided NiO-MoO3/γ-Al$_2$O$_3$ and NiO-WO$_3$/γ-Al$_2$O$_3$ Catalysts: Chemical Evidence for the Existence of Two Types of Catalytic Sites," *Catalysis Today* 4:117–131 (1988).

53. V. E. Arnold, "Lubricants Processing Strategies," Paper FL-97-111, presented at the Fuels and Lubes Meeting of the National Petroleum Refiners Association, Houston, Texas, November 6–7, 1997.

54. J. Scherzer and A. J. Gruia, *Hydrocracking Science and Technology* (New York: Marcel Dekker, 1996).

55. M. L. Poutsma, "Mechanistic Considerations of Hydrocarbon Transformations Catalyzed by Zeolites," in *Zeolite Chemistry and Catalysis*, J. A. Rabo, ed., Monograph 171 (Washington, DC: American Chemical Society, 1976).

56. A. P. Bolton, "Hydrocracking, Isomerization, and Other Industrial Processes," in *Zeolite Chemistry and Catalysis*, J. A. Rabo, ed., Monograph 171 (Washington, DC: American Chemical Society, 1976).

57. C. E. Langlois and R. F. Sullivan, "Chemistry of Hydrocracking," in *Refining Petroleum for Chemicals*, Advances in Chemistry Series vol. 97, R. F. Gould, ed., (Washington, DC: American Chemical Society, 1970).

58. N. Choudhary and D. N. Saral, "Hydrocracking: A Review," *Industrial and Engineering Chemistry Product Research and Development* 14(2):74–83 (1975).

59. S. P. Ahuja, M. L. Derrien, and J. F. Lepage, "Activity and Selectivity of Hydrotreating Catalysts," *Industrial and Engineering Chemistry Product Research and Development* 9(3):272–281 (1970).

60. G. M. Kramer, G. B. McVicker, and J. J. Ziemiak, "On the Question of Carbonium Ions as Intermediates Over Silica-Alumina and Acid Zeolites," *Journal of Catalysis* 92:355–363 (1985).

61. P. B. Venuto, "Organic Catalysis Over Zeolites: A Perspective on Reaction Paths Within Micropores," *Microporous Materials* 2:297–411 (1994).

62. T. F. Degnan, Jr., "Applications of Zeolites in Petroleum Refining," *Topics in Catalysis* 13:349–356 (2000).

63. H. Beuther, R. E. Donaldson, and A. M. Henke, "Hydrotreating to Produce High Viscosity Index Lubricating Oils," *Industrial and Engineering Chemistry Product Research and Development* 3(3):174–180 (1964).

64. R. F. Sullivan, C. J. Egan, G. E. Langlois, and R. P. Sieg, "A New Reaction That Occurs in the Hydrocracking of Certain Aromatic Hydrocarbons," *Journal of the American Chemical Society* 83:1156–1160 (1961).

65. C. J. Egan, G. E. Langlois, and R. J. Watts, "Selective Hydrocracking of C9- to C12-Alkylcyclohexanes on Acidic Catalysts. Evidence for the Paring Reaction," *Journal of the American Chemical Society* 84:1204–1212 (1962).

66. H. Pines and A. W. Shaw, "Isomerization of Saturated Hydrocarbons. XV. The Hydro-isomerization of Ethyl--C^{14}-cyclohexane and Ethyl--C^{14}-cyclohexane," *Journal of the American Chemical Society* 79:1474–1482 (1957).

67. E. Benazzi, L. Leite, N. Marchal-George, H. Toulhoat, and P. Raybaud, "New Insights into Parameters Controlling the Selectivity in Hydrocracking Reactions," *Journal of Catalysis* 217:376–387 (2003).

68. J. H. Roberts "Impact of Quality of Future Crude Stocks on Lube oils (Significance of VI in Measuring Quality)," Paper AM-85-21C, presented at the 1985 National Petroleum Refiners Association, San Antonio, Texas, March 24–26, 1985.

69. S. M. Jacob, R. J. Quann, E. Sanchez, and M. E. Wells, "Lube Oil Processing—1. Compositional Modeling Reduces Crude-Analysis Time, Predicts Yields," *Oil and Gas Journal* July 6:51–58 (1998).

70. S. M. Jacob, B. Gross, S. E. Voltz, and V. W. Weekman, Jr., "A Lumping and Reaction Scheme for Catalytic Cracking," *AIChE Journal* 22:701–713 (1976).

71. R. J. Quann and S. B. Jaffe, "Structure-Oriented Lumping: Describing the Chemistry of Complex Hydrocarbon Mixtures," *Industrial and Engineering Chemistry Research* 31:2483–2497 (1992).

72. R. J. Quann and S. B. Jaffe, "Building Useful Models of Complex Reaction Systems in Petroleum Refining," *Chemical Engineering Science* 51:1615–1635 (1996).

73. M. M. Boduszynski, "Composition of Heavy Petroleums. 2. Molecular Characterization," *Energy and Fuels* 2:597–613 (1988).

74. K. H. Altgelt and M. M. Boduszynski, *Composition and Analysis of Heavy Petroleum Fractions* (New York: Marcel Dekker, 1993).

75. H. P. Rønningsen and I. Skjevrak, "Characterization of North Sea Petroleum Fractions: Aromatic Ring Class Distribution," *Energy and Fuels* 4:608–626 (1990).

9 Urea Dewaxing and the BP Catalytic Process

9.1 INTRODUCTION

Dewaxing of paraffinic lube stocks is an essential step in the production of lubricants which will remain fluid and permit machinery to operate at winter temperatures. The next two chapters outline the technologies that have been developed for this purpose (except solvent dewaxing, which was discussed in Chapter 6). These processes reflect the historical development of dewaxing chemical knowledge during this century, first using chemical separation processes and more recently, chemical conversion.

This and the next chapter provide details of four processes for dewaxing:

- Urea dewaxing, which had a brief period of success in the 1950s to 1960s and which removed normal paraffins using shape selective clathrates.
- Catalytic dewaxing, developed by BP using a mordenite zeolite catalyst—the catalyst selectively cracked n-paraffins, but was limited in application to lighter stocks. The process does not appear to have been widely used.
- Catalytic dewaxing, developed in the 1970s by Mobil's immensely successful research program into zeolites, specifically ZSM-5 for this application, cracks paraffins and paraffinic groups in wax into light hydrocarbons and is applicable to the entire lube slate.
- Catalytic hydroisomerization, the most recent and commercially very successful technology was developed to isomerize paraffins and paraffinic groups to branched isomeric structures, thereby achieving two key objectives, reducing the pour point and increasing the viscosity index (VI). Yields are equivalent to or better than solvent dewaxing, but feedstocks must have low sulfur and nitrogen contents. Chevron and ExxonMobil are the sole licensors at the time of writing.

Prior to discussion of these processes, it is worthwhile understanding the composition of the petroleum wax to be separated or converted. That is the subject of the following section.

9.2 WAX COMPOSITION AND PROPERTIES

Petroleum wax is a mixture of simple and complex hydrocarbons that separate out as a crystalline or semicrystalline solid material when most petroleum distillates or residues are cooled, either alone or diluted by certain solvents. It is usually called "wax" or "slack wax" and it is much more than a collection of normal paraffins. The composition of wax depends on the crude source, the boiling range, and the temperature at which it is crystallized, and it becomes more complex with increasing molecular weight. The residue after removal of the wax is called the "oil" or "dewaxed oil" (DWO) depending on the context. The term "dewaxing" refers to the processes, either solvent or catalytic, for removing or altering wax molecules. The "waxes" that are physically obtained from solvent dewaxing and those that are chemically altered by catalytic dewaxing will be very similar in composition, but the extent and distribution of removal may not be the same.

The significant physical property distinguishing wax and oil is the pour point (pour point is defined as the lowest temperature at which a sample will flow; see Chapter 2); wax has a high pour point compared to oil because wax molecules have many of the properties of n-paraffins and these have high melting points. Pour points of some common substances are provided in Table 9.1.

The pour point of the oil after dewaxing depends on the amount of wax and the type remaining. The pour point is established in solvent dewaxing by the temperature at which the slack wax is filtered off and the solvent composition. Generally the pour points of paraffinic base stocks are in the range 0°C to −21°C and depend on how "severely" the dewax process is operated and the purposes of the lubricants. "Ultra low" pour point oils can be produced for specific applications. Distillates and residues from naphthenic crudes generally contain little wax and have naturally low pour points. Those from paraffinic crudes may contain 20% to 30% (or even higher) wax, have much higher pour points, and dewaxing is a must in the process to make usable lube base stocks. Figure 9.1 shows results from some Chevron[1] work on the variation in wax content and DWO VI that there can be in vacuum gas oils (VGOs) from a variety of crudes. Also, besides the y-axis, there are indicators of the economically applicable ranges of feed dewaxed VI for solvent refining and hydrocracking.

TABLE 9.1
Pour (Melting) Points of Some Common Materials

Sample	Pour or Melting Point or Range (°C)
Water	0
Petroleum waxes	50–80
Petroleum base stocks	−21 to 0
Antifreeze	−37
Engine oils	Approximately −40

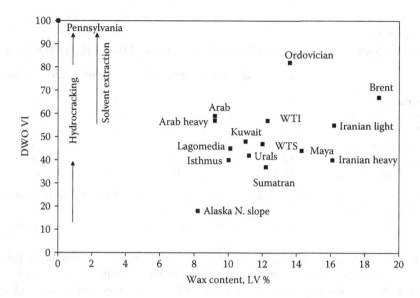

FIGURE 9.1 Wax content and VI of dewaxed oils for VGOs from various crudes. *Source:* M. W. Wilson, K. L. Eiden, T. A. Mueller, S. D. Chase, and G. W. Kraft, "Commercialization of ISODEWAXING™—A New Technology for Dewaxing to Manufacture High Quality Lube Base Stocks," Paper FL-94-112, presented at the National Fuels and Lubricants meeting of the National Petroleum Refiners Association, Houston, Texas, November 3–5, 1994. Figure copyrighted by Chevron Corporation and used with permission.

In solvent dewaxing, the wax filtered out is called slack wax and can contain as much as 20% oil. This slack wax can subsequently be "deoiled" in a further dewaxing step to produce a commercial grade wax that may be further purified (by clay treating or hydrofinishing), then blended with other waxes to give a final product. The oil from deoiling is called "foots" oil, a by-product whose n-paraffin content is low. Foots oil is mainly a mixture of isoparaffins and cycloparaffins. The oil content of wax is commonly measured by ASTM D721,[2] and other nonstandard methods are available[3,4] for measuring this and the wax content in waxy samples.[5,6] The wax content of a waxy lube sample is usually measured by performing a laboratory solvent dewaxing (with methyl ethyl ketone and toluene) to a specific oil pour point and corrected for any oil present. The oil pour point needs to be identified since wax content depends on how deep the dewaxing has been. Of course, wax yield increases as the oil pour point decreases, therefore it is normal to report wax yield, corrected for oil content, at the measured pour point of the oil.

The most easily identified and measurable components in wax are the n-paraffins, which extend up to about C_{45}.[7] Above C_{45} or thereabouts (about 1000°F), n-paraffin contents of heavy distillates or residues become quite small. The n-paraffin content decreases with increasing boiling point, and isoparaffins, cycloparaffins,

TABLE 9.2

Mass Spectrum Analyses of a Wax "A" and Some of Its Distillate Fractions

	Wax "A"	40%–45%	60%–65%	85%–90%	6.9% Residue
n-Paraffins	75.5	95.7	82.4	52.1	37.5
Branched paraffins	13.5	4.2	13.9	28.6	25.9
Monocyclic paraffins	10.2	0.1	3.6	17.7	30.5
Polycycloparaffins	0.5	0.0	0.0	1.3	4.4
Monocyclic aromatics	0.2	0.0	0.1	0.3	1.4
Aromatic cycloparaffins	0.0	0.0	0.0	0.0	0.2

Source: R. T. Edwards, "A New Look at Paraffin Waxes," *Petroleum Refiner* 36:180–187 (1957). With permission.

and aromatics with sufficient "paraffinic" character to be isolated together as part of the wax become more important. These distribution changes can be seen in Table 9.2, which gives the results from analyses by mass spectroscopy of a commercial wax and several of its distillate cuts (boiling ranges were unfortunately not given).[8] In this table, the decrease in n-paraffin levels and the increase in complexity with molecular weight and boiling range are quite noticeable. Figure 9.2 and Figure 9.3 show in more detail how the distributions of the normal and branched paraffins change between the 40% to 45% distillate cut and the 7% residue fraction from this wax. The overhead fraction (Figure 9.2) contains relatively few branched paraffins, whereas in the residue their concentration (Figure 9.3) has risen to almost equal those of the n-paraffins.

The variation in n-paraffin content can also be seen when n-paraffin levels are measured in distillation fractions of crudes; the two examples in Figure 9.4 and Figure 9.5 are from measurements on two paraffinic crudes, Sarir and Bomu.[9] Here, n-paraffin levels as a percent of the crude and as a percent of the slack wax obtained by solvent dewaxing decrease sharply with increased mid-boiling point.

Vacuum residues in nearly all crudes will have quite low n-paraffin levels. It should be noted that waxes from Fischer-Tropsch processes in all boiling ranges contain very little isoparaffins compared to petroleum waxes and they therefore have increased melting points and hardness.[10] These latter waxes also contain no sulfur components because of the process involved. Sulfur-containing molecules cannot be eliminated from petroleum waxes by deoiling or clay treating: hydrofinishing or hydrotreating is necessary.

These compositional changes are reflected in the physical properties of waxes. Paraffinic waxes, which are mainly linear C_{18} to C_{40}, form large crystals and are known as macrocrystalline or hard waxes; intermediate waxes have increased branching and range from C_{25} to C_{60} (there's obviously considerable overlap), while microcrystalline waxes have, as their name indicates, very small crystals and higher melting points, and possess very complex structures in the C_{25} to C_{85} range.[11] It can be surmised that many of the structures in microcrystalline waxes

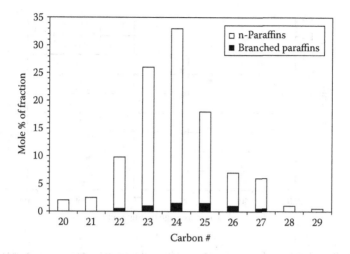

FIGURE 9.2 Branched and normal paraffin distributions in 40% to 45% overhead fraction from a wax sample.
Source: R. T. Edwards, "A New Look at Paraffin Waxes," *Petroleum Refiner* 36:180–187 (1957). With permission.

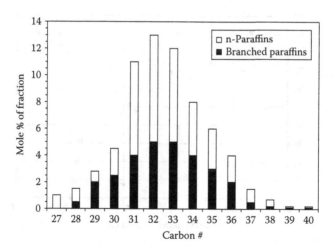

FIGURE 9.3 Branched and normal paraffin distributions of distillation residue from the wax sample.
Source: R. T. Edwards, "A New Look at Paraffin Waxes," *Petroleum Refiner* 36:180–187 (1957). With permission.

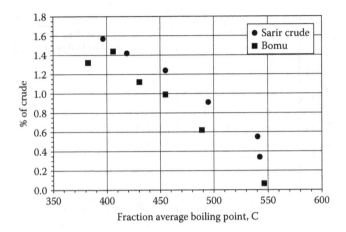

FIGURE 9.4 Variation of n-paraffin levels with boiling point in two crudes: Sarir (Libyan) and Bomu (Nigerian).
Source: J. V. Brunnock, "Separation and Distribution of Normal Paraffins from Petroleum Heavy Distillates by Molecular Sieve Adsorption and Gas Chromatography," *Analytical Chemistry* 38:1648–1652 (1966). With permission.

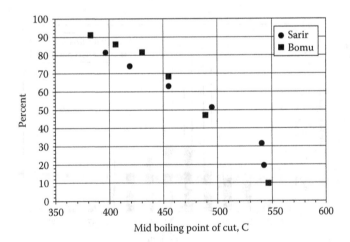

FIGURE 9.5 Variation of n-paraffin levels with boiling point in wax isolated from Sarir and Bomu crudes.
Source: J. V. Brunnock, "Separation and Distribution of Normal Paraffins from Petroleum Heavy Distillates by Molecular Sieve Adsorption and Gas Chromatography," *Analytical Chemistry* 38:1648–1652 (1966). With permission.

contain sufficient long chain or branched chain components attached to naphthene or aromatic rings to afford them enough "wax" chemical characteristics to separate with the n-paraffins. This is not a group that has been much explored from a compositional viewpoint, largely due to their complexity. Paraffinic waxes are employed in packaging and candle manufacturing, among others, while microcrystalline waxes, which have better adhesive properties, are used to impart improved flexibility and better low temperature properties.

Because n-paraffins can be identified and quantified by a number of techniques, these and some isoparaffins have come to be used as model compounds whose properties are used to develop, illustrate, and explain the chemical processes involved in dewaxing. One should always be mindful that the processes applied to lubes are "dewaxing" and not "de-n-paraffining" processes ("de-n-paraffining" processes do not work for heavy lube stocks and generally have not been commercially successful; see later sections in this chapter). While catalytic dewaxing of, for example, a waxy 100N may indeed largely remove n-paraffins, when it comes to a waxy 600N, most of the material targeted to be chemically altered or separated is not composed of n-paraffins any longer but structures that are much more complex in structure and which require greater flexibility in the operation of the catalyst pore structure.

9.3 UREA DEWAXING

In 1940, Friederich Bengen,[12] a German research chemist, accidentally found that urea and straight-chain organic molecules such as n-paraffins form insoluble molecular complexes known as adducts at room temperature. These complexes can subsequently be thermally decomposed to regenerate the n-paraffin and urea. This discovery led to the development of a new commercial process for dewaxing lubricating oil stocks, with a number of plants being built in the 1950s. Unfortunately experience showed that the technology is limited to light feedstocks whose wax is predominantly composed of n-paraffins, and a process that cannot be used with the entire base stock slate faces considerable economic challenges. In addition, operational difficulties were encountered in handling the large quantities of solid adduct. These problems limited use of this process to a few very specific applications and widespread use never developed. The chemistry is of interest, however, since it represents one of the first applications of shape selectivity in the petroleum industry, although its specific character is not one that has survived.

The type of adduct involved here was later termed a clathrate (from the Latin *clathratus*, meaning enclosed by the bars of a grating[13]) in which one material, in this case urea, forms a lattice which traps and holds a second type of molecule. Today this would be considered a subdivision of supramolecular chemistry.[13,14] The lattice formed has the shape and dimensions required to contain the molecule within the cage. In the case of urea, the cross section of the channel that urea can form is 5.5 Å × 4.4 Å, which is large enough to accommodate n-paraffins but insufficient for isoparaffins (6.0 Å) or aromatics (5.9 Å).[15] Figure 9.6 illustrates the

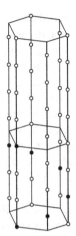

FIGURE 9.6 Lattice arrangement of the hexagonal system of urea complex.
Source: K. A. Kobe and W. G. Domask, "Extractive Crystallization—A New Separation Process. Part 1. Theoretical Basis," *Petroleum Refiner* 31(3):106–113 (1952). With permission.

spiral shape the urea molecules assume in winding around the adducted molecule, using the latter as a mandrel.[16] It should be noted that this cage forms only when n-alkanes or similar molecules and urea coexist in solution. Figure 9.7 shows the relationship between the urea channel size and the cross section of some common molecules.[15]

The ratio of the number of urea molecules to occluded paraffin ones was found to be approximately 0.7 urea molecules per carbon atom, which is the slope of the line in Figure 9.8.[15] Therefore the cage gets longer, as does the molecule inside, but the ends of the molecular structures do not affect the number of urea molecules needed.

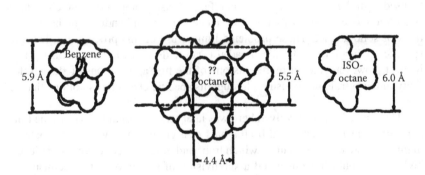

FIGURE 9.7 Relative sizes of urea adduct and specific hydrocarbons.
Source: T. H. Rogers, J. S. Brown, R. Diekman, and G. D. Kerns, "Urea Dewaxing Gets More Emphasis," *Petroleum Refiner* 36(5):217–220 (1957). With permission.

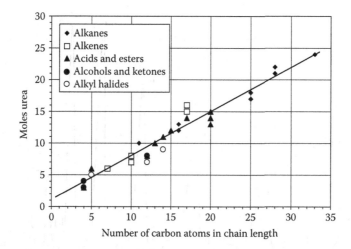

FIGURE 9.8 Moles of urea versus carbon chain length for alkanes, alkenes, and other long-chain molecules.
Source: T. H. Rogers, J. S. Brown, R. Diekman, and G. D. Kerns, "Urea Dewaxing Gets More Emphasis," *Petroleum Refiner* 36(5):217–220 (1957). With permission.

From the examples in Figure 9.8, it can be seen that these urea clathrates (also known as inclusion compounds, adducts, or channel or cage compounds; the process itself has been dubbed "extractive crystallization[17]") are not confined to n-alkanes, but are also formed with straight-chain olefins, alcohols, esters, ketones, halides, etc. An X-ray structure determination[18] first demonstrated the spiral hexagonal structure formed by the urea and the approximately 0.7 urea molecules per carbon of chain length.[15]

For n-paraffins, the shortest molecule that forms a urea adduct at 25°C is n-heptane, and n-paraffins up to n-C_{60} have been reported to react;[17] the time required for reaction increases with increasing molecular weight of the hydrocarbon.[15] Chain branching can prevent adduct formation, as expected for this size-/shape-specific reaction (e.g., 7-methyl tridecane, a C_{13} paraffin with a single methyl group in the middle, does not form an adduct). Neither does 3-ethyltetracosane, a molecule where the ethyl group is attached to the three position of a 24-carbon chain, in spite of leaving a 21-carbon straight unsubstituted chain available to form the adduct. However, if the unbroken chain is long enough[19] and the substituent is of the "right" size, formation of the urea adduct can take place (e.g., 1-phenyleicosane, 1-cyclohexyleicosane, and 1-cyclopentyl heneicosane with n-chain lengths of 20, 20, and 21 carbons all form adducts, but 2-phenyleicosane, 2-cyclohexyleicosane do not). Relative to these structures, the fact that 3-ethyltetracosane (with another C_{21} straight-chain part) does not form an adduct is surprising. For isoparaffins with chain length greater than C_{13}, where the branch is a single methyl group, adducts can form as seen by the appearance of small peaks corresponding to isoparaffins in the gas chromatograms of the

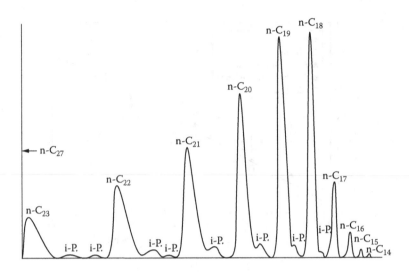

FIGURE 9.9 Gas chromatogram of paraffins obtained by urea dewaxing.
Source: A. Hoppe, "Dewaxing with Urea," in *Advances in Petroleum Chemistry and Refining*, vol. VIII (New York: Interscience Publishers, 1964), 193–234. With permission.

paraffins isolated (Figure 9.9).[20,21] These are presumably isomers with a single methyl group near the end of the chain and the methyl group is of small size. It is not apparent in these branched cases whether the channel formed by the urea molecules distorts to allow the substituent in or more likely the substituent stays outside and there is sufficient hydrocarbon within the channel to generate the attractive forces necessary to stabilize the overall structure.

It should be noted that thiourea, the sulfur analog of urea, also forms clathrates, but the larger size of the sulfur atom results in larger channels in the adducts which admit compact[22] molecules such as branched and cyclic hydrocarbons[19] and selectivity is no longer a feature. As a result, no commercial applications of thiourea to dewaxing have been developed.

Adduct formation and decomposition are equilibrium reactions. The position of the equilibrium is affected by external conditions (urea concentration and temperature mainly):

$$\text{Adduct} \leftrightarrow \text{n-Paraffin} + m \text{ Urea} \qquad (9.1)$$

where

$$m = 0.65*n + 1.5, \quad \text{for } (6 < n < 17),[22]$$

and n is the number of carbon atoms in the chain.

$$K = a_{HC}a_U{}^m/a_A$$

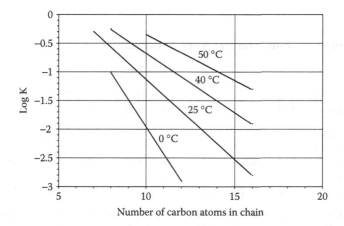

FIGURE 9.10 Influence of chain length on the equilibrium constants for n-paraffins and urea.

Source: O. Redlich, C. M. Gable, A. K. Dunlop, and R. W. Millar, "Addition Compounds of Urea and Organic Substances," *Journal of the American Chemical Society* 72:4153–4160 (1950). With permission.

$1/K$ is a measure of the stability of the adducts.[23] Adduct formation is favored by large m (excess urea) and by large n (long chain), so adducts of longer chain length are favored. Figure 9.10 shows the influence of chain length and temperature on the equilibrium constants for n-paraffins (these are all below the lube range).

These authors also found that linear relationships exist between chain length, the equilibrium constants, and the enthalpies of the adduct formation:

$$\log K = 2.20 - 0.403*m \text{ at } 25°C$$

and

$$\Delta H = -6.5 + 2.37*m \text{ kcal/mole (adduct formation is exothermic).}$$

The exothermicity of adduct formation means that the isolated adducts can be easily decomposed back to n-paraffins and urea by raising the temperature. Kobe and Damask have reviewed the theory and applications of this technology.[16,24,25]

That adduct formation is favored for higher molecular weight paraffins has also been seen in analytical method development by Marquart et al., where the use of urea was explored as a means to determine n-paraffin levels in heavy distillates.[21] In this work, the recoveries of n-paraffins were measured from blends

TABLE 9.3
Effect of Temperature on Recovery of n-Paraffins

n-Paraffin Content

Carbon number	8	9	10	12	14	17	21	30	32
Blend percent	0.50	0.48	0.51	0.53	0.55	0.54	0.56	0.25	0.34
By Adduction, Temperature °C									
At 25°C, recovery	0.04	0.06	0.07	0.38	0.43	0.52	0.58	0.25	0.32
At 0°C, recovery	0.08	0.12	0.28	0.50	0.54	0.54	0.58	0.23	0.32

Source: J. R. Marquart, G. B. Dellow, and E. R. Freitas, "Determination of Normal Paraffins in Petroleum Heavy Distillates by Urea Adduction and Gas Chromatography," *Analytical Chemistry* 40:1633–1637 (1968). With permission.

of n-paraffins in a dewaxed heavy gas oil. The n-paraffin levels were created by the addition of high purity n-paraffins (Table 9.3). These blends were subsequently treated with urea at both 0°C and 25°C, the adducts filtered off and decomposed, and the recoveries of the n-paraffins were determined. The results show that for n-paraffins of chain length greater than C_{17}, the temperature of adduct formation is unimportant and recoveries are 100% or close to that. Below C_{14}, low temperature gives better recoveries. However, these latter recoveries are far from complete. In terms of application to lubricating oil feedstocks, these results are not important for 100Ns or higher since they start at about C_{20}. For lighter stocks, temperatures close to 0°C would be beneficial.

When commercially dewaxing a lube slate, urea dewaxing should have its greatest effect on the pour point of lower viscosity cuts since adduct formation is selective for n-paraffins and the n-paraffin content increases as viscosity decreases. Marechal and de Radzitzky's work confirmed this when they dewaxed a 255°C to 550°C Middle East waxy distillate by solvent and urea methods to similar overall dewaxed pour points, 10°F and 0°F respectively.[26] The total dewaxed oils were then vacuum fractionated into 10 cuts each on which pour points and VIs were measured. From the chart of pour points versus average boiling point of the cuts (Figure 9.11) it can be seen that urea dewaxing has the greatest impact on the pour point of the lower boiling fractions and this pour point lowering is very much diminished above 400°C (750°F), where n-paraffin content decreases. Solvent dewaxing with methyl isobutyl ketone, on the other hand, gives a much flatter pour point profile and is more effective at higher boiling points. Solvent dewaxing is more selective for higher carbon number n-paraffins and it also removes both isoparaffins and more complex cycloparaffinic structures

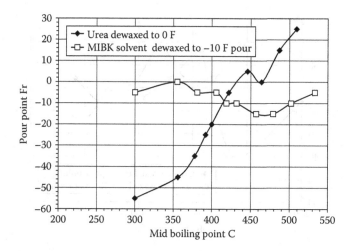

FIGURE 9.11 Urea and methyl isobutyl ketone dewaxing of waxy lube distillate: dependence of pour point on distillation range and dewaxing method.
Source: J. Marechal and P. de Radzitzky, "Some Aspects of Urea Dewaxing of Middle and Heavy Distillates," *Journal of the Institute of Petroleum* 46(434):33–45 (1960).

that contribute to pour point (400°C is within the boiling range of a 100N and therefore provides an upper marker as to the applicability of the process to lubes—beyond this, at least on a technical basis, it appears as if solvent dewaxing is favored).

The VI profile fits this selectivity for n-paraffins, with the lowest boiling fractions taking the greatest VI loss due to removal of the high VI n-paraffins, and above 375°C the VIs of the fractions are similar (Figure 9.12).

Figure 9.13 shows the effect of fractionating out the individual cuts from the waxy lube distillate and then dewaxing each fraction using urea. The pour points of the waxy fractions (undewaxed!) increase almost linearly with mid-boiling point, and the same is true of those of the urea-dewaxed cuts, but the curve for the latter lies about 60°F to 70°F lower. Closer examination shows that fractionation followed by dewaxing gives lower pour points (by about 20°F) for fractions with mid-boiling points less than 450°C.

These features were also observed, although perhaps not quite as clearly in the dewaxing by urea and by methyl ethyl ketone (MEK) of an Indian "washed Blue oil"[27] (Figure 9.14) in which the solid point (closely related to the pour point) for the fractions is plotted against midfraction percent. The conclusion from this work is the same, that urea is much more effective in reducing the pour or solidification point for low viscosity (low boiling point) fractions than is solvent dewaxing.

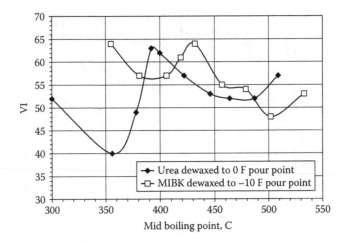

FIGURE 9.12 Urea and methyl isobutyl ketone dewaxing of waxy lube distillate: dependence of VI distribution on distillation range and dewaxing method.
Source: J. Marechal and P. de Radzitzky, "Some Aspects of Urea Dewaxing of Middle and Heavy Distillates," *Journal of the Institute of Petroleum* 46(434):33–45 (1960). With permission.

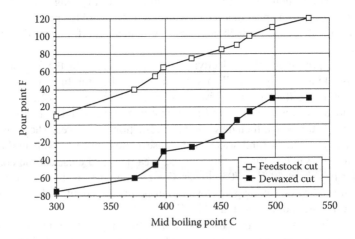

FIGURE 9.13 Pour points of distillation fractions versus distillation midpoint before and after urea dewaxing.
Source: J. Marechal and P. de Radzitzky, "Some Aspects of Urea Dewaxing of Middle and Heavy Distillates," *Journal of the Institute of Petroleum* 46(434):33–45 (1960). With permission.

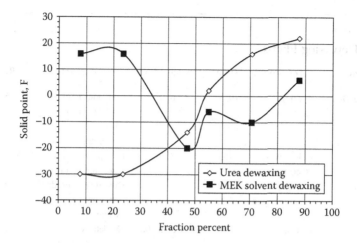

FIGURE 9.14 Urea and MEK dewaxing: solid point (°F) versus mid-volume percent after fractionation of the dewaxed oil.

Source: K. V. Gopalan, "Study of Urea-Reaction for Dewaxing of Washed Blue Oil from Indian Crude Oil," *Proceedings of the 4th World Petroleum Congress*, vol. V, Section IIIB, Paper 6, pp. 155–167 (1955). With permission.

9.4 UREA DEWAXING: COMMERCIAL APPLICATIONS

The foregoing shows that the commercial applications of urea dewaxing must lie with lighter waxy cuts because of its selectivity for n-paraffins and their location and distribution in waxy lube cuts. This is indeed the case and it has also been a significant limitation on the application of the technology. Lube refineries normally process a full slate of lubes for economic reasons and the urea technology's inability to handle this meant that its application has been confined to specific feeds.

While the initial discovery of this technology was made in 1940, development had to wait until after World War II, when there was much research activity in the period from 1950 to 1960. Process development was undertaken by the Edeleanu company of Frankfurt, Germany,[20,28,29] by Shell,[30] and by the Société Francaise des Pétroles (SFP)[31] with the objective of producing low pour oils. The technology is also claimed to be capable of manufacturing high purity (greater than 95%) n-paraffins.[29] Urea dewaxing plants reported to have been built are listed in Table 9.4.

In addition to dehazing, the Sonneborn white oil plant also reduced pour point from 4°C to 18°C and cloud point from +4°C to 10°C by removing 3 to 4% wax. This plant was unique in being a batch operation and being the first commercial use of this technology. Here, methanol was used as a promoter for adduct formation.

TABLE 9.4
Urea Dewaxing Plants

Company	Location	Date	Capacity, bpd	Products	Reference
Standard Oil	Whiting, Indiana	1955	650	Low pour point lubes (−70°F) for military use in the Arctic and other applications	15, 17, 41
Deutsche Erdoel AG	Heide, West Germany	1955	320	Low pour point diesel and spindle oil	17, 29
L. Sonneborn & Sons	Petrolia, Pennsylvania	1950		Dehazing to produce very low cloud point white oil	17, 41
Shell	Pernis, Holland		More than 200 tons/day	No details	29

A schematic for Standard Oil's Whiting, Indiana, plant is given in Figure 9.15. Solvent-dewaxed distillates of 0°F to 25°F pour points were judged as the most economic feeds to the urea plant for production of military Arctic oils. Examples of feed (feedstocks were given a preliminary solvent dewaxing) and product inspections from pilot plant studies prior to plant startup are given in Table 9.5. Product pour points ranging from −25°F to −55°F were achieved in yields of 90% to 96%. Yields decreased with increasing feed viscosity due to decreasing feed n-paraffin levels. This was an Edeleanu design in which methylene chloride was used to control the temperature of adduct formation by removing the heat through conversion of liquid to vapor.

In the Edeleanu process, waxy feed is mixed (Figure 9.15) in one of several reactors with urea and with recycled and fresh methylene chloride whose vaporization controls the exothermic reaction with urea. Prior to filtration, the adduct passes through a series of baffles to encourage complete reaction. After washing, the adduct is filtered from the oil, which proceeds to a stripper (to remove the methylene chloride) and then to product storage. The separate adduct stream from the filter is decomposed by steam at 75°C, then sent to a separator and subsequently stripped of solvent to yield the n-paraffins. The dewaxed oils disengage from the aqueous phase in a separator.

Details of pilot plant work undertaken to support process development by a number of companies have been published.[20,31,32] All are multistage units since the process steps involved are adduct formation, filtration, adduct decomposition, and solvent stripping. A successful process has to be capable of handling large solid quantities since the hexagonal crystalline form of the adduct is of greater volume than the tetragonal form of the urea itself. Examples of some feedstocks processed using the Edeleanu technology and their products are given in Table 9.6.

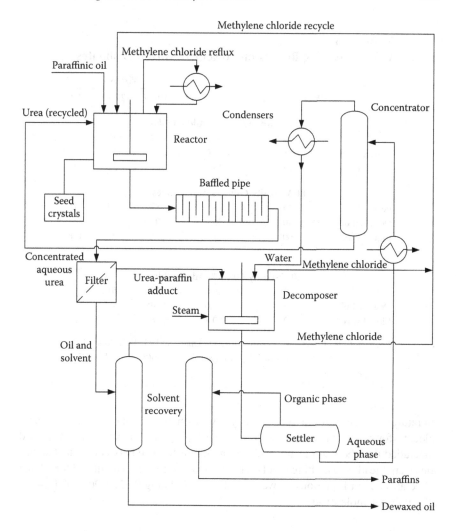

FIGURE 9.15 Schematic of urea dewaxing lube oil plant for Standard Oil in Whiting, Indiana.
Source: "Urea-Adduct Process Gains Ground in Lube-Oil Dewaxing," *Chemical Engineering* :114–118 (November, 1956). With permission.

The processes developed by Shell and IFP employed aqueous solutions of urea with the addition of other solvents or surface active materials, and the adduct decomposition step in each case again was to thermally reverse the equilibrium.

At the time of this writing, urea dewaxing has ceased to be a significant process in lube refining, overtaken by superior technologies. There is still interest in the chemistry, however, and it is conceivable that it may reemerge for other specific purposes. Papers on urea inclusion compounds regularly appear in the

TABLE 9.5
Urea Dewaxing Results on Solvent Dewaxed Distillates

	Pour Point, °F	Gravity, °API	Viscosity 100°F	210°F	VI
50 Viscosity Oil: 10% Adductables					
Feed	15	30.5	59	35	78
Dewaxed oil	−55	29.0	64	35	69
Liquid wax	75	46.1	44	32	139
80 Viscosity Oil: 7% Adductables					
Feed	35	28.6	82	37	75
Dewaxed oil	−40	26.9	92	38	61
Liquid wax	60	43.0	55	35	158
220 Viscosity Oil: 4% Adductables					
Feed	0	26.0	229	51	66
Dewaxed oil	−35	25.6	248	47	64
Liquid wax	80	40.8	68	38	167

Source: T. H. Rogers, J. S. Brown, R. Diekman, and G. D. Kerns, "Urea Dewaxing Gets More Emphasis," *Petroleum Refiner* 36(5):217–220 (1957). With permission.

literature; for example, a recent study[32] was on the thermodynamics of urea adducts formed with decane (C_{10}), dodecane (C_{12}), and hexadecane (C_{16}) and concluded that the main factor determining the difference between adduct stabilities were the alkane melting points. The free energy changes calculated for adduct association as in Equation 9.1 were small, the G_{as} being 0.95, 1.06, and 1.34 kJ, all being per mole of urea.

9.5 THE BP CATALYTIC DEWAXING PROCESS

In 1972 BP announced the development of a catalytic process for the production of very low pour point (less than −49°F, −45°C) base stocks.[33] These had been made traditionally from naphthenic oils, but a shortage of these and the ensuing high costs drove manufacturers to explore their production from waxy paraffinic stocks. The objective of BP's work was a more economical process than solvent dewaxing of paraffinic stocks to make base stocks with pour points in the less than −40°C to −57°C range, for transformer, refrigerator, automatic transmission, and hydraulic oils. This pour point range was one in which solvent dewaxing was and is particularly expensive.

BP found that it is possible to reduce pour points catalytically. A mordenite dual-function zeolite catalyst in the hydrogen form (to confer acidity and

TABLE 9.6

Feedstocks Dewaxed by the Edeleanu Urea Dewaxing Process

Feed:	Spindle Oil II	Spindle Oil O	Light Gas Oil	Light Gas Oil
Specific gravity at 15°C	0.894	0.871	0.817	0.833
Pour point, °C	33	13	−3	−22
Paraffin, wt. %	12	12	35	17
Boiling range, °C	414–480	300–400	240–330	210–320
Dewaxed with				
Urea solution, vol. %	70	90	200	140
Reaction temperature, °C	40	30	25	20
Dewaxed oil				
Yield, wt. %	85	83	68	84
Specific gravity at 15°C		0.890	0.839	0.849
Pour point, °C	−12	−20	−46	−63
Cloud point, °C		17		
Extract				
Yield, wt. %	15	17	32	16
Specific gravity at 15°C		0.808	0.776	0.774
Melting point, °C	55	35	12	6
n-Paraffins, wt. %	89.7	85	99	99

Source: A. Hoppe, "Dewaxing with Urea," in *Advances in Petroleum Chemistry and Refining*, vol. VIII (New York: Interscience Publishers, 1964), 193–234. With permission.

promote cracking) was discovered to effect selective hydrocracking of the normal and near-normal paraffins, which are in part responsible for the undesirable pour points.[34] This catalyst contained platinum to saturate olefins and reduce coke formation which otherwise would have shortened catalyst life. The process is conducted in a trickle bed reactor under hydrocracking conditions. A finishing step, either solvent extraction or hydrofinishing, was included in the process.

Zeolites are naturally occurring or synthetic minerals that have a porous structure due to their three-dimensional lattices. They are sometimes referred to as molecular sieves or ion exchange materials and possess highly structured three-dimensional frameworks. Their structure is based on silicate-alumina tetrahedral building blocks (Figure 9.16) and their catalytic properties are determined by their acidity and the dimensions and shapes of the pores, in turn controlled by the zeolite composition. The substitution of silicon by aluminum generates a charge imbalance that is corrected by cations; when these are protons acidity is affected. Zeolites have had a major impact on catalytic processes (e.g., as catalysts for fluidized catalytic cracker units [FCCUs] and fuel hydrocrackers), within the petroleum industry. The dimensions of the pores determine what molecules can

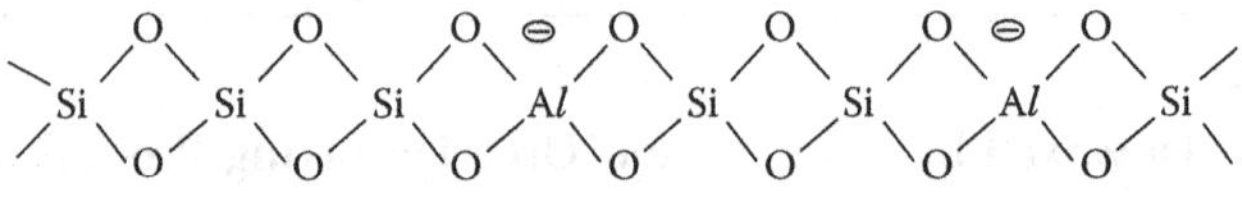

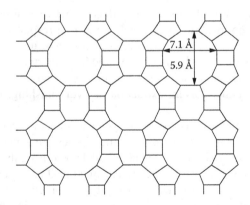

FIGURE 9.16 Two-dimensional representation of zeolite basic framework.
Source: K. Donaldson and C. R. Pout, "The Application of a Catalytic Dewaxing Process to the Production of Lubricating Oil Basestocks," presented at the Symposium on Advances in Distillate and Residue Technology, Division of Petroleum Chemistry, American Chemical Society meeting, New York, August 27–September 1, 1972. With permission.

enter, how they react within the catalyst, and what molecules can exit. They are frequently referred to as being "size selective" with respect to feed molecules.

Mordenite is a large pore zeolite with elliptical pores defined by 12 oxygen atoms and major and minor axes of 7.1 Å and 5.9 Å, respectively.[35] Figure 9.17 provides a schematic for mordenite's structure.[34] These dimensions mean that n- and isoparaffins can enter the pores where they react under the acidic conditions. However, these pore dimensions have also been interpreted to mean that larger molecules can also enter the pores[36] and react to form coke,[37] limiting catalyst life.

Pilot plant work by BP showed that the process is most applicable to lighter base stocks with TBP 95% points of less than 850°F, since the wax in these feeds is largely normal or near normal. In this respect the process bears similarities to urea dewaxing, but the urea process appeared to be more selective for n-paraffins. Like the urea process, high sulfur and nitrogen feeds can be dewaxed without problems since in this case the catalyst active sites are believed to come in contact only with paraffinic chains. The hydrocracking of the wax generates large

FIGURE 9.17 Schematic of mordenite structure.
Source: K. Donaldson and C. R. Pout, "The Application of a Catalytic Dewaxing Process to the Production of Lubricating Oil Basestocks," presented at the Symposium on Advances in Distillate and Residue Technology, Division of Petroleum Chemistry, American Chemical Society meeting, New York, August 27–September 1, 1972. With permission.

TABLE 9.7
Product Distribution in Weight Percent from Mordenite
Dewaxing of a Hydrogenated Middle East Waxy Distillate

Product by Boiling Range	Feed	Product from Platinum Hydrogen Mordenite
Methane		0.1
Ethane		0.5
Propane		9.7
Butane		19.2
Pentane to 82°C		16.9
82°C to 177°C	5.3	2.0
177°C to 232°C		1.3
232°C to 350°C	(32.7)	9.4
350°C to 371°C		3.7
Dewaxed residue, >371°C	62.0	37.2

Source: K. Donaldson and C. R. Pout, "The Application of a Catalytic Dewaxing Process to the Production of Lubricating Oil Basestocks," presented at the Symposium on Advances in Distillate and Residue Technology, Division of Petroleum Chemistry, American Chemical Society meeting, New York, August 27–September 1, 1972. With permission.

quantities of propane, butane, and pentane[34] in the weight ratios of 2:4:3 approximately (see Table 9.7).

Commercial operating conditions[38] were given as being within the following ranges:

- Hydrogen partial pressure: 300 to 1500 psi
- Temperature: 550°F to 750°F
- Liquid space velocity: 0.5 to 5.0 vol/vol/hr
- Gas recycle rate: 2000 to 5000 scf/bbl
- Hydrogen consumption: approximately 350 scf/bbl

Thus it was a relatively low pressure operation with most of the hydrogen consumption going to hydrocracking of selected C–C bonds of the n-paraffins.

Figure 9.18 is a schematic of BP's commercial demonstration unit[39] for their process, which can be seen to have the expected hardware for a process of this type, namely heat exchangers for the feed which is mixed with hydrogen prior to and after the furnace, a downflow trickle bed reactor followed by a separator, coolers, and a stripper system to remove naphtha and diesel and adjust the flash of the product.

Examples of the application of the process to naphthenic and paraffinic feeds are given in Table 9.8 and Table 9.9, respectively.[38,39] Even for the naphthenic

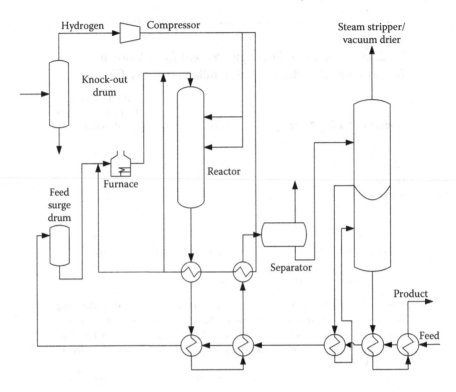

FIGURE 9.18 Schematic of BP's catalytic dewaxing demonstration unit.
Source: J. D. Hargrove, G. J. Elkes, and A. H. Richardson, "New Dewaxing Process Proven in Operations," *Oil and Gas Journal* 77:103–105 (1979). With permission.

TABLE 9.8
Naphthenic Feedstock Processing Summary: Commercial Operation for BP Catalytic Dewaxing

	Feedstock		Product
Viscosity Grade	Pour Point, °F	Boiling Range, 5%–95% °F	Pour Point, °F
250 SUS	−15	629–922	−45
500 SUS	+20	728–959	−20
1200 SUS	+20	786–1115	+10

Source: J. D. Hargrove, G. J. Elkes, and A. H. Richardson, "New Dewaxing Process Proven in Operations," *Oil and Gas Journal* 77:103–105 (1979). With permission.

TABLE 9.9
**Paraffinic Feedstock Processing Summary: Commercial
Operation for BP Catalytic Dewaxing**

Feedstock	Feedstock Pour Point, °F	Product Pour Point, °F
50N part dewaxed	+10	−55
75N part dewaxed	+15	−55
75N	+75	−60
100N part dewaxed	+15	−5
150N part dewaxed	+15	−5

Source: J. D. Hargrove, G. J. Elkes, and A. H. Richardson, "New Dewaxing
Process Proven in Operations," *Oil and Gas Journal* 77:103–105 (1979).
With permission.

feeds, particularly the 250N, the process can be seen to work quite well in spite
of the fact that naphthenic feeds do not contain substantial amounts of n-paraffins.
For the 500N and 1200N naphthenic feeds, the pour point reductions are modest
since n-paraffin content in these feeds is small, suggesting that part of the dewaxing here is due to removal of longer chains on polycyclic naphthenes which
the authors term "side chain stripping."[34] Polycyclic naphthenes will have difficulty entering the mordenite pores, but possibly some do or parts of the molecules
are able to enter and react, essentially to get chipped off.

In the case of paraffinic feeds, very low pour point products were most
economically obtained by using partially solvent dewaxed feedstocks (Table 9.9).
It can be seen as well that these are all relatively light lubes. The advantage cited
for this approach is that this can debottleneck the solvent dewaxing unit, where
throughput becomes very limited when attempting to produce very low pour
products.

Table 9.10[34] compares catalytic and solvent dewaxing of two waxy solvent
extracted paraffinic distillates, the first with a boiling range 360°C to 470°C and
the second 430°C to 540°C. As expected, it can be seen in both cases that catalytic
dewaxing results in lower VIs than solvent dewaxing since the high VI paraffins
are being selectively destroyed by the catalyst; solvent dewaxing retains many
lower molecular weight n-paraffins in the dewaxed oil, therefore the VI loss is
not as great. Overall, however, this VI loss is a serious impediment because of
the importance attached to the VI and its importance to low temperature viscosities. To achieve a VI close to that of solvent dewaxing, the feed must have a
higher VI, therefore it must be more deeply extracted and consequently there will
be a significant yield loss. The second consequence is that the base stock from
the catalytic dewaxing has a higher viscosity since it has lost the low-density
paraffins due to the greater selectivity of catalytic dewaxing for n- and isoparaffins. To maintain the same viscosity as the solvent dewaxed product, a lighter

TABLE 9.10

Comparison of Catalytic and Solvent Dewaxing for Paraffinic Solvent Raffinates

Inspection Data	360–470°C Waxy Raffinate			430–540°C Waxy Raffinate		
	Feed	BP Catalytically Dewaxed Product	Solvent Dewaxed Product	Feed	BP Catalytically Dewaxed Product	Solvent Dewaxed Product
Pour point, °C	32	−18	−15	46	−7	−12
Viscosity, cSt at 140°F	11.7	16.6	13.8	29.9	41.4	36.0
VI	118	85	98	93	62	79
Wax, wt. %	17.6	1.0	2.1	21.5	2.7	3.9
Nitrogen, ppm	57	94	—	—	—	—
Sulfur, wt. %	0.87	1.01	0.92	1.79	2.23	2.07
Yield on distillate, wt. %	68	47.6	53.7	84	54	68

Source: K. Donaldson and C. R. Pout, "The Application of a Catalytic Dewaxing Process to the Production of Lubricating Oil Basestocks," presented at the Symposium on Advances in Distillate and Residue Technology, Division of Petroleum Chemistry, American Chemical Society meeting, New York, August 27–September 1, 1972. With permission.

TABLE 9.11

Comparison of Solvent Dewaxing of Paraffinic Distillates with Catalytic Dewaxing Followed by Solvent Extraction

Inspection Data	360–470°C Waxy Raffinate			430–540°C Waxy Raffinate		
	Feed	Catalytically Dewaxed Product	Solvent Dewaxed Product	Feed	Catalytically Dewaxed Product	Solvent Dewaxed Product
Pour point, °C	29	−10	−15	41	−7	−12
Viscosity, cSt at 140°F	13.6	13.8	13.8	33.7	35.6	36.0
VI	85	98	98	75	74	79
Wax, wt. %	14.7	2.6	2.1	15.3	2.2	3.9
Yield on distillate, wt. %	100	50.0	53.7	100	50.0	68.0

Source: K. Donaldson and C. R. Pout, "The Application of a Catalytic Dewaxing Process to the Production of Lubricating Oil Basestocks," presented at the Symposium on Advances in Distillate and Residue Technology, Division of Petroleum Chemistry, American Chemical Society meeting, New York, August 27–September 1, 1972. With permission.

fractionated feed would have to be taken for catalytic dewaxing. While no densities are mentioned in this paper, undoubtedly the catalytically dewaxed products would also have higher densities due to loss of the low density paraffins.

BP found that the solvent extraction step used in traditional lubes manufacturing could be either before the catalytic dewax unit or downstream from it. They were able to solve the VI problem if solvent extraction followed catalytic dewaxing. In this configuration, the depth of extraction could be adjusted to produce base stock with the same viscosity and VI as solvent dewaxing—the extraction removes low VI, high viscosity material so both parameters are brought into line without further loss of yield (Table 9.11).

An alternative process arrangement in which hydrofinishing is employed instead of solvent extraction was also investigated. It was found that hydrofinishing prior to catalytic dewaxing was a satisfactory sequence, but when hydrofinishing came after catalytic dewaxing, the products were found to suffer from wax haze at temperatures well above the pour point.

Commercially, two plants used the BP process, one a converted 2000 bpd hydrotreater that came online in 1977,[39] and the second a grass-roots plant built in 1983, but apparently shutdown in 1986.[40] To the outside observer, the process seemed to suffer from the disadvantages shared with the urea process, namely that it could not handle the full scope of a refinery lube slate and incurred VI losses, and these may have been the reasons it (apparently) did not become more widespread.

REFERENCES

1. M. W. Wilson, K. L. Eiden, T. A. Mueller, S. D. Chase, and G. W. Kraft, "Commercialization of ISODEWAXING™—A New Technology for Dewaxing to Manufacture High Quality Lube Base Stocks," Paper FL-94-112, presented at the National Fuels and Lubricants meeting of the National Petroleum Refiners Association, Houston, Texas, November 3–5, 1994.

2. ASTM D721, "Standard Test Method for Oil Content of Petroleum Waxes," *ASTM Annual Book of Standards* (West Conshohocken, PA: American Society for Testing and Materials).

3. A. Crookell and A. Barker, "New Developments in the Measurements of Oil Content in Waxes Using Benchtop Pulsed NMR," Paper AM-98-49, presented at the Annual Meeting of the National Petroleum Refiners Association, , March 15–17, 1998.

4. M. Mafi, F. Yazdani, and F. Farhadi, "Determine Oil Content in Petroleum Waxes," *Hydrocarbon Processing* 85(6):95–97 (2006).

5. C. Giavarini, F. Pochetti, and C. Savu, "Determinazione delle paraffine negli oli minerali mediante la calorimetrica differenziale," *Rivista di Combustibili* 23:496–500 (1969).

6. F. Noel, "The Characterization of Lube Oils and Fuel Oils by DSC Analysis," *Journal of the Institute of Petroleum* 57(558):354–358 (1971).

7. ASTM D5442, "Standard Test Method for Analysis of Petroleum Waxes by Gas Chromatography," *ASTM Annual Book of Standards* (West Conshohocken, PA: American Society for Testing and Materials).

8. R. T. Edwards, "A New Look at Paraffin Waxes," *Petroleum Refiner* 36:180–187 (1957).

9. J. V. Brunnock, "Separation and Distribution of Normal Paraffins from Petroleum Heavy Distillates by Molecular Sieve Adsorption and Gas Chromatography," *Analytical Chemistry* 38:1648–1652 (1966).

10. E. A. Calcote, "Synthetic and Petroleum Paraffin Waxes: Complementary Tools in the Formulator's Toolbox," Paper LW-00-122, presented at the National Petroleum Refiners Association meeting, Houston, Texas, November 9–10, 2000.

11. D. Hess, "Microcrystalline Wax Under the Microscope," Paper LW-00-123, presented at the Lubricants and Waxes meeting of the National Petroleum Refiners Association, Houston, Texas, November 9–10, 2000.

12. F. Bengen, , German patent application, O.Z. 124438, 1940.

13. G. R. Desiraju, "Chemistry Beyond the Molecule," *Nature* 412:397 (2001).

14. J.-M. Lehn, "Steps Towards Complex Matter: Information, Self-Organization and Adaptation in Molecular and Supramolecular Systems," presented at the Nobel Centennial Symposia "Frontiers of Molecular Science," Orsundsbro, Sweden, December 4–7, 2001.

15. T. H. Rogers, J. S. Brown, R. Diekman, and G. D. Kerns, "Urea Dewaxing Gets More Emphasis," *Petroleum Refiner* 36(5):217–220 (1957).

16. K. A. Kobe and W. G. Domask, "Extractive Crystallization—A New Separation Process. Part 1. Theoretical Basis," *Petroleum Refiner* 31(3):106–113 (1952).

17. L. C. Fetterly, "Extractive Crystallization Today," *Petroleum Refiner* 36(7):145–152 (1957).

18. A. E. Smith, "The Crystal Structure of Urea-Hydrocarbon and Thiourea-Hydrocarbon Complexes," *Journal of Chemical Physics* 18:150–152 (1950).

19. R. W. Schiessler and D. Flitter, "Urea and Thiourea Adduction of C5-C42-Hydrocarbons," *Journal of the American Chemical Society* 74:1720–1723 (1952).

20. A. Hoppe, "Dewaxing with Urea," in *Advances in Petroleum Chemistry and Refining*, vol. VIII (New York: Interscience Publishers, 1964), 193–234.

21. J. R. Marquart, G. B. Dellow, and E. R. Freitas, "Determination of Normal Paraffins in Petroleum Heavy Distillates by Urea Adduction and Gas Chromatography," *Analytical Chemistry* 40:1633–1637 (1968).

22. O. Redlich, C. M. Gable, L. R. Beason, and R. W. Millar, "Addition Compounds of Thiourea," *Journal of the American Chemical Society* 72:4161–4162 (1950).

23. O. Redlich, C. M. Gable, A. K. Dunlop, and R. W. Millar, "Addition Compounds of Urea and Organic Substances," *Journal of the American Chemical Society* 72:4153–4160 (1950).

24. K. A. Kobe and W. G. Domask, "Extractive Crystallization—A New Separation Process. Part II. Thermodynamic Considerations," *Petroleum Refiner* 31(5):151–157 (1952).

25. K. A. Kobe and W. G. Domask, "Extractive Crystallization—A New Separation Process. Part III," *Petroleum Refiner* 31(7):125–129 (1952).

26. J. Marechal and P. de Radzitzky, "Some Aspects of Urea Dewaxing of Middle and Heavy Distillates," *Journal of the Institute of Petroleum* 46(434):33–45 (1960).

27. K. V. Gopalan, "Study of Urea-Reaction for Dewaxing of Washed Blue Oil from Indian Crude Oil," *Proceedings of the 4th World Petroleum Congress*, vol. V, Section IIIB, Paper 6, pp. 155–167 (1955).

28. H. Franz, "Urea Dewaxing Process Can Yield Normal Paraffins," *Hydrocarbon Processing* 44(9):183–184 (1965).

29. I. A. Hoppe and I. Franz, "Low Pour Oils Made by Urea Process," *Petroleum Refiner* 36(5):221–224 (1957).

30. L. N. Goldsbrough, "A Pilot Plant Employing a Novel Process for the Urea Extraction of Hydrocarbons," *Proceedings of the 4th World Petroleum Congress*, vol. V, Section IIIB, Paper 6, pp. 141–153 (1955).

31. A. Champagnat, J. Laugier, Y. Rollin, and C. Vernet, "La Cristallisation Extractive par L'Uree dans les Operations de Raffinage du Petrole," *Proceedings of the 4th World Petroleum Congress*, vol. V, Section IIIB, Paper 1, pp. 53–70 (1955).

32. M. A. White, "Origins of Thermodynamic Stability of Urea:Alkane Inclusion Compounds," *Canadian Journal of Chemistry* 76:1695–1698 (1998).

33. K. Donaldson and C. R. Pout, "The Application of a Catalytic Dewaxing Process to the Production of Lubricating Oil Basestocks," presented at the Symposium on Advances in Distillate and Residue Technology, Division of Petroleum Chemistry, American Chemical Society meeting, New York, August 27–September 1, 1972.

34. B. W. Burbridge, I. M. Keen, and M. K. Eyles, "Physical and Catalytic Properties of the Zeolite Mordenite," preprints of the 2nd International Conference on Molecular Sieves and Zeolites, pp. 400-409 (1970).

35. R. N. Bennett, G. J. Elkes, and G. J. Wanless, "New Process Produces Low-Pour Oils," *Oil and Gas Journal* 73(1):69–73 (1975).

36. M. P. Ramage, K. R. Graziani, and J. R. Katzer, "Science and Application of Catalytic Lube Oil Dewaxing," presented at the meeting of the Japan Petroleum Institute, Tokyo, October 27–28, 1986.

37. L. D. Rollman and D. E. Walsh, "Shape Selectivity and Carbon Formation in Zeolites," *Journal of Catalysis* 56:139–140 (1979).

38. J. D. Hargrove, G. J. Elkes, and A. H. Richardson, "New Dewaxing Process Proven in Operations," *Oil and Gas Journal* 77:103–105 (1979).

39. R. N. Bennett and G. J. Elkes, "Low Pour Oils from Paraffinic Crudes by the BP Catalytic Dewaxing Process," Paper F&L-74-52, presented at the National Fuels and Lubricants meeting, Houston, Texas, November 6–8, 1974.

40. A. Sequiera, Jr., *Lubricant Base Oil and Wax Processing* (New York: Marcel Dekker, 1994), 197.

41. , "Urea-Adduct Process Gains Ground in Lube-Oil Dewaxing," *Chemical Engineering*: 114–118 (November, 1956).

10 Dewaxing by Hydrocracking and Hydroisomerization

10.1 DEWAXING BY HYDROCRACKING

10.1.1 INTRODUCTION

Several years after BP's catalytic dewaxing process was announced in 1972, Mobil (now ExxonMobil) commercialized their own process at their Gravenchon (France) refinery. Like the BP methodology discussed in the previous chapter, the Mobil MLDW™ process reduces pour point by hydrocracking n-paraffins and similar molecules to gasoline and lighter range by-products and is unaffected by sulfur and nitrogen in the feed. For a number of years, this process became the predominant catalytic dewaxing process. In 1993 Chevron commercialized their ISODEWAXING™ technology that used an alternative chemical reaction available, namely selective isomerization of straight-chain paraffins to branched molecules with lower pour points. This was closely followed by the chemically similar ExxonMobil MSDW process, and these two technologies have come to dominate new developments in the dewaxing process picture because they retain the viscosity index (VI) and offer high yields. Their requirement for low sulfur and nitrogen feeds have not hindered their spread because of demand for group II and III base stocks.

10.1.2 MOBIL LUBE DEWAXING BY HYDROCRACKING

The Mobil process for catalytic dewaxing by selective hydrocracking of wax molecules arose from that company's development work on zeolites and the discovery of the remarkable selectivity exhibited by these catalysts some 20 years prior to first commercialization. In 1960 Weise and Frilette, of the then Socony Mobil Research and Development Laboratories,[1] reported that n-decane cracked readily to lighter paraffins over the sodium form of a zeolite known as "13X," whereas the bulkier molecules, α-pinene and isopropylbenzene, underwent no reaction (Figure 10.1).

Furthermore, the products from n-decane cracking were exclusively unbranched. In contrast, cracking the same feed over amorphous silica-alumina or the calcium form of the zeolite gave branched products (Table 10.1). This implied that not only was there selectivity for the shape of molecules that could access the reactive site, but also for those that departed. The eventual finding for

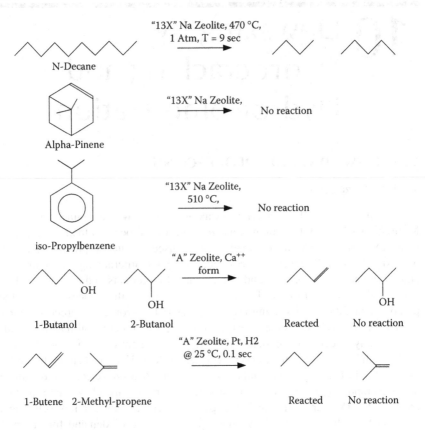

FIGURE 10.1 Key illustrations of zeolite shape selectivity in chemical reactions.

TABLE 10.1
n-Decane Cracking (3 Hr Operation, 470°C, τ = 9 sec, 1 atm)

Catalytic Solid	Percent Decane Converted	Iso-:n-Butane Ratio	Iso-:n-Pentane Ratio	Olefins in Cracked Products, wt. %
Silica-alumina	25	3.1	3.0	37
Na-aluminosilicate, X	32	0.0	0.0	62
Ca-aluminosilicate, X	39	1.1	2.4	30

Source: P. B. Weise and V. J. Frilette, "Intracrystalline and Molecular-Shape-Selective Catalysis by Zeolite Salts," *Journal of Physical Chemistry* 64:382 (1960). With permission.

our specific purposes was that waxy lubes, when reacted over certain zeolites, undergo selective destruction (hydrocracking) of n-, iso-, and other paraffins responsible for high pour points, and the process yields low pour point base stocks.

Zeolites, as previously described, are crystalline microporous solids with well-defined structures made up of interlocking microporous SiO_4 and AlO_4. Microporous means that the pores have dimensions of less than 20 Å, on the order of the size of many petroleum-related molecules, and their crystalline nature means that they have a narrow pore distribution (mesoporous materials have pore sizes between 20 Å and 500 Å, macroporous materials have pores larger than 500 Å). This combination of features not only restricts the size of molecules that can enter the pores, but also the dimensions of the transition state and of the molecules that can successfully leave. For these reasons, zeolites have been termed "shape selective."

The products from the cracking reactions mentioned above illustrate this feature. Other examples of this type of behavior in Figure 10.1 are the selective dehydration of 1-butanol in the presence of isobutyl alcohol and the selective hydrogenation of 1-butene in the presence of isobutene using a zeolite in which platinum had been incorporated within the crystalline cavity. Research on zeolite chemistry has led to the emergence of new industrial processes[2] that could not have been conceived of previously; for example, Mobil's methanol to gasoline process, isomerization to make p-xylene, and use of Linde's zeolite A to remove calcium and magnesium ions to convert hard water to soft. Catalytic dewaxing is based on various forms of Mobil's ZSM-5 catalyst, whose familiar three-dimensional pore structure is reproduced in Figure 10.2; a better depiction of the 10-membered ring structure (straight channels) is shown in Figure 10.3.

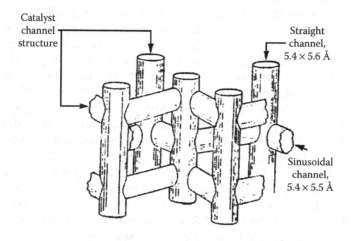

Catalyst channel structure

Straight channel, 5.4 × 5.6 Å

Sinusoidal channel, 5.4 × 5.5 Å

FIGURE 10.2 Schematic of channel structures within ZSM-5 zeolite.
Source: F. A. Smith and R. W. Bortz, "Applications Vary for Dewaxing Process over 10-Year Span," *Oil and Gas Journal* April 13, pp 51–55 (1990). With permission.

ZSM-5 skeletal diagram[1]

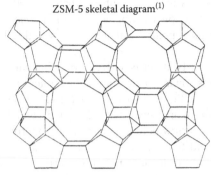

(1) Oxygen atoms not shown.

FIGURE 10.3 ZSM-5 skeletal diagram.
Source: R. G. Graven and J. R. Green, "Hydrodewaxing of Fuels and Lubricants Using ZSM-5 Type Catalysts," presented at the Australian Institute of Petroleum 1980 Congress, Sidney, Australia, September 14–17, 1980. With permission.

For convenience, microporous zeolites are further subdivided into those having small, medium, and large pores. The pore size is determined by the number of oxygen atoms in the rings or near rings that characterize their structure. Small pore zeolites have 8-membered rings, medium pore have 10, and large pore have 12-membered rings. The effect of pore size on what type of molecules can access the pores is seen in Figure 10.4,[3] where erionite is seen to have pores that are too small, mordenite too large, and ZSM-5 just right. Some specific examples of ring and pore sizes of zeolites are given in Table 10.2.[4] For the discussion here, the key zeolite to illustrate these chemical reactions is ZSM-5, originally developed by Mobil. By unwritten convention, the dewaxing process involving ZSM-5 and similar catalysts is referred to as "catalytic dewaxing," the first successful development of its type. In petroleum usage, the term "hydroisomerization" is employed in association with the processes using SAPO-11 (SAPO is a silica-alumina phosphate molecular sieve) and other catalysts central to Chevron's ISODEWAXING™ process and to ExxonMobil's MSDW™ and MWI™ processes, where paraffins are hydroisomerized rather than being hydrocracked. While use of "catalytic dewaxing" here would also be applicable, the distinction draws a convenient line between the two processes. In this section we'll focus on "catalytic dewaxing."

Pore size confers the ability to exclude molecules from reactive sites within the zeolite. Small pore zeolites can sorb (take up as "absorb") only n-paraffins, primary alcohols, or other straight-chain molecules, while medium pore ones, of which the prime example is ZSM-5, are accessible not only to n-paraffins, but also isoparaffins and some larger molecules as well. Large pore zeolites such as mordenite (12-membered rings) show poorer selectivity.

Two other features of zeolites are their silica:alumina ratio and acidity and these are closely related. The silica:alumina ratio can be varied since these are

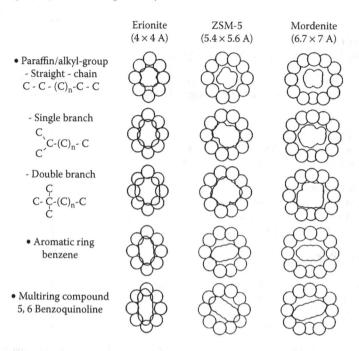

	Erionite (4 × 4 A)	ZSM-5 (5.4 × 5.6 A)	Mordenite (6.7 × 7 A)

• Paraffin/alkyl-group
- Straight - chain
C - C - (C)$_n$-C - C

- Single branch
C,
C-(C)$_n$- C
C

- Double branch
C
C- C-(C)$_n$-C
C

• Aromatic ring
benzene

• Multiring compound
5, 6 Benzoquinoline

FIGURE 10.4 Reactant shape selectivity for zeolite catalysts.
Source: M. P. Ramage, K. R. Graziani, and J. R. Katzer, "Science and Application of Catalytic Lube Oil Dewaxing," presented at the Japan Petroleum Institute meeting, Tokyo, Japan, October 27–28, 1986. With permission.

both tetrahedral atoms; when Al^{3+} is replaced by Si^{4+} an extra positive charge is created to maintain balance with the O–, and if this is provided by a proton, H^+, it confers acidity and the ability to crack or isomerize hydrocarbons.

In the case of the very versatile zeolite ZSM-5, the silica:alumina ratio is high, so that molecules that penetrate the pores experience a high acidity environment. This gives the interior environment a strong Bronsted activity and the ability to catalyze many reactions. Coupled with its shape-selective properties, ZSM-5 had the ability to become a catalyst of extraordinary breadth and did so.

Figure 10.5 shows the differences in feed selectivity that can occur due to pore size differences for erionite and ZSM-5 at several temperatures and for single temperature measurements, rare earth exchanged Y type (RE-Y) and for amorphous silica-alumina. In this study, first-order relative rates of cracking of n-hexane and 3-methylpentane were measured versus reactor temperature. The studies were performed either at 1 atm in argon or at 15 atm in a 15:1 hydrogen to hydrocarbon ratio. It can be seen that when the catalyst was erionite (a small pore zeolite), which nearly excludes all 3-methylpentane from its pores but admits n-hexane, the disappearance rate for n-hexane is much higher. In contrast for ZSM-5, with larger pore openings that can admit n- and isoparaffins and

TABLE 10.2
Key Properties of Molecular Sieves

Molecular Sieve Type	Ring Size of Channels	Pore Size, Largest Channel	Channel System Dimensionality
Small Pore			
Linde type A	8-8-8	4.1	3
Erionite	8-8-8	3.6 × 5.1	3
SAPO-34	8-8-8	3.8 × 3.8	3
Medium Pore			
ZSM-5	10-10-10	5.3 × 5.6	3
ZSM-11	10-10-10	5.3 × 5.4	3
SAPO-11	10	3.9 × 6.3	1
Large Pore			
Faujasite/X/Y	12-12-12	7.4	3
Beta	12-12	7.6 × 6.4	3
Mordenite	12-8	6.5 × 7.0	2
Linde type L	12	7.1	1
SAPO-5	12	7.3	1

Source: P. B. Venuto, "Organic Catalysis Over Zeolites: A Perspective on Reaction Paths within Micropores," *Microporous Materials* 2:297–411 (1994). With permission.

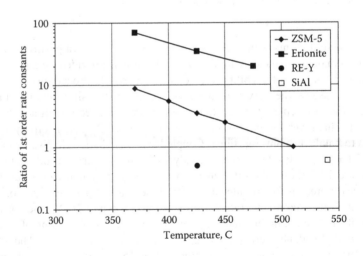

FIGURE 10.5 Shape selectivity between erionite and ZSM-5: relative rates of disappearance for n-hexane and 3-methylpentane at 1 and 5 atm.
Source: N. Y. Chen and W. E. Garwood, "Some Catalytic Properties of ZSM-5, a New Shape Selective Zeolite," *Journal of Catalysis* 52:453–458 (1978). With permission.

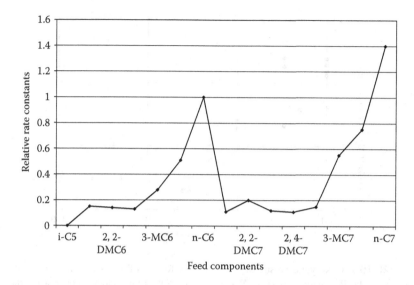

FIGURE 10.6 ZSM-5 catalyst: relative rate constant for reaction of paraffins at 370°C. *Source:* N. Y. Chen and W. E. Garwood, "Some Catalytic Properties of ZSM-5, a New Shape Selective Zeolite," *Journal of Catalysis* 52:453–458 (1978). With permission.

monocyclic hydrocarbons at significantly faster rates than those containing dimethyl-substituted or quaternary carbons, the relative rates are much smaller and more similar to those observed for the large pore catalyst RE-Y and amorphous silica-alumina surface.[5]

Figure 10.6 shows the relative disappearance rate constants at 35 atm and 340°C over ZSM-5 for a range of C_5, C_6, and C_7 paraffin isomers.[5,6] It can be seen that for the n- and isoparaffins, the n-isomers react more quickly and reactivity decreases as the chain gets shorter and as the branch point moves away from the end. As well, doubly branched isomers, either geminal (geminal substituents are two substituents on the same carbon) or otherwise, react very slowly. For effective dewaxing, it is important that the longer, more insoluble (in solvent dewaxing) and higher pour point n-alkanes react the quickest.

Further work from Mobil[7] showed that medium pore zeolites had a further advantage over those with large pores in that they experienced less coke formation (i.e., the coke make decreased as pore size decreased) (Figure 10.7). Coke formation reduces catalyst life, so low coke make is economically very important. The experimental study used a feed blend of n-hexane, 3-methylpentane, 2,3-dimethylbutane, benzene, and toluene. The authors measured coke make (grams of coke/100 g of catalyst) and pore size, the latter factor being estimated by the relative rate of disappearance of n-hexane to 3-methylpentane, this ratio being assumed to increase as pore size decreased. There is extensive scatter in the results due to the range of zeolites employed, but they do confirm the central thesis that small pores do reduce coke formation.

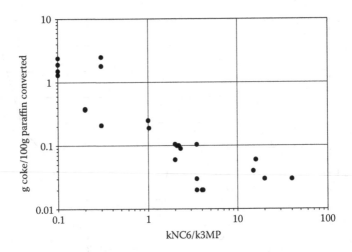

FIGURE 10.7 Coke yield versus shape selectivity for paraffin conversion.
Source: L. D. Rollmann and D. E. Walsh, "Shape Selectivity and Carbon Formation in Zeolites," *Journal of Catalysis* 56:139–140 (1979). With permission.

An additional effect of pore size on reaction course was seen during a study on zeolite cracking of paraffins in jet fuel,[8] using RE-X, a large pore system with 12-membered oxygen rings, and erionite and zeolite A, both having 8-membered rings. Using a jet fuel with n-hydrocarbons in the C_{11} to C_{16} range, the erionite and zeolite A catalysts reduced the pour point by 15°F at similar conversions, while the large pore nickel/RE-X catalyst increased the freeze point regardless of conversion. As n-paraffin number increases, erionite gives lower n-paraffin conversions (undesirable), whereas zeolite A was more capable of converting higher molecular weight n-paraffins. Both catalysts produced the same freeze point depressions, but in the erionite case this was due to greater conversion of nonnormal paraffins (see Table 10.3).

A final instructive example comes in Figure 10.8 from dewaxing studies using three very different catalysts—ZSM-5 (10-membered ring), erionite (8-membered ring), and a zeolite Y (12-membered oxygen ring opening).[9] The feed was a 10°C pour point, 315°C to 400°C vacuum gas oil (VGO) containing 12 wt. % n-paraffins. To assess the results, the product 315+°C pour point is plotted versus the percent conversion at 315°C (i.e., the yield loss). Best results will be points in the bottom left quartile (i.e., low pour point and high yield). As a yardstick of what just removing the n-paraffins would do, these were extracted with zeolite A and gave a 18°C pour point and a yield loss of 12% (this point is represented by the filled-in diamond shape). Best results among these catalysts were obtained using offretite, a 12-membered ring zeolite, followed by ZSM-5. It can be seen that neither erionite nor zeolite Y effect significant pour point reduction in spite of substantial conversion (top right quartile). This is due to the inability of erionite

TABLE 10.3
Comparison of Zeolite Catalysts for Freeze Point Lowering at 2000 psig, 27 H_2:1 Hydrocarbon

Zeolite Type	Ni/H Erionite	Pt/Ca-A		Ni/RE-X
Pore size	Small	Small		Large
Ring size	8	8	12	12
LHSV	30	0.5	2	20
Temperature, °F	750	800	480	650
Feedstock	A	B	A	A
Wt. % conversion to 358°F minus	18.2	13.5	12.3	78.8
n-Paraffins converted, wt. %	42.3	41.8	3.0	69.0
Non-normals converted, wt. %	10.2	6.1	11.6	82.0
Jet fuel yield, wt. %	81.8	86.5	87.7	21.2
Delta freeze point, °F	−15	−15	+10	+10

Source: N. Y. Chen and W. E. Garwood, "Selective Hydrocracking of n-Paraffins in Jet Fuels," *Industrial and Engineering Chemistry Process Design and Development* 19:315–318 (1978). With permission.

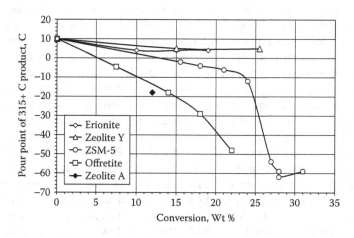

FIGURE 10.8 Zeolite dewaxing: comparison of catalyst behaviors: pour points of 315+°C versus weight percent conversion.
Source: N. Y. Chen, J. L. Schlenker, W. E. Garwood, and G. T. Kokotailo, "TMA-Offretite. Relationship Between Structural and Catalytic Properties," *Journal of Catalysis* 86:24–31 (1984). With permission.

to crack n-paraffins and zeolite Y showed no selectivity. Up to 25% conversion, ZSM-5 reduces the pour point in a near-linear fashion of about 0.5°C per 1 wt. % conversion and thereafter about 10°C per 1 wt. % conversion. Unfortunately, in this case there is no accompanying compositional analysis on the relative rates of disappearance of the paraffin types.

It is worth mentioning that the small pore Linde type 5 Å molecular sieves can be used as an analytical tool to determine n-paraffin content quantitatively up to at least C_{42}, since these n-paraffins are sorbed into the molecular sieve.[10-14] This method determines that number by difference (i.e., by weighing the material that is not sorbed). An alternative method already mentioned uses urea and claims a 95% accuracy and applicability to heavy gas oils.[15]

10.1.3 THE MLDW PROCESS: COMMERCIAL EXPERIENCE

The first commercial scale testing of the process was in France at Mobil's Gravenchon refinery in 1978 using waxy furfural raffinates from Middle East crudes.[16] This was followed in 1981 by a grass-roots plant in Mobil's Adelaide (Australia) refinery to dewax only bright stock, the slowest feedstock to dewax by the solvent route, resulting in their plant being debottlenecked and plant capacity was increased by 35%. A third Mobil plant came online at their Paulsborough (New Jersey) refinery in 1983 to produce a complete slate of base stocks.[17] Thus the MLDW process immediately became a serious contender to solvent dewaxing because it encompassed all viscosity grades, capital and operating costs were lower than for solvent dewaxing plants, very low pour specialty base stocks could be made at small additional cost, and there was the greater ease of operation associated with a continuous hydrotreating process. Overall lower dewaxed oil yields were said to be compensated by the much reduced operational costs and improved low temperature viscometrics.[18] By 1998 there were eight plants in operation.

The process employs two reactors, the first a dewaxing reactor and the second contains a hydrotreating catalyst to saturate any olefins produced. Figure 10.9 provides a schematic[3] and typical operating conditions[19] are given in Table 10.4. The catalysts can handle the sulfur and nitrogen levels in solvent refined stocks, so this process is applicable to virtually any waxy stock. In operation, dewaxing catalyst temperature depends on the feedstock, product pour point desired, feed nitrogen level, position in the catalyst cycle, and catalyst age. The dewaxing catalyst does undergo slow temporary deactivation and has to be hydrogen-reactivated at high temperatures (a hydrogen strip) at intervals the length of which depend on the feedstock. This period has shortened as improved versions of the catalyst—MLDW-1 (1981), MLDW-2 (1992), MLDW-3 (1993), and MLDW-4 (1996)—have been developed.[18] Not surprisingly, hydrocracked feedstocks exhibit very long cycles.

Two types of reactivations are mentioned,[20] a nonoxidative hydrogen-rich one at 15- to 100-day intervals,[21] and oxygen regeneration at longer intervals (6 to 12 months) to burn off coke buildup on the catalyst. Yields are given in Table 10.5 for commercial operations.

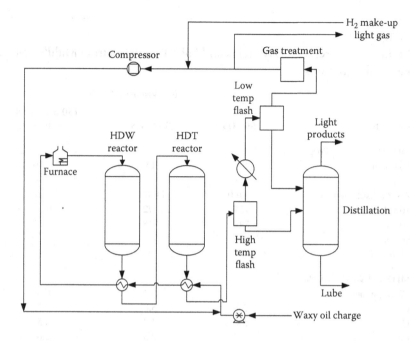

FIGURE 10.9 Schematic of MLDW lube dewaxing unit.
Source: M. P. Ramage, K. R. Graziani, and J. R. Katzer, "Science and Application of Catalytic Lube Oil Dewaxing," presented at the Japan Petroleum Institute meeting, Tokyo, Japan, October 27–28, 1986. With permission.

n-Paraffin hydrocracking leads to 4% to 10% by weight yields of light and heavy naphtha and the remaining by-products are butanes through pentanes with only trace amounts of methane and ethane. Hydrogen consumption decreases as viscosity increases and n-paraffin content of the feed decreases. Base stock yields are claimed to be as good as or better than from solvent dewaxing[21] for all grades

TABLE 10.4
Range of MLDW™ Operating Conditions

Reactor System	Fixed Bed
Type of feeds	Waxy solvent refined raffinates, waxy hydrocrackates, raw naphthenic distillates, soft waxes, deasphalted residuum
Space velocity	0.5–5
Reactor pressures	250–3000 psig
Hydrogen circulation	500–5000 scf/bbl
Hydrogen consumption	100–200 scf/bbl

Source: K. W. Smith, W. C. Starr, and N. Y. Chen, "New Process Dewaxes Lube Base Stocks," *Oil and Gas Journal* May 26:75–85 (1980). With permission.

TABLE 10.5
Initial Commercial MLDW Yields and Lube Properties from Middle East Furfural Raffinates

	Lube Viscosity Grade		
Characteristic	150 SUS	300 SUS	150 SUS Bright Stock
MLDW™ Charge Properties			
Pour point, °F	95	15	>130
ASTM distillation, °F	D1160	D1160	D1160
10 vol. %	764	772	913
50 vol. %	792	818	1012
90 vol. %	825	902	—
95 vol. %	835	921	—
MLDW™ Yields, wt. %			
C2 and lighter	0.6	0.3	0.6
C3	2.7	5.8	1.3
C4	5.4	4.7	4.5
C5	4.0	2.3	2.6
Light naphtha	7.0	3.4	3.5
Heavy naphtha	1.5	3.5	0.5
Lube product	79.1	80.0	87.1
Total[a]	100.3	100.2	100.1
Hydrogen consumption, scf/bbl	150	100	50
Lube Properties			
Pour point, °F	5	−50	15
Flash point, °F	430	435	600
VI	94	—	94

[a] Weight percent yields exceed 100% due to hydrogen incorporation in the products.

Source: K. W. Smith, W. C. Starr, and N. Y. Chen, "A New Process for Dewaxing Lube Basestocks: Mobil Lube Dewaxing," *Proceedings of the American Petroleum Institute Meeting, Refining Department* 59:151 (1980). With permission.

(see Table 10.6), but in a number of cases with the lighter lubes, catalytic dewaxing gives lower yields.

In fact, ZSM-5-type catalysts work best for heavier feedstocks where there is little VI or yield penalty due to the almost complete absence of n-paraffins. There dewaxing is believed to take place by shortening of long chains on both naphthene and aromatic molecules.[6] For the heavier stocks, overall compositional changes compared to solvent dewaxing are therefore relatively small (see Table 10.7[22] for a comparison of base stocks solvent and catalytically dewaxed to the same

TABLE 10.6
Comparison of MLDW™ and Solvent Dewaxing Yields

Feed	Yield on Charge, vol. %	
	MLDW™	Solvent Dewaxing
Very light neutral	78	78
Light neutral	79	78
Heavy neutral	82	77
Bright stock	90	75

Source: R. G. Graven and J. R. Green, "Hydrodewaxing of Fuels and Lubricants Using ZSM-5 Type Catalysts," presented at the Australian Institute of Petroleum 1980 Congress, Sidney, Australia, September 14–17, 1980. With permission.

pour points). But, as in so many things, the devil is in the details; in spite of this apparent similarity, the low temperature properties of catalytically dewaxed base stocks are much superior.

The effect of feed n-paraffin content can be seen from Taylor and MacCormack's (Texaco) excellent study[23] comparing the compositional differences

TABLE 10.7
Hydrocarbon Compositions of Feeds and Products in Solvent and Catalytic Dewaxing

	Light Neutral				Heavy Neutral				Bright Stock			
	Feed	Wax	SD	CD	Feed	Wax	SD	CD	Feed	Wax	SD	CD
Weight percent												
Paraffins	37.0	77.6	25.2	21.0	23	23	18	14	16	26	14	13
n-Paraffins	15.0	68.0	0.9	0.2	2.5	15	0	0	<0.2	<2	—	—
Monocycloparaffin s	15.3	16.4	14.3	16.5	15	37	15	17	14	21	12	14
Polycycloparaffins	24.9	2.8	34.0	34.7	24	25	24	27	23	10	24	26
Aromatics	23.0	3.2	26.4	27.9	38	15	43	43	47	43	50	47
Number of branches	3.1	0.1	3.5	4.1	4.8	1.1	5.6	5.7	5.5	3.1	6.7	6.7
Pour point, °C	—	—	–6	–6	—	—	–6	–6	—	—	–3.5	–3.5
Yield, vol. %	—	—	78	78	—	—	73	83	—	—	76	90
VI			108	98			95	89			95	95

CD, catalytically dewaxed oil; SD, solvent dewaxed oil.

Source: M. P. Ramage, K. R. Graziani, and J. R. Katzer, "Science and Application of Catalytic Lube Oil Dewaxing," presented at the Japan Petroleum Institute meeting, Tokyo, Japan, October 27–28, 1986. With permission.

TABLE 10.8
Comparison of Normal Paraffin Contents with Dewaxed Oil Yields and VI

Raffinate	n-Paraffin Content, wt. %	Yield Penalty Relative to Solvent Dewaxing, wt. %	VI Penalty Relative to Solvent Dewaxing
100N	12	6–8	10–13
320N	6	2–3	6–7
850N	4	0	4–5

Source: R. J. Taylor and A. J. MacCormack, "Study of Solvent and Catalytic Lube Oil Dewaxing by Analysis of Feedstocks and Products," *Industrial and Engineering Chemistry Research* 31:1731–1738 (1992). With permission.

between solvent and catalytic dewaxing of a range of feedstocks. Their results (Table 10.8) highlight the importance of n-paraffin levels in determining yields and VI penalty from catalytic dewaxing relative to solvent dewaxing. It can be seen in Table 10.8 that changes in yield accurately reflect changes in n-paraffin levels and that the VI penalty follows this trend as well. In the case of lube hydrocracking plants, hydrocracking severity has to be increased (with its accompanying yield loss) to offset this VI loss. For solvent refining units, the corollary is that extraction severity has to be increased, again with a yield loss. In addition, catalytic dewaxing leads to higher viscosity products (due to n-paraffin loss), and to keep viscosities unchanged, distillation cuts have to be altered (i.e., mid-boiling points moved lower), which will impact volatility unless improvements are made to the fractionation tower to produce sharper cuts.

Figure 10.10 and Figure 10.11 show that for heavy neutral feed, yield and VI changes are quite small, being about 0.1 vol. %/°F pour point change and about 0.2 VI/°F pour point change, and these features deteriorate very little as the catalyst ages and undergoes a number of activations.

10.1.4 CHEVRON BY HYDROCRACKING DEWAXING

In 1983, Chevron's Richmond Lube Oil Project (RLOP), just outside San Francisco, came online producing 9000 bpd light, medium, and heavy lubes via an all-hydroprocessing route.[24,25] The plant schematic is shown in Figure 10.12 and is described in more detail in Chapter 7.

The light (100N) and medium (240N) streams were catalytically dewaxed using what is probably a ZSM-5-type catalyst, since it is described as selectively cracking the wax molecules (mostly n-paraffins) to propane, butane, and light gasoline. Dewaxing is accompanied by a decrease in VI and an increase in viscosity compared to solvent dewaxing. Chevron continued with solvent dewaxing for the heavy stream (500N) since it was their lowest cost option (this situation changed when they later employed their hydroisomerization catalyst, which was used for all waxy streams). All dewaxed products were hydrofinished.

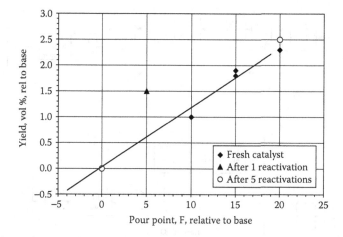

FIGURE 10.10 Mobil lube dewaxing of heavy neutral raffinate: yield versus lube pour point (°F).
Source: K. W. Smith, W. C. Starr, and N. Y. Chen, "New Process Dewaxes Lube Base Stocks," *Oil and Gas Journal* May 26:75–85 (1980). With permission.

In spite of the fact that Chevron chose not to use their dewaxing catalyst for their heavy stream, in an interesting application they did employ the same (or very similar) catalyst to dehaze the heavy dewaxed oil when pinholes or tears in the filter fabric caused leakage of wax into the solvent dewaxed oil. The refinery's

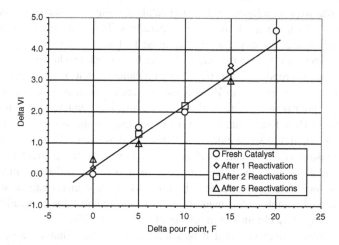

FIGURE 10.11 Mobil lube dewaxing: heavy neutral raffinate—dewaxed lube oil VI versus pour point (°F).
Source: K. W. Smith, W. C. Starr, and N. Y. Chen, "New Process Dewaxes Lube Base Stocks," *Oil and Gas Journal* May 26:75–85 (1980). With permission.

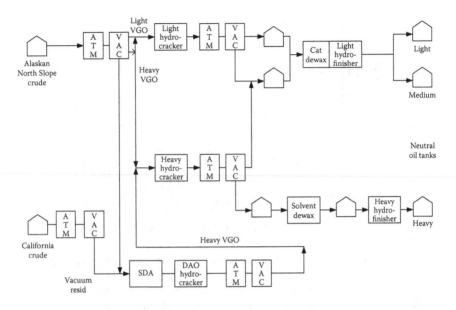

FIGURE 10.12 Schematic of Chevron all-hydroprocessing RLOP plant.
Source: J. A. Zakarian and J. N. Ziemer, "Catalytic Dehazing of Heavy Lube Oil: A Case History," *Energy Progress* 8:109–112 (1988). With permission.

problem with haze turned out to be an intermittent one, the worst sort to deal with, and necessitated either downgrading affected product or its reprocessing. Their solution[26]—conceived, pilot planted, and implemented over an eight-month period—was to place a layer of dewaxing catalyst at the top of the heavy hydrofinisher reactor, to hydrocrack any normal paraffins. This was successful, in spite of the high space velocity that this catalyst had to operate at. There was no negative effect reported on the quality of the base stock produced.

Chevron also reported[27] kinetic studies on catalytic dewaxing of several waxy streams (see Table 10.9) with ZSM-5 using a technique that gave the activation energies for pour point reduction whose values turned out to have some interesting implications. They found that E_A's were dependent on the stock being dewaxed (Figure 10.13), and to some extent the degree of dewaxing. These E_A's decreased with increasing viscosity and their average values bear an apparent near-linear converse relationship to the percent n-paraffins in the wax, and more importantly, the percent non-normals in the wax (Figure 10.14). Their investigations with partially dewaxed samples of a medium neutral showed that the n-paraffins were virtually completely eliminated in all samples (Table 10.10).

The authors concluded that it was not the n-paraffins that were rate determining, but rather it was the nonnormal paraffins composed of iso- and branched paraffins and naphthenes with chains attached. The rate determining steps were therefore the catalytic modification of these more highly branched structures, with E_A's decreasing as molecular weight increased. The conversion of the n-paraffins

TABLE 10.9
Properties of Hydrocracked Lube Feedstocks

	Light Neutral	Medium Neutral 1	Medium Neutral 2	Heavy Neutral
Pour point, °F	80	95	100	135
Viscosity, cSt at 100°C	3.71	6.79	5.63	10.52
Solvent dewaxing				
Dewaxed oil pour point, °F	5	10	10	10
Wax content, wt. %	12.3	9.28	14.4	25.3
n-Paraffins in wax, wt. %	71.3	56.3	42.3	24.0

Source: D. J. O'Rear and B. K. Lok, "Kinetics of Dewaxing Neutral Oils Over ZSM-5," *Industrial and Engineering Chemistry Research* 30:1105–1110 (1991). With permission.

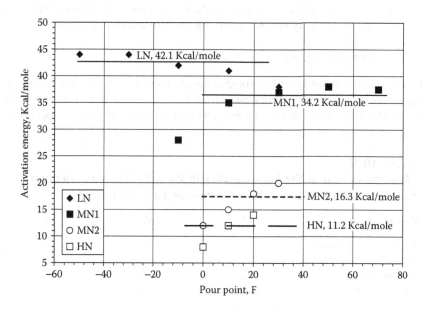

FIGURE 10.13 Variation in activation energy for lube dewaxing versus product pour point.
Source: D. J. O'Rear and B. K. Lok, "Kinetics of Dewaxing Neutral Oils Over ZSM-5," *Industrial and Engineering Chemistry Research* 30:1105–1110 (1991). With permission.

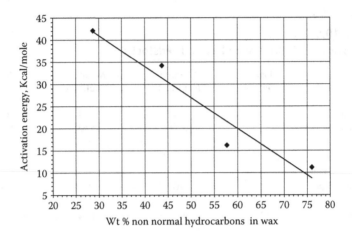

FIGURE 10.14 ZSM-5 catalytic dewaxing of waxy hydrocrackates: activation energies versus percent non-normals.
Source: D. J. O'Rear and B. K. Lok, "Kinetics of Dewaxing Neutral Oils Over ZSM-5," *Industrial and Engineering Chemistry Research* 30:1105–1110 (1991). With permission.

was very rapid in comparison. The authors were not able to distinguish to their satisfaction whether these differences in the reactivities of non-n-paraffins were due to inherent reactivity differences or due to diffusion characteristics.

10.1.5 FURTHER STUDIES

Ramage et al.[3] investigated the compositional changes in a light neutral where the n-paraffins are a substantial portion of the wax. Their results are given in Figure 10.15 and show that both normal and total paraffins (both n- and isoparaffins)

TABLE 10.10
Characterization of the Residual Wax in Partially Dewaxed Medium Neutral, Pour Point 100°F, Wax Content at 10°F, 14.4 wt. %

Pour point, °F	65	45	15
Batch solvent dewaxing			
Dewaxed oil pour point, °F	15	10	0
Wax content, wt. %	5.3	3.5	0.4
n-Paraffin in wax, %	<2.0	<2.0	<2.0
Conversion, wt. %			
n-Paraffins	>97	>98	>99
Non-n-Paraffins	3	36	93

Source: D. J. O'Rear and B. K. Lok, "Kinetics of Dewaxing Neutral Oils Over ZSM-5," *Industrial and Engineering Chemistry Research* 30:1105–1110 (1991). With permission.

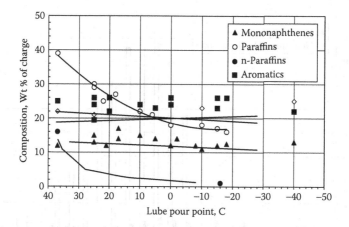

FIGURE 10.15 Mobil ZSM-5 catalytic dewaxing of a light neutral: compositional changes. *Source:* M. P. Ramage, K. R. Graziani, and J. R. Katzer, "Science and Application of Catalytic Lube Oil Dewaxing," presented at the Japan Petroleum Institute meeting, Tokyo, Japan, October 27–28, 1986. With permission.

show steep declines as the pour point decreases from its initial value, with the n-paraffin content falling the faster of the two. In this case the mononaphthenes show a small change and the polynaphthenes and aromatics show little if any.

In the Texaco study by Taylor and McCormack[22] on the differences between solvent and catalytic dewaxing of a light neutral, analyses of the waxes (Figure 10.16

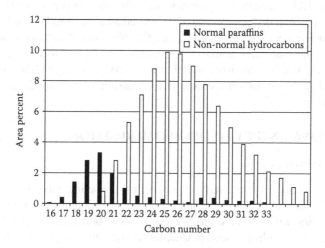

FIGURE 10.16 Normal and non-normal paraffins remaining after solvent dewaxing of a waxy 100N.
Source: R. J. Taylor and A. J. MacCormack, "Study of Solvent and Catalytic Lube Oil Dewaxing by Analysis of Feedstocks and Products," *Industrial and Engineering Chemistry Research* 31:1731–1738 (1992). With permission.

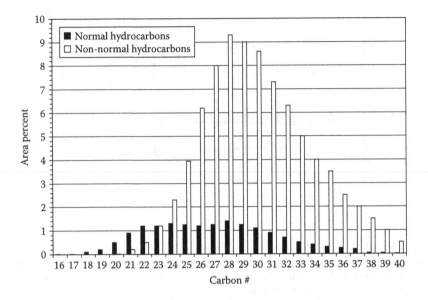

FIGURE 10.17 Normal and non-normal hydrocarbons in wax remaining after CD dewaxing of a light neutral.
Source: R. J. Taylor and A. J. MacCormack, "Study of Solvent and Catalytic Lube Oil Dewaxing by Analysis of Feedstocks and Products," *Industrial and Engineering Chemistry Research* 31:1731–1738 (1992). With permission.

and Figure 10.17) remaining in the dewaxed oils (SD SD wax and SD CD wax) showed that solvent dewaxing essentially removed all n-paraffins beyond C_{24}, but did leave a group of n-paraffins between C_{16} and C_{23}. Non-normal distribution was nearly "normal," with a maximum at about C_{26}. In contrast, catalytic dewaxing leaves a more uniform (but low) distribution of n-paraffins between C_{20} and C_{32}, and the non-normals are shifted to higher carbon number, with a maximum at about C_{28} and residual nonnormals to C_{40}.

10.2 DEWAXING BY HYDROISOMERIZATION

10.2.1 INTRODUCTION

Hydroisomerization is the catalytic process for dewaxing waxy lubes and conversion of waxes to high VI base stocks by isomerization of n-paraffin structures to isoparaffins with one or more branches. These branches are usually methyl branches. We have already seen in Chapters 2 and 3 that iso-paraffins have lower pour points than n-paraffins and can have quite high VIs if the branches are close to the chain ends. Hydroisomerization is distinguished from catalytic dewaxing via ZSM-5-type catalysts in that the latter cracks n-paraffin structures to C_3 to C_8 molecules (Figure 10.18), whereas the former causes isomerization and has

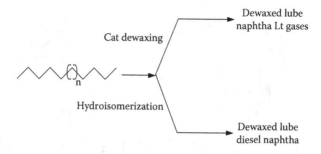

FIGURE 10.18 Chemical pathways for catalytic dewaxing and hydroisomerization.

the distinct advantage that VI is retained in the dewaxing step. Isomerization is not the sole reaction pathway, since some cracking does occur, producing high cetane diesel and some naphtha.

10.2.2 COMMERCIAL DEWAXING BY HYDROISOMERIZATION

Dewaxing by isomerization was first commercialized by Chevron[28] in 1993 at their Richmond, California, refinery using their proprietary ISODEWAXING™ technology employing a zeolite catalyst (SAPO-11) with a noble metal as a hydrogenation component. Subsequently the ExxonMobil MSDW[29] process was announced, employing a medium pore (10-ring) zeolite.[30] This was installed in their Jurong, Singapore, hydrocracking lubes plant in 1997.[18] Both processes are used with waxy streams containing low levels of sulfur and nitrogen to avoid poisoning the noble metal incorporated in the catalyst. There is yet no technology to isomerize wax in waxy solvent refined stocks that have not been severely hydrotreated to reduce sulfur and nitrogen to very low levels. Both Chevron and ExxonMobil processes employ hydrofinishing reactors after dewaxing. Figure 10.19 is a schematic of the Chevron process,[31] and that of the ExxonMobil process is similar. As a consequence of diesel formation, fractionation of the dewaxer/hydrofinisher reactor product is required to separate byproducts from base stock and establish base stock volatility. ExxonMobil also offers a specific process—Mobil Wax Isomerization (MWI)[32]—for conversion of waxes to 140+ VI lubes. At the time of this writing, Chevron and ExxonMobil are the only two companies offering this type of dewaxing technology for lubricants production.

The advantages of hydroisomerization include[27]

- Capital costs for building a hydroisomerization plant are considerably less than for a solvent dewaxing plant.
- Operating costs and emissions are reduced.

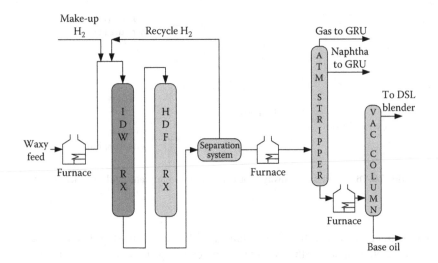

FIGURE 10.19 Schematic of Chevron hydroisomerization dewaxing process.
Source: K. R. Krishna, A. Rainis, P. J. Marcantonio, J. F. Mayer, J. A. Biscardi, and S. I. Zones, "Next Generation Isodewaxing® and Hydrofinishing Technology for Production of High Quality Base Oils," Paper LW-02-128, presented at the Lubricants and Waxes meeting, National Petroleum Refiners Association, Houston, Texas, November 14–15, 2002. Figure copyrighted by Chevron Corporation and used with permission.

- Hydroisomerization is applicable to the full slate of products from 40N to bright stock.
- Base stock yields are equal or better compared to either solvent dewaxing or catalytic dewaxing.
- Hydroisomerization produces base stocks with VIs higher than for solvent or catalytic dewaxing. Product viscosity is similar to that of the waxy feed.
- When coupled to a lubes hydrocracker, for the same VI as produced by solvent dewaxing, hydroisomerization allows hydrocracker severity to be reduced, giving higher yields of base stocks.
- Product formulation costs are reduced and product performance is enhanced by the higher paraffinicity of the products.
- Hydroisomerization of wax feeds produces group III and group III+ products.
- Higher yields of group III base oils from hydrocracker bottoms are achieved.
- Higher VI base stocks reduce volatility.
- Hydroisomerization produces high value middle distillates as by-products.

By 2004, this technology had penetrated so far that Chevron[33] had licensed some 150,000 bpd in design capacity, while ExxonMobil's MSDW catalyst was in use in eight units with lube production capacity of more than 100,000 bpd.[30]

TABLE 10.11
Hydrocracker/Dewaxing Arab Light VGO to Make Finished 150N with 100 VI

	Hydrocracker Feed Rate, bpd	Hydrocracker Yield, vol. % Feed	Dewaxer Feed Rate, bpd	Dewaxer Yield, vol. %	Base Oil Yields, bpd
Catalytic dewaxing	10,000	52	5200	81	4200
Solvent dewaxing	10,000	63	6300	84	5300
Isodewaxing	10,000	72	7200	89	6400

Using isodewaxing to make group II base stocks allows the refiner to lower processing severity in the hydrocracker and substantially increase yields.

Source: W. Qureshi, L. Howell, C.-W. Hung, and J. Xiao, "Isodewaxing—Improving Refining Economics," *Petroleum Technology Quarterly* Summer:17–23 (1996). Table copyrighted by Chevron Corporation and used with permission.

Table 10.11 shows the difference that dewaxing process selection can make in the yields of 150N base stock of 100 VI from Arab light VGO in a lubes hydrocracking plant.[34] In this instance, catalytic dewaxing gives the lowest overall base oil yield from a 10,000 bpd hydrocracker feed, since the VI loss in dewaxing necessitates a high VI feed to the dewaxing unit and therefore high severity (and low yield) in the hydrocracker. Solvent dewaxing produces a better overall base stock yield than catalytic dewaxing since there is less wax and VI loss. Finally, of these three options, hydroisomerization results in the highest base oil yield since a significant part of the n-paraffins are converted to high VI isoparaffins and severity in the hydrocracker can be reduced.

By-products from an early ISODEWAXING™ catalyst are compared in Table 10.12. They can be seen to be largely diesel with some naphtha and obviously no wax by-product is produced from either of the catalytic modes.

High wax yields by solvent dewaxing mean low oil yields. Base stock yields from ISODEWAXING™ also decrease with increasing wax yields, but the slope is not as steep as for solvent dewaxing. Consequently the greatest yield benefit of hydroisomerization relative to solvent dewaxing, at least as practiced by Chevron's catalyst, is with feeds of high wax content (Figure 10.20).

This can also be seen from the results in Table 10.13, where at one extreme a deoiled wax gives a 66% yield of base stock and at the other an Alaskan North Slope hydrocrackate with 10% wax by solvent dewaxing gives a 92% yield of base stock by ISODEWAXING™.[35] In between is the high wax 150N of South American crude origin which gave a 14% yield increase over solvent dewaxing.

Note that the two side-by-side comparisons are at the same pour points and viscosities, with the greater yield benefit being for the high wax South American

TABLE 10.12
By-Products of Dewaxing Processes: Distribution of Nonlube Products after Hydrofinshing ANS Hydrocrackate

	ISODEWAXING™	Solvent Dewaxing	Conventional Catalytic Dewaxing
C_1–C_3, wt. %	1	1	11
C4	5	2	22
Naphtha	22	0	67
Mid-distillate	72	0	0
Wax, wt. %	0	97	0
Total by-products	100	100	100
Mid-distillate properties			
Jet smoke point, mm	34	N/A	N/A
Diesel cetane index	73	N/A	N/A

Source: W. Qureshi, L. Howell, C.-W. Hung, and J. Xiao, "Isodewaxing—Improving Refining Economics," *Petroleum Technology Quarterly* Summer:17–23 (1996). Table copyrighted by Chevron Corporation and used with permission.

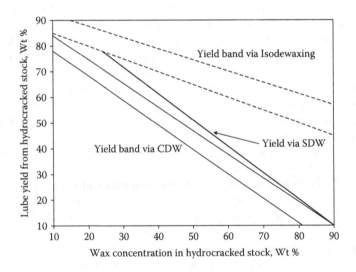

FIGURE 10.20 Lube yield comparison for hydrocracked stocks and different dewaxing methods.
Source: M. W. Wilson, K. L. Eiden, T. A. Mueller, S. D Case, and G. W. Kraft, "Commercialization of Isodewaxing—A New Technology for Dewaxing to Manufacture High-Quality Lube Basestocks," Paper FL-94-112, presented at the National Fuels and Lubricants meeting, National Petroleum Refiners Association, Houston, Texas, November 30, 1994. Figure copyrighted by Chevron Corporation and used with permission.

TABLE 10.13

Properties and Yields of Dewaxed Lube Base Oils from Solvent and ISODEWAXING™

Feed Source	Hydrocracked Alaskan North Slope VGO		Hydrocracked South American VGO		Deoiled Slack Wax from Solvent Refined West Texas VGO
Dewaxing process	IDW	SDW	IDW	SDW	IDW
Pour point, °C	−12	−12	−15	−15	−15
Yield, LV %	92	90	84	68	66
Viscosity, cSt at 100°C	4.016	4.163	5.304	5.380	3.938
VI	96	87	134	131	155
Volatility, LV %, % < 371°C, ASTM D2887	20	17	5.0	4.1	5.0

Source: S. J. Miller, J. Xiao, and J. M. Rosenbaum, "Application of Isodewaxing, a New Wax Isomerization Process for Lubes and Fuels," *Science and Technology in Catalysis*:379–382 (1994). With permission.

hydrocrackate and the greatest VI increase for the low wax feed. As expected, the ISODEWAXED™ product has a greater paraffin content (Table 10.14) than the solvent dewaxed oil, which must lose paraffins by removal during the dewaxing step. In this particular example, finished base stock composition by mass spectrometry is almost identical to that of the waxy feed.

Since development of their technologies, both Chevron[33,35] and ExxonMobil[36-38] have worked on refinements of the catalysts to give higher base stock yields and VIs (i.e., increase the efficiency of isomerization versus cracking). Figure 10.21 and Figure 10.22 illustrate these improvements for the ExxonMobil catalysts MSDW-1 and MSDW-2 with a waxy light neutral feed. It can be seen that the slopes of the yield and VI changes with pour point are fairly similar for the two catalysts and for solvent dewaxing, but the catalytic benefits are quite evident as are the improvements within that group.

The results from similar work by Chevron on their own next-generation catalysts can be seen in Table 10.15.[33] Of these data, the results with bright stock stand out from the rest in terms of their remarkable yield improvements, but the VI improvements are also very noteworthy.

10.2.3 POUR POINTS, VI, AND PARAFFIN STRUCTURE

To briefly refresh the memory with data from Chapters 2 and 3, Figure 10.23 shows that the melting points (used as proxies for pour points) of the mono-methyl isomers of C_{10} to C_{20} paraffins depend on the position of the methyl group[39] and that they

TABLE 10.14
Compound Types in Dewaxed Alaskan North Slope Medium Neutral Oil

	Feed	Solvent Dewaxed	Chevron ISODEWAX™ Catalyst
Pour point, °C	+ 42	−15	−15
VI		120	121

Saturate Compound Types, LV % of Total, ASTM D2786			
Paraffin	33.7	29.6	34.4
1-ring	34.1	34.5	35.4
2-ring	16.3	16.7	18.4
3-ring	6.6	6.5	6.7
4+-ring	3.1	3.3	3.0
Monoaromatics	0.6	0.6	0.3

Source: S. J. Miller, M. A. Shippey and G. M. Masada, "Advances in Lube Base Oil Manufacture by Catalytic Hydroprocessing," Paper FL-92-109 presented at the National Fuels and Lubricants Meeting, National Petroleum Refiners Association, Houston, Texas, November 5–6, 1992. Table copyrighted by Chevron Corporation and used with permission.

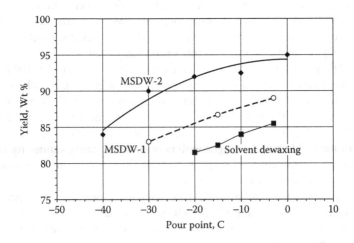

FIGURE 10.21 ExxonMobil dewaxing technology: comparison of light neutral base stock yields versus pour point for solvent dewaxing and MSDW-1 and MSDW-2 catalysts. *Source:* M. Daage, "Baseoil Production and Processing," available at http://www.prod.exxonmobil.com/refiningtechnologies/pdf/base_oil_refining_lubes_daage_france070601.pdf. With permission.

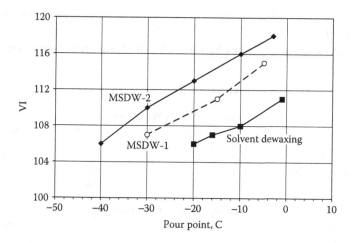

FIGURE 10.22 ExxonMobil technology: base stock VI versus pour point for solvent dewaxing and MSDW-1 and MSDW-2 catalysts.
Source: M. Daage, "Baseoil Production and Processing," available at http://www.prod.exxonmobil.com/refiningtechnologies/pdf/base_oil_refining_lubes_daage_france070601.pdf. With permission.

decline significantly as the methyl group moves toward the middle of the chain. For example, altering the carbon architecture for the C_{20} paraffin by changing the structure from the normal paraffin eicosane to the 2-methyl isomer, a 20°C decrease in melting point is achieved. These examples are below the carbon number range for lube base stocks, but we can assume with reasonable safety that the behavior here is broadly representative of paraffins in the lube range.

However, offsetting these reductions in pour points is the parallel decrease in VI, which can be seen in Figure 10.24,[40] for substituted C_{20} and C_{22} hydrocarbons, and this is very steep as the substituent's position moves away from the end of the hydrocarbon chain. The substituents in these examples are n-butyl, cyclohexyl, and phenyl and we have assumed that there is a parallel curve to these for methyl-substituted cases. Therefore a balance is needed between pour point reduction and VI and the position where the methyl group is created.

Specific commercial process parameters—hydrogen pressure, space velocity, catalyst temperatures—have not been published. One Chevron patent[41] gives preferred conditions as: pressure 200 to 3000 psig and LHSV 0.2 to 10, with conditions for a specific example as 1 LHSV charge rate and 2200 psig pressure, 8000 scf/bbl recycle hydrogen. Model compound studies by Chevron include 1000 psig pressure, WHSV 2.8, 16 moles H_2/mole of feed, and temperatures of 330°C to 400°C. ExxonMobil cites hydrogen consumption as being 100 to 400 scf/bbl,[42] together with pressure as "high" and space velocity as "low."[43]

TABLE 10.15
Chevron Dewaxing Studies by Solvent and Using Hydroisomerization Catalysts

	Solvent Dewaxed	ICR 408	ICR 418
Group III: 100N Waxy Feed			
Base oil yield	Base	Base + 5.5%	Base + 11%
Pour point, °C	−15	−16	−15
Viscosity, cSt at 100°C	4.2	4.1	4.1
VI	129	130	131
Group II: Waxy 150N Feed			
Base oil yield	90	91	93.5
Pour point, °C	−11	−12	−15
Viscosity, cSt at 100°C	5.3	5.4	5.3
VI	104	105	107
Waxy 500N Feed			
Base oil yield	80	92	94
Pour point, °C	−18	−18	−20
Viscosity, cSt at 100°C	11.1	10.6	10.5
VI	106	111	113

Bright Stock A		
	Solvent Dewaxed	ICR 408/ICR 418
Base oil yield	48	91
Pour point, °C	−20	−19
Viscosity, cSt at 100°C	30.4	27.8
VI	106	114

Bright Stock B		
	Solvent Dewaxed	ICR 408	ICR 408
Base oil yield	33.5	95.5	93.4
Pour point, °C	−15	−13	−22
Viscosity, cSt at 100°C	30.9	29.4	29.1
VI	104	116	114

Source: K. R. Krishna, A. Rainis, P. J. Marcantonio, J. F. Mayer, J. A. Biscardi, and S. I. Zones, "Next Generation Isodewaxing® and Hydrofinishing Technology for Production of High Quality Base Oils," Paper LW-02-128, presented at the Lubricants and Waxes meeting, National Petroleum Refiners Association, Houston, Texas, November 14–15, 2002. Table copyrighted by Chevron Corporation and used with permission.

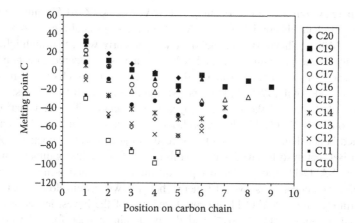

FIGURE 10.23 Dependence of the melting points of isomeric methyl substituted alkanes on the position on the chain of the methyl group.
Source: K. J. Burch and E. G. Whitehead, "Melting Points of Alkanes," *Journal of Chemical and Engineering Data* 49:858–863 (2004).

10.2.4 HYDROISOMERIZATION: MODEL COMPOUND STUDIES

A number of studies on model compounds such as n-octane, n-hexadecane, and n-C$_{24}$ were reported from Chevron Research and Technology[44–47] using their successful SAPO-11 catalyst. SAPO-11 has an intermediate pore size with a one-dimen-

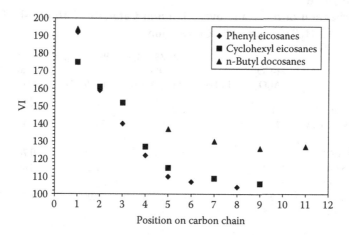

FIGURE 10.24 Dependence of the viscosity index of substituted n-paraffins on the position of the substituent.
Source: R. T. Sanderson, "Viscosity-Temperature Characteristics of Hydrocarbons," *Industrial and Engineering Chemistry* 41:368 (1949).

sional pore system with a free opening of 6.7 Å × 4.4 Å. (ZSM-5's dimensions for the straight channels are 5.3 Å × 5.6 Å and for the sinusoidal are 5.1 Å × 5.5 Å).

S. J. Miller (Chevron) published results from early work that highlighted the selectivity of the platinum form of SAPO-11 catalyst compared to a number of others. These others were amorphous silica-alumina, from which one would expect little or no selectivity, ZSM-5, HY, and Na-Beta zeolites. All the catalysts carried 1 wt. % platinum and the feed employed was n-octane. He found that at 30% conversion, only SAPO-11, the amorphous silica-alumina, and the HY catalysts exhibited better than 94% selectivity for feed isomerization to isooctanes. ZSM-5 and Na-Beta catalysts behaved poorly in this regard. Selectivity for dimethylhexanes was low. SAPO-11 also produced equal quantities of 2- and 3-methyl heptanes, whereas the other catalysts favored 3-methyl heptane, with a ratio close to that favored by thermodynamics. SAPO-11 also produced one of the lowest levels of doubly-branched hexanes (Table 10.16[46]) and the predominant ones formed were those separated by more than one carbon—only minor amounts of the less thermally stable (bond breaking here can produce tertiary carbonium ions) geminal-dimethyl (2,2 and 3,3-) ones were formed. Noble metal presence was a key to success since replacement of the hydrogenation metal platinum by palladium did not alter the isomeri-zation selectivity much, but replacement by nickel led to very poor isomerization.

With the larger n-hexadecane (n-C_{16}) molecule, just below the start of the lube carbon number range of about C_{20}, as feed, SAPO-11 was found to be clearly superior to the other catalysts selected since at 94% conversion, selectivity for isomerization was about 85% versus about 70% for the Pt-SiO_2-Al_2O_3 catalyst

TABLE 10.16
Isomerization of n-Octane over Platinum Catalysts at 1000 psig, 2.8 WHSV, 16 H_2/HC, and 30% Conversion

Catalyst	Pt-SiO_2-Al_2O_3	Pt-HY	Pt-ZSM-5, 80 SiO_2/Al_2O_3	Pt-ZSM-5, 650 SiO_2/Al_2O_3	Pt-Na-BETA	Pt-SAPO-11
Temperature, °C	371	257	260	343	367	331
iso-C8 selectivity, wt. %	96.4	96.8	56.5	58.4	74.3	94.8
2MC7/3MC7	0.67	0.71	1.54	0.88	0.70	1.07
C3 + C5/C4, mole ratio	0.95	0.64	2.1	1.2	0.68	1.0
Iso-C4/n-C4	0.96	3.5	1.1	0.96	1.7	0.92
DMC6 selectivity, wt. %	8.5	12	1.8	5.6	10	2.3

Source: S. J. Miller, "New Molecular Sieve Process for Lube Dewaxing by Wax Isomerization," Paper presented at the Symposium on New Catalytic Chemistry Utilizing Molecular Sieves, 206th National Meeting of the American Chemical Society, Aug 23–27, 1993. Table copyrighted by the Chevron Corporation and used with permission.

TABLE 10.17
Isomerization of Hexadecane over Platinum Catalysts at 1000 psig, 3.1 WHSV, and 30 H_2/HC

Catalyst	HY	ZSM, 80 SiO_2-Al_2O_3	ZSM-5, 650 SiO_2-Al_2O_3		SiO_2-Al_2O_3	SAPO-11
Conversion, wt. %	60	95	70	93	94	94
Temperature, °C	246	260	349	368	361	340
Iso-C_{16} selectivity, wt. %	59.7	1.0	2.8	2.5	69.2	84.7

Source: S. J. Miller, "New Molecular Sieve Process for Lube Dewaxing by Wax Isomerization," Paper presented at the Symposium on New Catalytic Chemistry Utilizing Molecular Sieves, 206th National Meeting of the American Chemical Society, Aug 23–27, 1993. Table copyrighted by the Chevron Corporation and used with permission.

(Table 10.17). From a separate experiment at low pressure it was concluded that the stability of SAPO-11 was much superior to that of the Pt-SiO_2-Al_2O_3 catalyst.

Product distribution[45] from n-hexadecane at 96% conversion over SAPO-11 produced 53% monomethyl pentadecane isomers, 30% dimethyl tetradecanes, and a pour point of −51°C, whereas over Pt-SiO_2-Al_2O_3 there was just 22% monomethyl products, nearly 40% dimethyl ones, and a pour point of −28°C (Table 10.18).

TABLE 10.18
Isomerization of Hexadecane at 1000 psig, 3.1 WHSV, 30 H_2/HC, and 96% Conversion

Catalyst	Pt-SAPO-11	Pt-SiO_2-Al_2O_3
Temperature, °C	340	360
Isomerization selectivity, wt. %	85	64
C_{16} product composition, wt. %		
2M-C_{15}	7.7	3.3
3M-C_{15}	8.1	3.2
4M-C_{15}	7.5	3.1
5M-C_{15}	7.1	3.2
6M-C_{15}	22.9 (includes 6M)	2.7
7- + 8-M-C_{15}		6.1
Total M-C_{15}	53.3	21.6
NC_{16}	4.7	6.0
DM-C_{14}	29.8	37.8
Other C_{16}	12.2	34.6
Pour point, °C	−51	−28

Source: S. J. Miller, "New Molecular Sieve Process for Lube Dewaxing by Wax Isomerization," Microporous Materials 2:439–449 (1994). With permission.

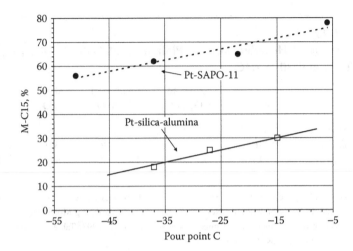

FIGURE 10.25 Hydroisomerization of n-hexadecane over Pt-SAPO-11 and Pt-silica-alumina catalysts: monosubstituted methyl pentadecane contents versus pour point.
Source: S. J. Miller, "Studies on Wax Isomerization for Lubes and Fuels," *Studies in Surface Science and Catalysis* 84:2319–2326 (1994). With permission.

The SAPO-11 catalyst was clearly superior, and among the monomethyl penta-decanes, the distribution of the methyl substituent along the carbon chain did not appear to favor any particular position.

Furthermore, it was found for this isomerization that SAPO-11 gave the higher content of the mono-substituted methyl-C_{15} isomers over a range of pour points (Figure 10.25) and the number of branches was minimized with this catalyst (Figure 10.26). Miller's concept of the isomerization process is that the 2- isomer is produced first, which leads to the other single methyl isomers by methyl migration down the chain. The monomethyl isomers can then subsequently react further catalytically to form 2-, 3-, and 3+ branched isomers.

Also studied was a comparison of the products from Pt-SAPO-11 and Pt-silica-alumina, with the higher molecular weight (right in the lube range) n-C_{24} as feed at 99% conversion in which Pt-SAPO-11 gave a significantly higher lube (316+°C) yield and VI, although the pour point was not quite as good (Table 10.19).

Isomer ratios (Table 10.20) for the isomerization on n-C_{24} by Isodewaxing to −15°C and by the Pt-silica-alumina catalyst to +22°C, then solvent dewaxed to −15°C, show that from the Pt-SAPO-11 catalyst, the 4-methyl isomer amounts to about 50% more than the 2-methyl case and from the Pt-silicon-alumina, the 4-methyl constitutes are three times as much as the 2-methyl. In the case of n-octane, Pt-SAPO-11 gave about equal quantities of the 2- and 3-isomers, and a similar situation occurred with n-hexadecane, where the 2-, 3-, 4-, and 5- isomer yields were just about equal.

Taylor and Petty of Texaco (now part of Chevron) examined Pd-SAPO-11 together with a number of other palladium-containing zeolites.[48] ZSM-5, SAPO-11,

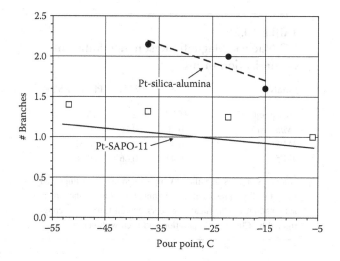

FIGURE 10.26 Hydroisomerization of n-hexadecane over Pt-SAPO-11 and Pt-silica-alumina catalysts: pour point versus the average number of branches.
Source: S. J. Miller, "Studies on Wax Isomerization for Lubes and Fuels," *Studies in Surface Science and Catalysis* 84:2319–2326 (1994). With permission.

TABLE 10.19
Yields in Isomerization of n-C$_{24}$ over Pt-SAPO-11 and Pt-Silica-Alumina

Catalyst	Pt-SAPO-11	Pt-Silica-Alumina
n-C$_{24}$ conversion, wt. %	99.4	99.1
Yields, wt. %		
C1–C2	0	0
C3	0.1	0.3
C4	0.6	0.8
C5–82°C	2.3	3.2
82–177°C	2.5	5.1
177–316°C	8.4	23.6
316+°C	86.1	67.0
316+°C inspections		
Pour point, °C	−35	−43
Viscosity at 100°C	2.49	2.41
VI	126	112

Source: S. J. Miller, "Wax Isomerization for Improved Lube Quality," presented at the annual meeting of the American Institute of Chemical Engineers, March 1998. Table copyrighted by Chevron Corporation and used with permission.

TABLE 10.20
^{13}C Nuclear Magnetic Resonance Ratios in Isomerization of n-C$_{24}$

Catalyst	IDW	Pt/S-A	Pt/S-A + SDW
Pour point, °C	−15	+22	−15
2M/5+-M	0.30	0.26	0.22
3M/5+-M	0.33	0.33	0.31
4M/5+-M	0.44	0.66	0.66

Source: Source: S. J. Miller, "Wax Isomerization for Improved Lube Quality," presented at the annual meeting of the American Institute of Chemical Engineers, March 1998. Table copyrighted by Chevron Corporation and used with permission.

and Beta are all medium-sized pore catalysts and have 10-membered rings, while mordenite, USY, and SDUSY have 12-membered rings and, as previously mentioned, are considered large pore catalysts. Their investigations using n-hexadecane concluded that of these catalysts, only Pd-SAPO-11 was able to give high yields of liquid product and minimized cracking reactions. Figure 10.27 shows a plot of the percent C$_{16}$ isomerized versus the percent C$_{16}$ converted and it can be seen that all the other catalysts fall short of the Pd-SAPO-11 at high conversion.

To gauge the selectivity of the catalysts ability to distinguish between n-paraffins and isoparaffins in the feed, runs were performed with a 50/50 blend of n-hexadecane/iso-C$_{19}$ (2,6,10,14-tetramethylpentadecane). Figure 10.28, a plot of

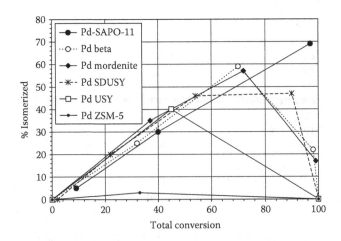

FIGURE 10.27 Study on isomerization of n-C$_{16}$ over lead zeolites.
Source: R. J. Taylor and R. H. Petty, "Selective Hydroisomerization of Long Chain Normal Paraffins," *Applied Catalysis A: General* 119:121–138 (1994). With permission.

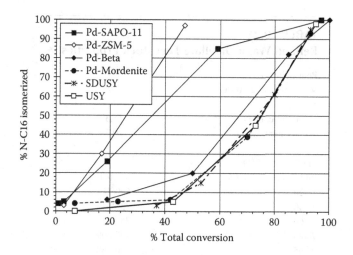

FIGURE 10.28 Study on hydroisomerization of an n-C16 and iso-C19 blend.
Source: R. J. Taylor and R. H. Petty, "Selective Hydroisomerization of Long Chain Normal Paraffins," *Applied Catalysis A: General* 119:121–138 (1994). With permission.

percent isomerized C_{16} product formed versus percent total conversion, shows that only the two intermediate pore size catalysts show selectivity for reacting with the n-C_{16} in the feed. The other large pore catalysts convert the more reactive iso-C_{19} molecules first by cracking to smaller molecules.

However, in spite of the promising features discovered for this form of SAPO-11, when applied to an actual waxy feed, the authors found the catalyst to be uncompetitive in terms of yield with solvent dewaxing (no VI information was published). The feed in this case was a nominal 4 cSt refined waxy distillate (Table 10.21) with 21% wax and 15% n-paraffins. The experimental results were disappointing with this form of the SAPO-11 catalyst in that yields were consistently poorer (Figure 10.29) than for solvent dewaxing and indeed were similar to those from the ZSM-5 employed.

10.2.5 ExxonMobil MWI Process

ExxonMobil developed two hydroprocessing technologies for isomerization of slack waxes and similar very high wax streams to highly paraffinic 140+ VI base stocks. These were used at their Fawley, England, refinery and illustrate aspects of the development of catalyst technology.

The first process developed employed an amorphous catalyst reputedly with high fluoride levels to increase acidity and bring about partial hydroisomerization. Since wax conversion was incomplete, the final step was solvent dewaxing to remove unconverted wax and complete achievement of the target pour point. A preliminary hydrotreatment step was used to remove sulfur and nitrogen.

TABLE 10.21
Refined Waxy Distillate Feedstock Properties

Pour point, °F	80
Viscosity, cSt at 100°C	3.808
VI	111
n-Paraffin content, wt. %	15.3
Wax content, wt. %	20.7
SimDis, wt. %, °F	
Initial boiling point/5	605/670
10/20	690/713
30/40	730/745
50/60	759/774
70/80	790/809
90/95	840/868
Final boiling point	950

Source: R. J. Taylor and R. H. Petty, "Selective Hydroisomerization of Long Chain Normal Paraffins," *Applied Catalysis A: General* 119:121–138 (1994). With permission.

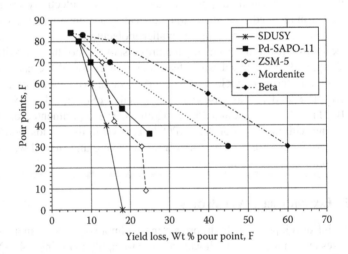

FIGURE 10.29 Pour points versus yield loss for solvent dewaxing and using five hydro-isomerization catalysts.
Source: R. J. Taylor and R. H. Petty, "Selective Hydroisomerization of Long Chain Normal Paraffins," *Applied Catalysis A: General* 119:121–138 (1994). With permission.

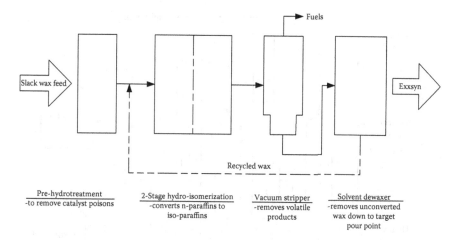

FIGURE 10.30 Schematic of ExxonMobil's initial plant at Fawley, England, for wax isomerization.

Source: T. T. Releford and K. J. Ball, "Exxon's New Synthetic Basestocks—Exxsyn," Paper FL-93-117, presented at the National Fuels and Lubricants meeting, Houston, Texas, November 4–5, 1993. With permission.

Figure 10.30 is a simplified schematic of this process[49] to produce an Exxsyn 6 base stock.

In 2003 the process was adapted to use the much more effective zeolite hydroisomerization catalysts MWI-1 and MWI-2; the initial change was to replace the solvent dewax unit by a catalytic dewax one employing MWI-1 catalyst. Subsequently a variation on their MSDW catalyst replaced the amorphous one, and since isomerization was now complete, no final wax removal step was needed any longer. This modified schematic[50] is shown in Figure 10.31, and it retains the cleanup unit at the front, after which there is an interstage removal of any H_2S and NH_3, followed by the isomerization reactor, and an exchanger to adjust temperature prior to the noble metal hydrofinisher unit (MAXSAT). Hydrogen consumption is given as between 200 and 400 scf/bbl, producing base stocks of 50% to 70% yield with VIs between 130 and 160.[42] The products from light and heavy waxes are termed Visom 4 and 6. Properties of these are given in Table 10.22.

To put the foregoing in some perspective, wax isomerization has had a long history, beginning with work using such acid catalysts as aluminum chloride (Friedel-Crafts-type catalyst) in the presence of a "cracking suppressor," such as decahydronaphthalene or methylcyclohexane, and in the absence of any hydrogen.[51] By this means, and after a dewaxing step to remove unconverted wax, base stocks with VIs in the range of 130 to 160 (ASTM D567) were obtained in rather poor yields (15% to 30%). For example, pure $n-C_{25}H_{52}$ gave a product with a viscosity at 210°F of 2.88 cSt, pour point of −10°C, and VI of 130.

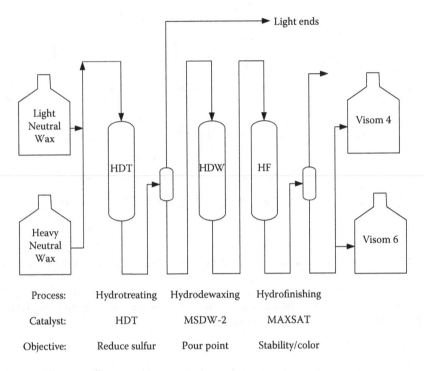

FIGURE 10.31 Schematic of ExxonMobil's updated Fawley, England, plant for wax isomerization to base stocks.
Source: W. B. Genetti, A. B. Gorshteyn, A. Ravella, T. L. Hilbert, J. E. Gallagher, C. L. Baker, S. A. Tabak, and I. A. Cody, "Process Options for High Quality Base Stocks," presented at the 3rd Russian Refining Technical Conference, Moscow, Russia, September 25–26, 2003. With permission.

Further work by Shell[52] found that isomerization could be accomplished in a hydrotreating step using platinum on alumina catalyst, preferred conditions being temperatures in the range of 375°C to 490°C and pressures of 300 to 1000 psig. While VIs were comparable to those obtained previously, yields of oil after dewaxing and fractionation were still disappointing. More detailed studies[53] with other noble or nonnoble catalysts of this type showed that oil yields could reach about 60%, but VIs fell sharply with the extent of the reaction, decreasing from an initial 145 to about 100 at 70% conversion to cracked products. As reaction progressed, initial high VI products must undergo further isomerization to both lower VI products and cracked material. More recent work[54] with platinum on silica-alumina with a high silica content showed similar yield variation with wax conversion and high VIs (greater than 140), but a solvent dewaxing step was still necessary.

TABLE 10.22
Products from ExxonMobil Wax to Lubes Process, MWI

	Visom 4	Visom 6	Exxsyn 4	MWI 650+°F	MWI-2 Feed	A	B	C
Feed								
Viscosity at 100°C				4.2	8.4			
Percent oil				8				
Product								
Yield				64	100	65	50	35
Viscosity at 100°C	3.9-4.1	6.4-6.8	4.0	3.8		6.8	6.5	6.3
VI	135-140	144	140	147		168	158	146
Noack, mass %		15.2				10.7	13	15
Pour point, °C	−15	−8	−18			−21	−39	<−65
Cloud point, °C						+8	11	33
CCS at 35°C	1600	8500						

Source: A. Sapre, "ExxonMobil Advanced Technology: A Key to Clean Fuels and Premium Lubricants," presented at the 6th European Fuels Conference, Paris, France, March 2005. T.E. Helton, T.F. Degnan, Jr., D.N. Mazzone, M.P. McGuiness, T.L. Hilbert and R.C. Dougherty, "Catalytic Hydroprocessing a Good Alternative to Solvent Processing," *Oil & Gas J.,* July 20, pp. 58–67 (1998). With permission.

REFERENCES

1. P. B. Weise and V. J. Frilette, "Intracrystalline and Molecular-Shape-Selective Catalysis by Zeolite Salts," *Journal of Physical Chemistry* 64:382 (1960).
2. J. D. Sherman, "Synthetic Zeolites and Other Microporous Oxide Molecular Sieves," *Proceedings of the National Academy of Science USA* 36:3471–3478 (1999).
3. M. P. Ramage, K. R. Graziani, and J. R. Katzer, "Science and Application of Catalytic Lube Oil Dewaxing," presented at the Japan Petroleum Institute meeting, Tokyo, Japan, October 27–28, 1986.
4. P. B. Venuto, "Organic Catalysis Over Zeolites: A Perspective on Reaction Paths within Micropores," *Microporous Materials* 2:297–411 (1994).
5. N. Y. Chen and W. E. Garwood, "Some Catalytic Properties of ZSM-5, a New Shape Selective Zeolite," *Journal of Catalysis* 52:453–458 (1978).
6. J. J. Wise and J. R. Katzer, "Catalytic Dewaxing in Petroleum Processing," presented at the American Chemical Society annual meeting, April 13–18, 1986.
7. L. D. Rollmann and D. E. Walsh, "Shape Selectivity and Carbon Formation in Zeolites," *Journal of Catalysis* 56:139–140 (1979).

8. N. Y. Chen and W. E. Garwood, "Selective Hydrocracking of n-Paraffins in Jet Fuels," *Industrial and Engineering Chemistry Process Design and Development* 19:315–318 (1978).

9. N. Y. Chen, J. L. Schlenker, W. E. Garwood, and G. T. Kokotailo, "TMA-Offretite. Relationship Between Structural and Catalytic Properties," *Journal of Catalysis* 86:24–31 (1984).

10. N. Y. Chen and S. J. Lucki, "Determination of n-Paraffins in Gas Oils by Molecular Sieve Adsorption," *Analytical Chemistry* 42:508–510 (1970).

11. J. G. O'Connor and M. S. Norris, "Molecular Sieve Adsorption, Application to Hydrocarbon Type Analysis," *Analytical Chemistry* 32:701–706 (1960).

12. J. G. O'Connor, F. H. Burow, and M.S. Norris, "Determination of Normal Paraffins in C_{20} to C_{32} Paraffin Waxes by Molecular Sieve Adsorption," *Analytical Chemistry* 34(1):82–85 (1962).

13. J. V. Brunnock, "Separation and Distribution of Normal Paraffins from Petroleum Heavy Distillates by Molecular Sieve Adsorption and Gas Chromatography," *Analytical Chemistry* 38:1648–1652 (1966).

14. J. V. Mortimer and L. A. Luke, "The Determination of Normal Paraffins in Petroleum Products," *Analytica Chimica Acta* 38:119–126 (1967).

15. J. R. Marquart, G. B. Dellow, and E. R. Freitas, "Determination of Normal Paraffins in Petroleum Heavy Distillates by Urea Adduction and Gas Chromatography," *Analytical Chemistry* 40:1633–1637 (1968).

16. F. A. Smith, "Mobil Lube Oil Dewaxing (MLDW) Technology," presented at the Texaco Lubricating Oil Symposium, May 18, 1982.

17. C. N. Rowe and J. A. Murphy, "Low-Temperature Performance Advantages for Oils Using Hydrodewaxed Base Stocks," Paper 831715, presented at the Fuels and Lubricants meeting, San Francisco, October 31–November 3, 1983.

18. T. E. Helton, T. F. Degnan, Jr., D. N. Mazzone, M. P. McGuiness, T. L Hilbert, and R. C. Dougherty, "Catalytic Hydroprocessing a Good Alternative to Solvent Processing," *Oil and Gas Journal* July 20:58–67 (1998).

19. K. W. Smith, W. C. Starr, and N. Y. Chen, "New Process Dewaxes Lube Base Stocks," *Oil and Gas Journal* May 26:75–85 (1980).

20. K. W. Smith, W. C. Starr, and N. Y. Chen, "A New Process for Dewaxing Lube Basestocks: Mobil Lube Dewaxing," *Proceedings of the American Petroleum Institute Meeting, Refining Department* 59:151 (1980).

21. R. G. Graven and J. R. Green, "Hydrodewaxing of Fuels and Lubricants Using ZSM-5 Type Catalysts," presented at the Australian Institute of Petroleum 1980 Congress, Sidney, Australia, September 14–17, 1980.

22. A. Sequeira, Jr., *Lubricant Base Oil and Wax Processing* (New York: Marcel Dekker, 1994).

23. R. J. Taylor and A. J. MacCormack, "Study of Solvent and Catalytic Lube Oil Dewaxing by Analysis of Feedstocks and Products," *Industrial and Engineering Chemistry Research* 31:1731–1738 (1992).

24. T. R. Farrell and J. A. Zakarian, "Lube Facility Makes High-Quality Lube Oil from Low-Quality Feed," *Oil and Gas Journal* May 19:47–51 (1986).

25. J. A. Zakarian, R. J. Robson, and T. R. Farrell, "All-Hydroprocessing Route for High-Viscosity Index Lubes," *Energy Progress* 7:59–64 (1987).

26. J. A. Zakarian and J. N. Ziemer, "Catalytic Dehazing of Heavy Lube Oil: A Case History," *Energy Progress* 8:109–112 (1988).

27. D. J. O'Rear and B. K. Lok, "Kinetics of Dewaxing Neutral Oils Over ZSM-5," *Industrial and Engineering Chemistry Research* 30:1105–1110 (1991).

28. M. W. Wilson, K. L. Eiden, T. A. Mueller, S. D. Case, and G. W. Kraft, "Commercialization of Isodewaxing—A New Technology for Dewaxing to Manufacture High-Quality Lube Basestocks," Paper FL-94-112, presented at the National Fuels and Lubricants meeting, National Petroleum Refiners Association, Houston, Texas, November 30, 1994.

29. C. H. Baker, M. P. McGuiness, "Mobil Lube Dewaxing Technologies," Paper AM-95-56, presented at the annual meeting of the National Petroleum Refiners Association, San Francisco, March 19–21, 1995.

30. A. Sapre, "ExxonMobil Advanced Technology: A Key to Clean Fuels and Premium Lubricants," presented at the 5th European Fuels Conference, Paris, France, March 15–17, 2004.

31. K. R. Krishna, A. Rainis, P. J. Marcantonio, J. F. Mayer, J. A. Biscardi, and S. I. Zones, "Next Generation Isodewaxing® and Hydrofinishing Technology for Production of High Quality Base Oils," Paper LW-02-128, presented at the Lubricants and Waxes meeting, National Petroleum Refiners Association, Houston, Texas, November 14–15, 2002.

32. ExxonMobil Research and Engineering, MWI™—Wax Isomerization Technology, brochure (Fairfax, VA: ExxonMobil).

33. J. Mayer, D. Brossard, K. Krishna, and B. Srinivasan, "The All-Hydroprocessing Route. Group II and Group III Base Oils: One Company's Experience," Paper AM-04-68, presented at the annual meeting of the National Petroleum Refiners Association, San Antonio, Texas, March 21–23, 2004.

34. W. Qureshi, L. Howell, C.-W. Hung, and J. Xiao, "Isodewaxing—Improving Refining Economics," *Petroleum Technology Quarterly* Summer:17–23 (1996).

35. S. J. Miller, J. Xiao, and J. M. Rosenbaum, "Application of Isodewaxing, a New Wax Isomerization Process for Lubes and Fuels," *Science and Technology in Catalysis*:379-382 (1994).

36. J. E. Gallagher, Jr., I. A. Cody, S. A. Tabak, R. G. Wuest, A. A. Claxton, L. Loke, and C. T. Tan, "New ExxonMobil Process Technology for Producing Lube Basestocks," paper presented at the Asia Pacific Refining Technology Conference, Kuala Lumpur, Malaysia, March 9, 2000.

37. A. Ravella, "Manufacturing High Quality Basestocks. Chemical Reaction Engineering IX: Meeting the Challenges for New Technology," presented at, Quebec City, Quebec, Canada, June 29–July 4, 2003.

38. M. Daage, "Baseoil Production and Processing," available at http://www.prod. exonmobil.com/refiningtechnologies/pdf/base_oil_refining_lubes_daage_france 070601. pdf.

39. K. J. Burch and E. G. Whitehead, "Melting Points of Alkanes," *Journal of Chemical and Engineering Data* 49:858–863 (2004).

40. R. T. Sanderson, "Viscosity-Temperature Characteristics of Hydrocarbons," *Industrial and Engineering Chemistry* 41:368 (1949).

41. S. J. Miller, "Catalytic Dewaxing Process Using a Silicoaluminophosphate Molecular Sieve," U.S. Patent 4,859,311.

42. A. Sapre, "ExxonMobil Advanced Technology: A Key to Clean Fuels and Premium Lubricants," presented at the 6th European Fuels Conference, Paris, France, March 2005.

43. P. Kamienski, "Technology for High Quality Basestocks and Finished Waxes," presented at the 7th Annual Roundtable—Central and Eastern European Refining and Petrochemicals, Prague, Czech Republic, October 19, 2004.

44. S. J. Miller, "New Molecular Sieve Process for Lube Dewaxing by Wax Isomerization," presented at the Symposium on New Catalytic Chemistry Utilizing Molecular Sieves, Division of Petroleum Chemistry, 206th National Meeting of the American Chemical Society, August 22–27, 1993.

45. S. J. Miller, "Studies on Wax Isomerization for Lubes and Fuels," *Studies in Surface Science and Catalysis* 84:2319–2326 (1994).

46. S. J. Miller, "New Molecular Sieve Process for Lube Dewaxing by Wax Isomerization," *Microporous Materials* 2:439–449 (1994).

47. S. J. Miller, "Wax Isomerization for Improved Lube Quality," presented at the annual meeting of the American Institute of Chemical Engineers, March 1998.

48. R. J. Taylor and R. H. Petty, "Selective Hydroisomerization of Long Chain Normal Paraffins," *Applied Catalysis A: General* 119:121–138 (1994).

49. T. T. Releford and K. J. Ball, "Exxon's New Synthetic Basestocks—Exxsyn," Paper FL-93-117, presented at the National Fuels and Lubricants meeting, Houston, Texas, November 4–5, 1993.

50. W. B. Genetti, A. B. Gorshteyn, A. Ravella, T. L. Hilbert, J. E. Gallagher, C. L. Baker, S. A. Tabak, and I. A. Cody, "Process Options for High Quality Base Stocks," presented at the 3rd Russian Refining Technical Conference, Moscow, Russia, September 25–26, 2003.

51. R. J. Moore and B. S. Greensfelder, "Hydrocarbon Conversion," U.S. Patent 2,475,358.

52. G. M. Good, J. W. Gibson, and B. S. Greensfelder, "Isomerization of Parrafin Wax," U.S. Patents 2,668,866 and 2,668,790.

53. F. Breimer, H. I. Waterman, and A. B. R. Weber, "Hydroisomerization of Paraffin Wax," *Journal of the Institute of Petroleum* 43(407):297–306 (1956).

54. V. Calemma, S. Peratello, C. Perego, A. Moggi, and R. Giardino, "Hydroisomerization of Slack Wax Over Pt/Amorphous SiO_2-Al_2O_3 Catalyst to Produce Very High Viscosity Index Lubricating Base Oils," presented at Worldwide Perspectives on the Manufacture, Characterization and Application of Lubricant Base Oils: IV, Division of Petroleum Chemistry, 218th National Meeting of the American Chemical Society, New Orleans, Louisiana, August 22–26, 1999.

11 Technical and Food Grade White Oils and Highly Refined Paraffins

11.1 WHITE OILS

11.1.1 INTRODUCTION

The term "white oil" refers to highly refined distillate fractions in the lubes boiling range whose water white color (and therefore the "white" descriptor) is due to the almost complete absence of aromatics as well as sulfur- and nitrogen-containing compounds. White mineral oils, also known as "paraffin oil," "liquid paraffin," and "white mineral oil," are liquids at room temperature and are predominantly mixtures of isoparaffins and naphthenes with lesser amounts of n-paraffins. White oils are manufactured for use in agriculture and the chemicals and plastics, textiles, food, pharmaceuticals, and personal care and cosmetics industries, and their purity is regulated in most countries. The manufacturing objective is to produce oils of high purity and low toxicity with the composition, as mentioned above, being almost entirely saturated hydrocarbons. Toxicity specifications require polynuclear aromatic hydrocarbons (PAHs) to be at very low levels.

White mineral oils were first developed by a Russian chemist, J. Markownikoff, and the first plant for their manufacture was set up in Riga, Latvia, around 1885. When European supplies to the United States were cut off during World War I, the L. Sonneborn Company was the first U.S. company to begin to manufacture them and used Pennsylvanian crude.[1] This was later followed by the Pennsylvania Refining Company (Penreco) and many others.[2,3] The major North American manufacturers now are Sonneborn, Lyondell-Citgo, Penreco, and Petro-Canada.[4]

White oils are of either "technical" or "food/medicinal" grade, with the food/medicinal grade having tighter specifications and therefore requiring more stringent processing. Technical grade white oils are employed as components of nonfood articles intended for use in contact with food[5] (e.g., in food machinery lubricants) and in the United States are governed by Food and Drug Administration (FDA) regulations (21 CFR 178.3620(b)). For technical grade white oils, color must be better than 20 on the Saybolt scale (ASTM D156), however, most technical grade material made today is +30, the same as food grade.[3] To control levels of polynuclear aromatics (PNAs), ultraviolet (UV) absorbance measured on dimethyl sulfoxide extracts must be lower than the values in Table 11.1.

TABLE 11.1
U.S. FDA UV Absorbance Limits for Technical
Grade White Oils

Wavelength Range	Maximum Absorbance per Centimeter of Path Length
280–289 nm	4.0
290–299 nm	3.3
300–329 nm	2.3
330–350 nm	0.8

Source: "Indirect Food Additives: Adjuvants, Production Aids and Sanitizers. Mineral Oil," 21 CFR 178.3620. With permission.

Food/medicinal grade specifications (21 CFR 172.878) are designed such that products meeting these specifications can be safely used in food and pharmaceuticals.[6] The specifications control PNA levels by the UV absorption limits given in Table 11.2 and by the carbonizable substances test (ASTM D565). These are discussed in more detail later in the chapter, and these are the difficult specifications to meet.

Food/medicinal oils are frequently referred to as meeting United States Pharmacopeia (USP) or National Formulary (NF) specifications, usually written as "meets USP/NF specifications." USP and NF specifications differ only in specific gravity and viscosity. USP oils must have specific gravities between 0.845 and 0.905 at 25°C and have viscosities greater than 34.5 cSt at 40°C. NF oils must have viscosities less than 33.5 cSt at 40°C and must have densities between 0.818 and 0.880 at 25°C. Further details on specifications are provided later.

In addition, food grade white oils must satisfy the following:

- The product must be water white, with a Saybolt color of +30.
- There must be no taste or odor.
- The oil must be neutral, being neither acidic or basic.

TABLE 11.2
U.S. FDA UV Absorbance Limits for Food/Medicinal Grade White Oils

Wavelength Range	Maximum Absorbance Per Centimeter of Path Length
260–350 nm	0.1

Source: "Food Additives Permitted for Direct Addition to Food for Human Consumption. White Mineral Oil," 21 CFR 172.878. With permission.

- Sulfur compounds must be sufficiently low to pass the lead oxide test.
- The oil must pass the solid paraffin test at 0°C (this is essentially a cloud point test) by remaining clear at this temperature. This is established in the dewaxing step.

These all attest to the stringency of the process itself and also that of the downstream handling, storage, and shipping procedures.

11.1.2 MANUFACTURE BY ACID TREATMENT

The traditional method of manufacturing is by a separation method (like most traditional methods) with a solvent extracted and dewaxed base oil as the usual feedstock. By reaction of either a paraffinic or a naphthenic base oil with oleum, aromatics are converted to the corresponding sulfonic acids and separate out. This may require several mixing steps to complete reaction of all the aromatics. Paraffins and naphthenes are essentially unaffected by the acid treatment.[7] After separation of the sulfonic acids, the remaining oil, now free of aromatics, is neutralized with caustic, washed with alcohol, and finally treated with bauxite or clay. Figure 11.1 provides a schematic for this type of process.

White oils, containing no significant levels of aromatics or sulfur-containing compounds, have no antioxidant capability and will undergo oxidation to peroxides and hydroperoxides readily and at the same time show no visible signs of chemical change. Therefore storage conditions should be maintained at ambient temperatures. Because of this instability, addition of an antioxidant such as butylated hydroxytoluene (BHT) or vitamin E is permitted in some jurisdictions.

White oils are referred to as naphthenic or paraffinic depending on their crude source. Alternatively it has been suggested that they may be defined by the percent type of carbon atoms, with paraffinic white oils having 50% or more paraffinic carbons and naphthenic oils having at least 40%.[8]

11.1.3 HYDROTREATMENT PROCESSES

11.1.3.1 Introduction

The traditional acid treatment process is still employed but suffers from poor yields, the economic disadvantages of being a batch process, and the use of oleum, a hazardous chemical. Furthermore, the process generates waste materials (acid sludge, caustic, clay) whose disposal has become increasingly difficult and expensive. The poor yields are due to this being a "carbon-rejection" process in which the aromatic hydrocarbons are removed during the process rather than being converted. Licensors of hydrotreating processes (e.g., Exxon,[9,10] Gulf,[11] [now Chevron], IFP,[12] Atlantic Richfield[13] [developers of the Duotreat process; now Lyondell], and BASF[14-16]) recognized that hydrotreatment could convert aromatics to naphthenes with only minor amounts of cracking and not only improve the process yields but also provide a continuous process with no by-products requiring landfill disposal. The hydrotreatment processes generally retained dewaxed

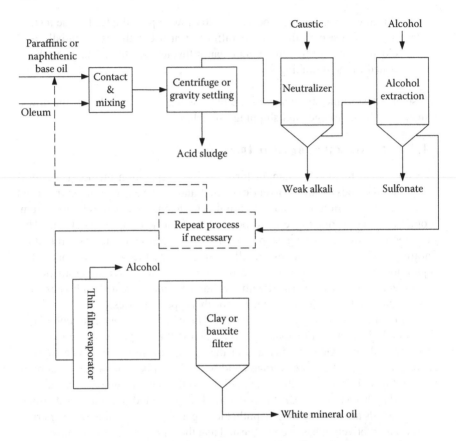

FIGURE 11.1 Acid treating process schematic for white oil production.
Source: J. Weeks, "The White Oil Industry in North America," Paper LW-98-131, presented at the Lubricants and Waxes meeting, National Petroleum Refiners Association, Houston, Texas, November 13–14, 1998. With permission.

solvent refined base stocks as feedstocks to minimize changes to the overall process. (It should be noted that food grade white oils can also be made from poly-alpha-olefins,[17,18] where the needs of specific applications justify the additional cost.)

The hydrotreatment processes developed to convert solvent refined lube base stocks to food grade white oils have all involved two stages, since current catalyst technology cannot reduce the high feed aromatics to near extinction in a single stage. Figure 11.2 is the flow scheme for the BASF two-stage process and can be taken as generally representative of most hydrotreatment processes for food grade white oils. The first stage performs preliminary aromatics hydrotreatment together with saturation of nearly all nitrogen and sulfur compounds to meet technical grade specifications and prepare the feed to meet the requirements for stable second-stage catalyst operation which removes remaining aromatics.

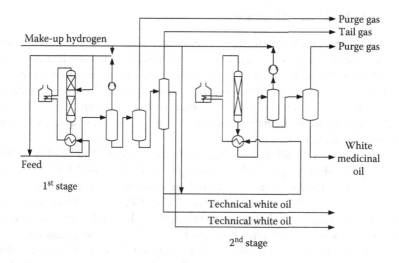

FIGURE 11.2 BASF process schematic for two-stage hydroprocessing route to food grade white oil manufacturing.
Source: W. Himmel, T. Anstock, R. Spahl, and K. Kussner, "White Oils and Fully Refined Paraffins," *Erdol und Kohle* 39:408–414 (1986). With permission.

Some white oil manufacturers have invested the capital necessary to construct both first- and second-stage units in their plants, completely replacing the old technology, while others (e.g., Penreco[3]) found that introduction of one stage only best suited their business. In a case where a first-stage hydrotreater is employed, the benefits predicted were reduced load on the acid extraction unit, less acid consumption and correspondingly less acid sludge produced, and improved yields.[11]

This process development work for hydrotreated white oils took place in the 1960s through the 1980s, when hydroprocessing was a relatively new technology. It should be recognized that, with the proliferation of high pressure catalytic lube units due to demand for group II and III base stocks, many more plants have the hardware necessary for manufacturing white oils than previously (e.g., SK Corporation in South Korea includes white oils among the products it manufactures, together with group II and III base oils, from a fuels hydro-cracker unit).[19,20] The relative absence of n-paraffins in these white oils due to dewaxing by hydroisomerization has been claimed to be an advantage. Toxicological issues with respect to n-paraffins in white oils are being investigated[21–23] and catalytic dewaxing by hydroisomerization has been proposed as a solution.[24] The product from SK's other Korean refiner, S-Oil, also meets food grade white oil quality.[25]

In these cases, white oil is an additional product to the base stocks. Of course, there is much more to white oil than the ability to produce it. Maintaining quality through product storage and delivery is not a simple undertaking.

11.1.3.2 First-Stage Operation

As in the traditional acid extraction process, the feedstock is generally dewaxed solvent refined base stock, since levels of the aromatics, polynuclear aromatics, and nitrogen and sulfur compounds are already reduced relative to a straight-run gas oil. This facilitates hydroprocessing by lightening the load on the catalysts and extending their lives. Equally important is that this is an already dewaxed feed, so the white oil producer does not have to bear the capital costs of crude fractionation and dewax units.

The first stage in all these processes has been reported to employ sulfided base metal catalysts (e.g., NiO/MoO on Al_2O_3 for BASF; CoO/MoO, NiO/MoO, or NiO/WO for the Exxon process) to reduce aromatics to the low levels that the second-stage unit can handle and essentially eliminate poisons (nitrogen, sulfur) that will affect the second-stage catalyst's activity. In BASF's process, the first stage operates at a high temperature (300°C to 380°C)[16] and pressure (8 to 15 MPa) and low enough space velocity (0.1 to 1.0 weight hourly) to achieve these objectives. These catalysts are tolerant of the nitrogen and sulfur in the feed. Since the needs of this stage are hydrodearomatization (HDA), hydrodesulfurization (HDS), and hydrodenitrification (HDN) only, any hydrocracking will cause yield loss. Therefore a nonacidic catalyst will normally be used for this to reduce cracking, and while cracking is minimized, some inevitably will occur.

From Figure 11.2 it can be seen that product from the first-stage high pressure separator flows via a high pressure separator to a low pressure separator where the product gases (e.g., hydrogen sulfide, ammonia, and light hydrocarbons) are disengaged. Liquid product is fractionated to separate low aromatic distillates—which will be water white or nearly so and contain only trace quantities of sulfur and nitrogen—from the bottoms stream and establish the viscosity and volatility of the white oil. Product may be sent to storage at this point for sale as a technical white oil or further processed in the second-stage hydrofinisher unit to higher value food/medicinal grade product. White oil papers, which include inspections on feeds to and products from first-stage white oil units, generally indicate that pour point increases by 2°F to 5°F through this stage. This might be expected from the poorer solvency of the technical grade product with most of the aromatics in the feed saturated.

11.1.3.3 Second-Stage Operation

The second stage is said to operate "cold" (i.e., greater than 150°C but less than 340°C) and universally employs a very active hydrogenating catalyst (e.g., a noble metal such as platinum or lead or a Raney nickel-type catalyst) whose purpose is to hydrogenate remaining aromatics, particularly polyaromatics. The "cold" operation is to keep the aromatic saturation temperature in the region of kinetic control, particularly for polyaromatics. At higher temperatures, thermodynamic control can take over and cause reversible formation of polyaromatics from three-ring and higher naphthenes. This eventuality would cause the product to fail specifications for polynuclear aromatics levels. If the reactor temperature is too low, the product may also fail specifications due to kinetic failure (i.e., insufficient removal of PNAs).

Process operating conditions for the BASF second-stage unit are given as a 120°C to 300°C reactor temperature, 10 to 20 MPa hydrogen partial pressure, and 0.1 weight hourly space velocity.[16] In the case of both stages, increased hydrogen partial pressures will obviously assist in meeting product specifications more easily.

Feedstock to the second-stage catalyst is essentially nitrogen- and sulfur-free to maintain catalyst activity and long life and maximize the hydrogenative activity. For this reason, the hydrogen systems for the first- and second-stage catalysts have to be isolated from each other—the first-stage recycle gas may have sufficient levels of hydrogen sulfide to affect the activity of the second-stage catalyst. In addition, operation of the first stage to a fractionator bottom sulfur specification will extend the life of the second-stage catalyst.

Newly developed second-stage catalysts which increase their ability to resist either sulfur spikes in the second-stage feed or overall higher sulfur levels will greatly enhance the economics of the process. Criterion[26] developed a more poison-resistant catalyst that in one plant extended the catalyst cycle length to 3.5 years from 6 months attained by the previous nickel catalyst.

The second-stage catalyst must also have essentially no acidity to prevent cracking of the feedstock and generation of light ends. It is normal to have a stripper at the end of the process to remove small amounts of light ends formed and correct the product flash point, but any further cracking would lead to unnecessary yield loss.

It is appropriate to point out here that since catalyst technology is a rapidly advancing art, it is to be expected that more recently developed catalysts for both first- and second-stage reactors will significantly outperform any mentioned here.

11.1.3.4   Products

To address any concerns that there might be chemical differences between white oils produced by the acid process and hydrotreatment, the mass spectra of Lyondell Duotreat products were compared with those from acid treatment.[8] The authors concluded that there was indeed little difference at the same viscosity level (Table 11.3). White oils made by acid treating can have higher sulfur levels than those that are produced by hydrotreating.[27]

Table 11.4, a comparison of the mass spectra of white oils produced from lube hydrocracking and SK's fuels hydrocracking process for lubes, which entails severe hydrocracking followed by hydroisomerization and hydrofinishing, shows higher paraffin (presumably essentially all isoparaffins) in the SK product[28] compared with the hydrocracked material and lower polycyclic naphthene content. The SK product will also obviously have higher VIs (which is not among the white oil specifications).

TABLE 11.3
Mass Spectrometry Analyses of Food Grade Oils Produced
by Acid Treatment and by the Duotreat Process

	Acid Treated	Duotreat
Viscosity, SUS at 100°F	70	70
Composition, wt. %		
Paraffins	39.3	36.3
Naphthenes		
1-ring	28.2	32.1
2-ring	15.9	19.3
3-ring	7.9	7.9
4-ring	6.8	4.4
5-ring	1.9	0.0
6-ring	0.0	0.0
Aromatics by clay gel analysis	0.0	0.0

Source: H. C. Moyer and M. K. Rausch, "Duotreat Oils: Hydrogenated Technical and Food Grade White Oils," *Proceedings of the American Petroleum Institute, Division of Refining* 49:863–876 (1969).

TABLE 11.4
Properties of White Oil (70N) from UCO Lube Process Products

	UCO Process Product Based	Lube Hydrocracking Based	Remarks
Acid treating test			
Yield, vol. %	96	89	By 3% Oleum + 5% EtOH
Compositional analysis			
Paraffins, wt. %	41.7	30.6	ASTM D2549 and D2786
1-ring naphthenes	26.3	25.6	
2-ring naphthenes	18.0	19.6	
3-ring naphthenes	13.8	11.7	
4-ring naphthenes	0.2	8.0	
5-ring naphthenes	0.0	3.4	
6-ring naphthenes	0.0	1.1	

Source: W. S. Moon, Y. R. Cho, C. B. Yoon, and Y. M. Park, "VHVI Base Oils from Fuels Hydrocracker Bottoms," presented at the Oil and Gas Producers Conference, June 1998. With permission.

TABLE 11.5
Feedstocks and Products from the BASF White Oil Process

	Neutral Oil			Heavy Machine Oil			Dewaxed Hydrocrackate		
		Products			Products			Products	
	Feed	Technical	Food	Feed	Technical	Food	Feed	Technical	Food
Specific gravity	0.877	0.866	0.866	0.905	0.895	0.893	0.890	0.870	0.878
Viscosity at 20°C	319	220	221	940	542	545	203	187	189
Flash point, °C	258	258	256	262	250	262	246	248	248
Sulfur, ppm	1500	3		12000	4		10	3	
Aromatics, wt. %	12	Trace	Trace	28	Trace	Trace	9.1	Trace	Trace

Source: P. J. Polanek, D. J. Artrip, and G. Kons, "Specialties by Catalytic Hydrogenation: White Oils and Fully Refined Paraffins," Paper FL-96-112, presented at the National Fuels and Lubricants meeting, National Petroleum Refiners Association, Houston, Texas, November 7–8, 1996. With permission.

Table 11.5 shows the properties of technical and food/medicinal grade oils produced by the BASF white oil process from three different feeds, all measured at the same flash point. It can be seen that feed sulfur is essentially eliminated at less than 10 ppm, as are aromatics in both technical and food/medicinal grade oils. Viscosities of the technical and food grade oils are significantly reduced from those of the feeds due to the combination of aromatic saturation and molecular weight lowering from hydrocracking in the first stage. There is no change in viscosity (therefore no cracking) between the technical and food/medicinal grades.

The combination of first-stage catalysts and the process conditions (reactor temperature, hydrogen pressure, and space velocity) employed allowed licensors to claim substantial success in producing technical grade white oil in a single stage (e.g., the Duotreat process exceeds technical grade's UV specification by a factor of about 10 when processing raw feedstocks) (see Table 11.6). When processing solvent extracted feedstocks, technical grade color of +30 Saybolt was obtained and UV results are claimed to be even lower.[13]

The chemistry of the second-stage unit is basically a competition between the rates of aromatic saturation and the reverse reaction—loss of hydrogen with formation of aromatics. This particularly applies to the polyaromatics:

$$ArH + nH_2 \leftrightarrow Naphthene.$$

When reactor temperature increases sufficiently the rate of the reverse reaction becomes competitive with that of the forward saturation reaction, and with further temperature increases, the reverse reaction becomes predominant and polyaromatic levels cause product failure in either the carbonizable substances test, the UV test, or both.

TABLE 11.6
Comparison of Duotreat Product with Technical Grade White Oil Specifications

	Technical Grade White Oil Specification	Duotreat Technical Grade White Oil Typical Results
Saybolt color, minimum	+20	+28
UV absorbance limit		
FDA 121.2589(b)		
Maximum absorbance/cm		
280–289 nm	4.0 maximum	0.6
290–299 nm	3.3 maximum	0.4
300–329 nm	2.3 maximum	0.3
330–350 nm	0.8 maximum	0.1

Source: M. K. Rausch and G. E. Tollefson, "Process Makes High-Grade White Oil," *Oil and Gas Journal* 71(7):84–86 (1973). With permission.

In an interesting investigation of these aspects by Chevron,[27] in their patent on the use of sulfided NiO/MoO, platinum, and palladium catalysts for white oil hydrofinishing, they described use of a reactor whose top-half and bottom-half temperatures could be independently controlled. Product quality was measured by the percent transmittance of orange light through the acid layer of the carbonizables test with greater than about 93% transmittance being required for a pass. This transmittance was plotted against the temperature of the outlet half of the reactor, with the results for the several types of catalysts being summarized in Figure 11.3. Note that the nickel/molybdenum catalyst used an inlet temperature of 650°F, while the platinum and lead catalysts used inlet temperatures of 500°F. Higher temperatures in the top half of the reactor will attain maximum monoaromatic saturation.

Of the three catalysts employed, only the palladium one was successful at outlet temperatures of about 400°F, this catalyst resulting in a passing product at 0.5 LHSV, and the fail at 0.25 LHSV was probably a bad data point, with the curve indicating that product quality declined as outlet temperature was increased from 350°F to more than 500°F. The platinum catalyst was not successful in meeting the product specification, but came close, and had the same overall shaped curve. For the sulfided NiO/MoO catalyst product quality, product quality never came close to passing, even using the high initial temperatures in the reactor. Product quality improved from 275°F to 350°F, where it achieved its best quality, although still far short of a pass and then quality declined as temperature was further increased to 500°F and presumably thermodynamics took over, and indeed product carbonizables content was eventually

Hydrogenation of hydrofined raffinate at 2000 PSIG, except as noted

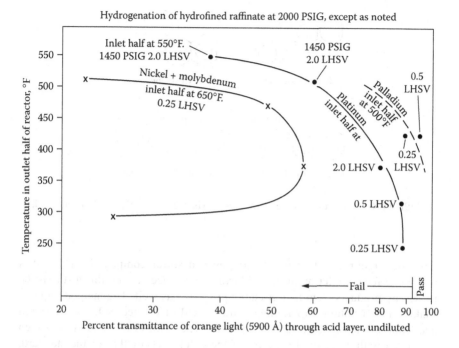

FIGURE 11.3 Summary of Chevron study on white oil hydrotreatment using noble metal and sulfided base metal catalysts: reactor outlet temperatures versus product quality. *Source:* M. L. Diringer and C. R. Hare, "Two-Stage Hydrotreatment for White Oil Manufacture," U.S. Patent 3,340,181. With permission.

poorer than that of the feed. The sulfided catalyst just did not have the kinetic capability necessary at low temperatures and this led to it succumbing easily to thermodynamics. In contrast, the noble metal catalysts either met or just failed to meet the kinetic criteria, and in the palladium case was able to achieve product quality over a small temperature range. This patent also reports that even brief contact of the exiting oil with the reactor steel internals at 450°F or greater caused degradation of the product, although the reactor inlet may go as high as 550°F.

11.1.3.5 Product Specifications for Polynuclear Aromatics

The key toxicity specifications required by FDA food/medicinal requirements are the readily carbonizable substances test (ASTM D565) and the UV test for polynuclear aromatics (ASTM D2269).

Readily carbonizable substances are those that form colored compounds with fuming sulfuric acid. These carbonizable compounds include polynuclear

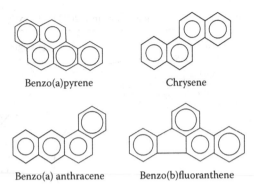

Benzo(a)pyrene   Chrysene

Benzo(a) anthracene   Benzo(b)fluoranthene

FIGURE 11.4 Structures of some of the 30+ PAHs used in FDA's development work.

aromatics, olefins, and aromatic nitrogen and sulfur compounds. The test is really one for completion of the acid extraction process,[29] but this test must be passed by products from hydrotreating processes as well. The method requires shaking equal volumes of fuming sulfuric acid of defined SO_3 concentration and white oil at 100°C for a specified time. The color of the acid layer is then compared with a standard solution. If the color exceeds that of the standard, then the sample has failed the FDA specification. This is not an easy test to perform on a consistent basis because its success depends on judgment of color by the human eye. The German DAB IX test is similar, but uses a different acid strength.

The UV specification addresses polynuclear aromatic levels specifically. This test and the associated dimethyl sulfoxide (DMSO) extraction procedure arose from work performed by the FDA in the 1960s to regulate food additives and their potential contamination with trace amounts of carcinogenic PAHs (3+ aromatic rings). Some examples of these compounds are given in Figure 11.4. The objective of those studies was to place an upper limit for their content in food and develop a method for measuring their level.

Polyaromatic compounds absorb UV radiation more intensively at more than 260 nm than compounds with one or two aromatic rings. This absorbance is proportional to their concentration, therefore a measurement method is available. However, some preconcentration is required before reliable concentration estimates can be made. One of the difficulties in concentrating these materials was to avoid interference from other substances. Haenni et al.[30] found that in using selective extraction by acetonitrile, dimethylformamide (DMF), and DMSO, the latter two gave the highest extraction efficiency per pass, but the DMF extract also contained quantities of "extraneous background material" that affected measurement. The procedure with DMSO was therefore pursued and developed into a standardized method (ASTM D2269[31]) in which the UV spectrum of the DMSO extract is recorded and absorbances in the 260 to 350 nm range confirm that they

TABLE 11.7
Polynuclear Aromatics in Synthetic White Oils

	Concentration (ppb by Weight)		
PNA	2 cSt	4 cSt	6 cSt
Benzo(a)Pyrene	<0.1	<0.1	<0.1
Pyrene	<0.1	<0.1	<0.1
Benzo(a)anthracene	<0.1	<0.1	<0.1

Source: R. D. Galli, B. L. Cupples, and R. E. Rutherford, "A New Synthetic Food Grade White Oil," *Journal of the American Society of Lubrication Engineers* 36:365–372 (1982). With permission.

are below the maximum allowed.[32] This maximum corresponds to approximately 0.3 ppm PAH.

In the case of white oils from poly-alpha-olefins, the original work showed that the usual tests for PNAs, namely, the carbonizable and FDA UV tests, were passed by the products. Further analyses for specific PNAs, benzo[a]pyrene, pyrene, and benzo(a)anthracene showed the levels of these contaminants to be less than 0.1 ppb (see Table 11.7).

Table 11.8 contrasts the compositional differences between white mineral oils from petroleum sources and synthetic white oil, the latter being 100% branched

TABLE 11.8
Comparison of Conventional White Mineral Oil with Synthetic White Oils

	White Mineral Oil				Synthetic White Oils		
Property	Naphthenic Oils		Paraffinic Oils		2 cSt	4 cSt	6 cSt
Viscosity, cSt at 100°F	20.5	75.5	8.85	13.06	521	17.90	33.82
Composition							
Paraffins	13.9	11.8	49.2	42.2	100.0	100.0	100.0
1-ring naphthenes	23.1	22.0	24.0	26.3	—	—	—
2-ring naphthenes	21.8	20.3	13.1	15.0	—	—	—
3-ring naphthenes	16.2	19.6	7.3	7.7	—	—	—
4-ring naphthenes	14.0	14.3	5.5	6.4	—	—	—
5-ring naphthenes	7.6	8.0	0.8	2.3	—	—	—
6-ring naphthenes	3.4	2.8	0.1	0.1	—	—	—
Number of carbons/molecule	22	30	21	23	20	31	37

Source: R. D. Galli, B. L. Cupples, and R. E. Rutherford, "A New Synthetic Food Grade White Oil," *Journal of the American Society of Lubrication Engineers* 36:365–372 (1982). With permission.

paraffins.[17] These synthetic white oils have lower volatility and significantly lower pour points than corresponding ones of direct petroleum origin.

11.2 REFINED WAXES

Solvent dewaxing produces an initial wax, known as slack wax, that contains substantial quantities of oil, up to 20% by volume. A second treatment of the wax, essentially another "dewaxing" step called deoiling, produces essentially oil-free wax and, as a by-product, "footes oil" consisting of low melting point paraffins and naphthenes. Deoiled wax from hydrocrackates will contain only parts per million quantities of nitrogen and sulfur compounds. From solvent refined oils, the level of these impurities in deoiled wax will necessarily be higher. In both deoiled wax cases, further treatment is necessary to meet food grade standards.

Feedstocks to dewaxing units are generally waxy distillates intended for lube base stock production, but dewaxing to produce wax may be performed on crude distillates if the wax content is high enough.

Wax has traditionally been refined by acid/clay treating, which reduces the amount of aromatics together with sulfur and nitrogen heterocycles, improves color, and reduces odor. Hydrotreating provides an economically attractive alternative for making food grade waxes, just as has been the case for white oils.

For paraffinic (macrocrystalline) waxes, predominantly n-paraffins, the compounds that require hydrotreatment are mostly oil components present in a few percent or less in deoiled wax. In the case of microcrystalline waxes, which are largely naphthenic, polar compounds will be more prevalent. In both cases, feed quality is higher than for white oils. Therefore a single catalytic hydrotreatment stage has usually been required, employing a catalyst and reaction conditions such that neither cracking nor isomerization of n-paraffins occurs that will affect wax properties such as viscosity, melting point, and penetration. Figure 11.5 is a schematic of the BASF process for food grade waxes. Table 11.9 shows the range of wax feeds that can be handled by IFP's process.[12]

BASF,[15,16] ExxonMobil,[33] IFP,[12] Gulf (Chevron),[11] and Lyondell[34] all offer licensed processes for refining of waxes in yields of 99% or better. Recent ExxonMobil technology employs two reactors.[35] The products from these processes meet 21 CFR 172.886 for food/medicinal grade waxes. The analytical procedure for this test specification extracts a sample of wax with DMSO/phosphoric acid, with subsequent measurement of UV absorbances in specific wavelength ranges after extraction into isooctane.[36] The limits to be met are shown in Table 11.10.

Table 11.11 shows inspection results on feedstock and fully refined paraffins from the BASF process for four feedstocks. The process improves color, removes

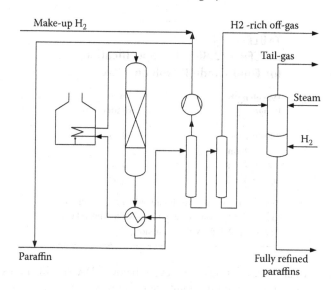

FIGURE 11.5 Schematic of the BASF process for food grade waxes.
Source: W. Himmel, T. Anstock, R. Spahl, and K. Kussner, "White Oils and Fully Refined Paraffins," *Erdol und Kohle* 39:408–414 (1986). With permission.

TABLE 11.9
IFP Hydrotreatment of Waxes: Typical Feedstocks

Type	Melting Point, °C	Percent Oil	Pen., 1/10 mm	n-Paraffins, wt. %	ASTM Color	Sulfur, ppm
Macrocrystalline	50–52	0.2	18	82	0.5–1	30–50
	52–54	0.2	87	87	0.5–1	50–200
	58–60	0.2	17	70	0.5–2.5	100–400
	63–65	0.4	15	65	—	300–600
Slack Wax	60–64[a]	9	>100	—	—	—
	60–64[a]	20	>100	—	—	—
Microcrystalline	75–80[a]	05–2.0	20–30	20	3–8	3000–8000
	75–80[a]	2–4	40–60	20	3–8	3000–8000

[a] Drop point.

Source: A. Billon, J. P. Peries, M. Lafforgoe, and J. Rossarie, "Improvements in Waxes and Special Oil Refining," *Proceedings of the American Petroleum Institute, Division of Refining* 59:168–177 (1980). With permission.

TABLE 11.10
21 CFR 172.886: UV Specification
for Food Grade Petroleum Wax

Wavelength Range	Maximum UV Absorbance/cm
280–289 nm	0.15
290–299 nm	0.12
300–359 nm	0.08
360–400 nm	0.02

Source: "Food Additives Permitted for Direct Addition to Food for Human Consumption. White Mineral Oil," 21 CFR 172.878. With permission.

essentially all sulfur, and brings UV levels below FDA specifications without affecting specific gravity, melting point, and viscosity.

The BASF[15] and IFP[12] processes are reported to use sulfided NiO/MoO catalysts, and catalysts of this general sulfided base metal type can be expected to be employed by all licensors. While noble metal or nickel catalysts result in low reactor temperatures, their use is unlikely in this application. Sulfur levels in the feeds make for short catalyst life unless they are from dewaxing a hydrocrackate.

A recent paper reported on wax hydrotreatment over Ni-W/Al$_2$O$_3$ catalysts at hydrogen pressures ranging from 40 to 160 kg/cm^2 (570 to 2300 psi) and temperatures ranging from 290°C to 370°C.[37] The feed was a microcrystalline wax from solvent dewaxing of a heavy neutral stream and was described as having high sulfur, nitrogen, and aromatic levels, although measured numbers were not given. The feed Saybolt color was 16. Reactions were tracked using Saybolt color (ASTM D156), the carbonizables test (ASTM D612), and oil content as a measure of the extent of isomerization to branched paraffins. Four catalysts, with different nickel/tungsten contents were assessed. The work showed, as might be expected, that high pressures and higher temperatures lead to better Saybolt (+30 is the target) and carbonizables results and these were also accompanied by slightly higher oil levels which would have affected penetration results. Interestingly, higher pressures appeared to lead to decreased paraffin cracking.

An obvious new source of highly refined waxes is from Fischer-Tropsch wax—this type of wax is already on the market from Shell's Bintulu plant in Malaysia, and it undoubtedly will be upgraded, if it has not been already, to meet FDA standards and equivalent ones from other countries. It will be unique in its properties because of the near-complete absence of isoparaffins and the heavy waxes will contain no naphthene components that give microcrystalline waxes their properties.

TABLE 11.11
Inspection Results of Feedstocks and Products from BASF's Wax Refining Process

| | Macrocrystalline | | | | Microcrystalline | | | |
| | A | | B | | C | | D | |
	Feed	Product	Feed	Product	Feed	Product	Feed	Product
Specific gravity at 70°C	0.778	0.776	0.782	0.782	0.805[a]	0.805[a]	0.815	0.814
Melting point, °C	57	57	62	62	69	69	55	55
Viscosity, cSt at 100°C	4.1	4.1	5.1	5.1	16.8	16.9	13.1	13.2
Percent oil	0.25	0.28	0.40	0.44	2.5	2.6	5.6	5.8
Sulfur, ppm	38	<2	360	3	550	7	2500	15
Color, ASTM D1500	<0.5		1.5		6		4	
Color, Saybolt		+30		+30		+19		+24
DAB UV								
275 nm (maximum 0.60)		0.072		0.052				
295 nm (maximum 0.30)		0.008		0.021				
310 nm (maximum 0.10)		0.004		0.010				
FDA UV								
280–289 nm (maximum 0.15)						0.036		0.043
290–299 nm (maximum 0.12)						0.067		0.076
300–359 nm (maximum 0.08)						0.064		0.066
360–400 nm (maximum 0.02)						0.017		0.012

[a] At 100°C.

Source: W. Himmel, T. Anstock, R. Spahl, and K. Kussner, "White Oils and Fully Refined Paraffins," *Erdol und Kohle* 39:408–414 (1986). With permission.

REFERENCES

1. E. Meyer, *White Mineral Oil and Petrolatum* (Brooklyn, New York: Chemical Publishing Company, 1950).
2. D. Schramm, "The USP Petrolatum Industry in North America," Paper LW-01-125, presented at the Lubricants and Wax meeting, National Petroleum Refiners Association, Houston, Texas, November 8–9, 2001.
3. J. Weeks, "The White Oil Industry in North America," Paper LW-98-131, presented at the Lubricants and Waxes meeting, National Petroleum Refiners Association, Houston, Texas, November 13–14, 1998.
4. L. Tocci "Awash with Challenges," *Lubes 'N' Greases* 11(7):32 (2005).
5. "Indirect Food Additives: Adjuvants, Production Aids and Sanitizers. Mineral Oil," 21 CFR 178.3620.
6. "Food Additives Permitted for Direct Addition to Food for Human Consumption. White Mineral Oil," 21 CFR 172.878.
7. V. A. Kalichevsky and B. A. Stagner, *Chemical Refining of Petroleum: The Action of Various Refining Agents and Chemicals on Petroleum and Its Products* (New York: Reinhold Publishing, 1942), chap. 2.
8. H. C. Moyer and M. K. Rausch, "Duotreat Oils: Hydrogenated Technical and Food Grade White Oils," *Proceedings of the American Petroleum Institute, Division of Refining* 49:863–876 (1969).
9. J. B. Gilbert, C. Olavssen, and C. H. Holder, "Hydroprocessing for White Oils," *Chemical Engineering* 82(19):87–89 (1975).
10. J. Lecomte, J. B. Gilbert, C. Olavssen, and C. H. Holder, "For White Oil Purity: Hydrogenate," *Hydrocarbon Processing*, April:157–159 (1977).
11. H. C. Murphy, Jr., R. P. Nejak, and J. R. Strom, "High Pressure Hydrogenation—Route to Specialty Products," *Proceedings of the American Petroleum Institute, Division of Refining* 49:877–904 (1969).
12. A. Billon, J. P. Peries, M. Lafforgoe, and J. Rossarie, "Improvements in Waxes and Special Oil Refining," *Proceedings of the American Petroleum Institute, Division of Refining* 59:168–177 (1980).
13. M. K. Rausch and G. E. Tollefson, "Process Makes High-Grade White Oil," *Oil and Gas Journal* 71(7):84–86 (1973).
14. E. F. Gallei and M. Schwarzmann, "The BASF Process for Preparation of Technical and Food- or Medicinal-Grade White Oils by Catalytic Hydrogenation," presented at the Congress of Large Chemical Plants, Antwerpen, Holland, 1982.
15. W. Himmel, T. Anstock, R. Spahl, and K. Kussner, "White Oils and Fully Refined Paraffins," *Erdol und Kohle* 39:408–414 (1986).
16. P. J. Polanek, D. J. Artrip, and G. Kons, "Specialties by Catalytic Hydrogenation: White Oils and Fully Refined Paraffins," Paper FL-96-112, presented at the National Fuels and Lubricants meeting, National Petroleum Refiners Association, Houston, Texas, November 7–8, 1996.
17. R. D. Galli, B. L. Cupples, and R. E. Rutherford, "A New Synthetic Food Grade White Oil," *Journal of the American Society of Lubrication Engineers* 36:365–372 (1982).
18. D. R. Holmes-Smith, "Process for Producing Isoparaffinic White Oil," World Patent WO9108276.

19. H. Y. Sung, S. H. Kwon, and J. P. Andre, "VHVI Base Oils and White Oils from Fuels Hydrocracker Bottoms," presented at the Asia Fuels and Lubricants Conference, Singapore, January 25–28, 2000.

20. W.-S. Moon, Y.-R. Cho, and J.-S. Chun, "Application of High Quality (Group II, III) Base Oils to Specialty Lubricants," presented at the 6th Annual Fuels and Lubes Asia Conference, Singapore, January 28, 2000.

21. R. A. Barter, "API Research Program on Waxes and White Oils," Paper FL-95-119, presented at the National Fuels and Lubricants meeting, National Petroleum Refiners Association, Houston, Texas, November 2–3, 1995.

22. L. E. Twerdok, "Update on the American Petroleum Institute White Oils and Waxes Research Program," Paper LW-99-133, presented at the Lubricants and Waxes meeting, National Petroleum Refiners Association, Houston, Texas, November 11–12, 1999.

23. L. E. Twerdok, "Food Grade White Oils and Waxes—Update on Recent Research and Regulatory Review," Paper LW-02-130, presented at the Lubricants and Waxes Meeting, National Petroleum Refiners Association, Houston, Texas, November 14–15, 2002.

24. B. A. Narloch, M. A. Shippey, and M. W. Wilson, "Process for Preparing White Oil Containing a High Proportion of Isoparaffins," U.S. Patent 5,453,176.

25. S-Oil and ExxonMobil, "Successful Conversion of a Fuels Hydrocracker to Group III Lube Production," presented at the Asia Refining and Technology Conference (ARTC), 7th Annual Meeting and Reliability Conference, Singapore, April 2004.

26. G. L. Everett and A. Suchanek, "Lubricant Oil Production: The Proper Marriage of Process and Catalyst Technologies," Paper AM-96-37, presented at the annual meeting of the National Petroleum Refiners Association, San Antonio, Texas, March 17–19, 1996.

27. M. L. Diringer and C. R. Hare, "Two-Stage Hydrotreatment for White Oil Manufacture," U.S. Patent 3,340,181.

28. W. S. Moon, Y. R. Cho, C. B. Yoon, and Y. M. Park, "VHVI Base Oils from Fuels Hydrocracker Bottoms," presented at the Oil and Gas Producers Conference, , June 1998.

29. H. Schindler and R. Rhodes, "Where Should White Oils be Used in Industry," Paper FL-68-57, presented at the Fuels and Lubricants meeting, National Petroleum Refiners Association, New York, September 11–12, 1968.

30. E. O. Haenni, J. W. Howard, and F. L. Joe, "Dimethyl Sulfoxide: A Superior Analytical Extraction Solvent for Polynuclear Hydrocarbons and for Some Highly Chlorinated Hydrocarbons," *Journal of the Association of Official Analytical Chemists* 45:67–70 (1962).

31. ASTM D2269, "Standard Test Method for Evaluation of White Mineral Oils by Ultraviolet Absorption," *ASTM Annual Book of Standards*, vol. 05.01 (West Conshohocken, PA: American Society for Testing and Materials).

32. E. O. Haenni, F. L. Joe, Jr., J. W. Howard, and R. L. Liebel, "Food Additives. A More Sensitive and Selective Ultraviolet Absorption Criterion for Mineral Oils," *Journal of the Association of Official Analytical Chemists* 45:59–66 (1962).

33. J. B. Gilbert and R. Kartzmark, "Advances in the Hydrogen Treating of Lubricating Oils and Waxes," *Proceedings of the 7th World Petroleum Congress* 4:193–205 (1967).

34. "Wax Hydrotreating Process," brochure (Houston, Texas: Lyondell).

35. P. Kamienski, "Technology for High Quality Base Stocks & Finished Waxes," presented at the Central and Eastern European Refining and Petrochemicals Conference—7th Annual Roundtable, Prague, October 19, 2004.

36. "Petroleum Wax," 21 CFR 172.886.

37. J. Sanchez, M. F. Tallafigo, M. A. Gilarranz, and F. Rodriguez, "Refining Heavy Neutral Oil Paraffin by Catalytic Hydrotreatment Over Ni-W/Al$_2$O$_3$ Catalysts," *Energy and Fuels* 20:245–249, 2006.

12 Base Stocks from Fischer-Tropsch Wax and the Gas to Liquids Process

12.1 THE FISCHER-TROPSCH PROCESS

It is anticipated that by 2007 to 2008, significant quantities of base stocks will be produced from Fischer-Tropsch waxes by cracking and isomerization. These will be of very high quality, made by isomerization of wax, and therefore highly paraffinic and are expected to be very competitive with the synthetic poly-alpha-olefins (PAOs) in cost, viscosity index (VI), volatility, and performance. The gas-to-liquids (GTL) plants currently planned in Qatar are expected to produce more than 48,000 bpd[1] of base stocks—about 5% of the world's current production and about 24% of the current volume of group II + III petroleum-based base stocks.[2] These volumes and quality may revolutionize the industry. The plants will be driven by fuels (gasoline and diesel) production, with the lubes part piggybacking at the end of the process. Fischer-Tropsch wax is unique in being more than 99% n-paraffins, and the lubes will also be unique compositionally in being totally long-chain paraffins with mainly methyl branches and essentially no aromatics or naphthenes.

The Fischer-Tropsch process catalytically converts synthesis gas to hydrocarbons and was developed by Franz Fischer and Hans Tropsch in Germany in the 1920s.[3] The reaction is basically the hydrogenation of CO to $-CH_2$ units which simultaneously oligomerize to straight-chain molecules from C_2 to more than C_{100} and even as far as C_{200}.[4] The products contain no sulfur, nitrogen, or oxygen. The CO is produced from methane. The process was employed in Germany during World War II for the production of fuels and lubricants and more recently in South Africa, where Sasol produces 135,000 bpd of fuels from coal. The commercial developments taking place now are (1) because of the desire to convert "stranded" low cost natural gas into income-producing fuels rather than flaring, (2) because the current prices for crude oil make the capital and operating expenditures manageable, and (3) because the fuel quality will be extremely high, with diesel cetane in the 70 to 80 range.

The chemical basis is

- Synthesis gas (syngas) formation: $CH_4 + \frac{1}{2}O_2 \rightarrow 2H_2 + CO$
- Fischer-Tropsch reaction: $(2n + 1)H_2 + nCO \rightarrow C_nH_{2n+2} + nH_2O$
- Conversion to lubes: waxy FT product $\rightarrow$ dewaxed lubes

The FT process produces a mixture of hydrocarbon gases, gasoline, diesel ("green diesel"), and wax in the lube range, depending on conditions and the catalyst employed. The German plants used both iron- and cobalt-based catalysts,[5] as did the Sasol plants, the latter producing only fuels. Shell built the first modern GTL plant with an output of about 1500 bpd of lubes range wax[6,7] at Bintulu in Malaysia in 1993, and a total capacity of 12,500 bpd of liquids.[3] Sasol's plant in Sasolburg, South Africa, produces a similar product. Shell's next project is a 140,000 bpd refinery (Pearl) in Qatar, slated to come online in 2008 or 2009, with a 9600 bpd base oil plant.[6] In addition, Sasol Chevron and Qatar have announced a GTL plant whose lube plant (Oryx) will produce 8000 bpd of base oils,[8] while ExxonMobil plans to build a GTL joint venture with Qatar which will have an associated lube plant with a capacity of 30,800 bpd.[9] Finally, Chevron and Sasol announced[10] the joint Escravos GTL plant in Nigeria, which will produce 34,000 bpd of diesel and naphtha, but there was no mention of heavier wax or lubes production. A generalized schematic for a GTL plant is shown in Figure 12.1 and involves an oxygen plant, the syngas unit, where natural gas is oxidized to CO and hydrogen, which react in the Fischer-Tropsch unit, and a fractionation tower, where the products are separated. The 650+°F wax bottoms are catalytically hydrocracked, isomerized, and catalytically dewaxed, and then the product is fractionated to produce base stocks.

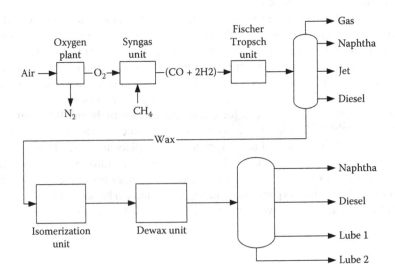

FIGURE 12.1 GTL lubes production from natural gas: general process schematic.

Several types of reactors have been and are being employed for the FT reaction, including fixed bed, fluid bed, and slurry reactors. Slurry beds, in which the catalyst is suspended in the heavy liquid product from the reaction, were part of the early work by Fischer and Tropsch,[11] and became an area that was neglected for quite a while. Interest was restored by commercial developments: for example, by ExxonMobil's Advanced Gas Conversion for the 21st Century slurry-based process (AGC-21),[12] and Sasol's slurry phase distillate process (SSPD),[13] which will be used in their proposed Qatar and Nigerian developments. Shell's Bintulu, Malaysia, plant employs a fixed bed multitube reactor, and they will continue to use this type in the Qatar project. Other companies actively developing FT processes include BP,[13] Syntroleum,[14] Rentech,[15] and ConocoPhillips.[3]

Catalyst technology has shifted from iron as the active metal to cobalt and it has been suggested that all new plants in the immediate future will have cobalt catalysts, preferred because of their inherent stability, excellent activity, compatibility with slurry reactor operations, and ability to make waxy products as well as diesel and gasoline.[16]

Gas to liquids operating conditions employ relatively low temperatures (200°C to 250°C) and pressures (20 to 30 bar).

12.2 PRODUCT DISTRIBUTIONS

Product distributions[13] depend on the catalyst and operating conditions and can be generalized as

- Naphtha, 15 to 25 vol. %
- Middle distillates (jet fuel, diesel), 65 to 85 vol. %
- Lubes/wax, 0 to 30 vol. %

High naphtha and light olefin yields are favored by operation at high temperatures. The naphtha becomes excellent feed for naphtha crackers to make ethylene, the diesel is a green diesel with a cetane number of about 70, and obviously zero sulfur and aromatics, therefore it is an excellent blending component for poorer quality counterparts. The high molecular weight wax can either be cracked to fuels products or isomerized to reduce the pour point and produce lube base stocks.

Product distribution from the Fischer-Tropsch unit is generally regarded as being approximated by the Anderson-Schulz-Flory equation:

$$W_n = n(1 - \alpha)^2 \alpha^{(n-1)}$$

where W_n is the weight fraction of a chain with n carbons, and α is the chain growth probability factor for insertion of a further carbon. A typical distribution curve for products as a function of α is shown in Figure 12.2.

The term α is defined by the equation

$$\alpha = k_p/(k_p + k_t)$$

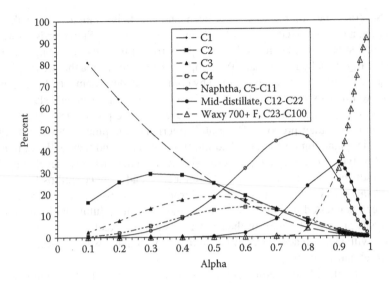

FIGURE 12.2 Anderson-Schulz-Flory product distribution from Fischer-Tropsch synthesis.

where k_p and k_t are the rates of propagation and termination, and is therefore the ratio of chain propagation to propagation plus termination. In this mechanism, CO and hydrogen are adsorbed on the catalyst surface where they react, and chain extension takes place by insertion of CH_2 units, one unit at a time. The two rates are considered to be independent of chain length (Figure 12.3). In practice, in many cases α is, in fact, not constant. Product distributions are affected by the catalyst and reactor operating conditions, with low temperatures and high pressures favoring formation of higher molecular weight products.[17] It can be seen from Figure 12.2 that formation of 700+°F product is favored by high values (greater than 0.9) of alpha.

12.3 BASE STOCK PROPERTIES

The purity of this feedstock to make lubes, essentially a homologous series of n-paraffins—a near-impossible dream to petroleum chemists—is almost unique in the lubes industry; the only match being the olefin feeds used to make synthetic lubes. Thus GTL lubes will almost certainly be regarded as "synthetics" (after all, the key process is called the Fischer-Tropsch synthesis!), and the n-paraffin feed structure means that in principle the structures of the base stocks are up to the ingenuity of development chemists and engineers. GTL lubes will have very different structures from "conventional" synthetics, which have a "star"-type structure (Figure 12.4). This means that their low temperature properties are not as good (multiple methyl substituents versus a single long substituent).

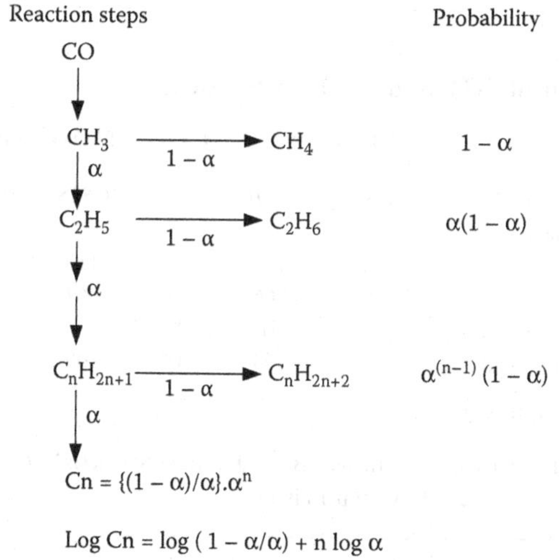

Reaction steps   Probability

$$CO$$

$$CH_3 \xrightarrow{1-\alpha} CH_4 \qquad 1-\alpha$$

$$\downarrow \alpha$$

$$C_2H_5 \xrightarrow{1-\alpha} C_2H_6 \qquad \alpha(1-\alpha)$$

$$\downarrow \alpha$$

$$C_nH_{2n+1} \xrightarrow{1-\alpha} C_nH_{2n+2} \qquad \alpha^{(n-1)}(1-\alpha)$$

$$\downarrow \alpha$$

$$C_n = \{(1-\alpha)/\alpha\}.\alpha^n$$

$$Log\ C_n = log\ (1-\alpha/\alpha) + n\ log\ \alpha$$

FIGURE 12.3 Growth of paraffin chains in the Fischer-Tropsch synthesis and Schulz-Flory kinetics.
Source: J. Eilers, S. A. Posthuma, and S. T. Hie, "The Shell Middle Distillate Synthesis Process (SMDS)," *Catalysis Letters* 7:253–270 (1990). With permission.

Gas to liquids base stock structures will therefore be founded on linear overall structures with methyl groups at intervals and perhaps a few longer branches. In the future it may be possible to synthesize base stocks with a predetermined number of branches of known length with those branches at predetermined positions along the chain, providing the detailed chemistry needed to give the optimum quality for a particular application.

Gas to liquids base stocks offer low pour points, low volatility, and very high VIs, and are comparable in their properties to those of PAOs (Table 12.1).

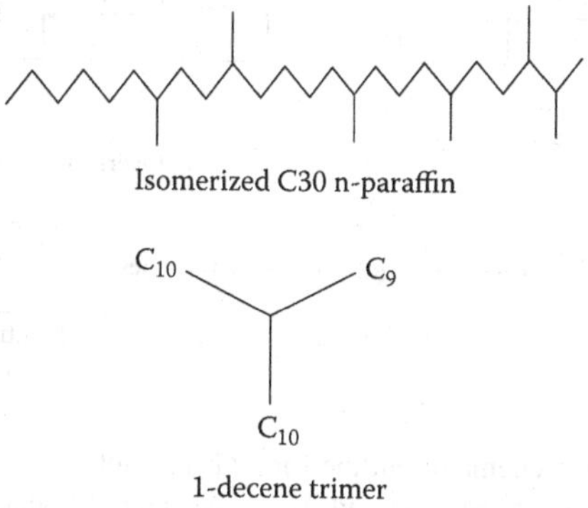

Isomerized C30 n-paraffin

$$C_{10} \qquad C_9$$

$$C_{10}$$

1-decene trimer

FIGURE 12.4 Comparison of GTL and synthetic C_{30} lubricant molecules.

TABLE 12.1
Comparison of GTL and PAO Base Stocks

Type	GTL-3	PAO-3	GTL-5	PAO-5	GTL-7	PAO-7
Density at 15°C, kg/L	0.805	0.801	0.818	0.815	0.820	0.831
Viscosity, cSt at 100°C	2.7	2.7	4.5	4.6	7.0	7.0
VI	117	114	144	132	147	134
Pour point, °C	−57[a]	−66	−39[a]	−67	−39[a]	−54
Noack volatility, wt. %	34	51	8	13	2	5
Composition, wt. % alkanes	100	100	100	100	100	100

[a] Includes 0.1 wt. % ppd.

Source: H. E. Henderson, "Performance Beyond Current Synthesis," *Hydrocarbon Engineering* August:13–17 (2002). With permission.

12.4 GTL PROCESSES

A possible schematic for the lubes part of an ExxonMobil GTL plant, based on one of their patents, is illustrated in Figure 12.5[18] and features

(A) The FT 650°F to 700+°F wax is first hydrotreated to clean up the waxy feed by removing any trace nitrogen, sulfur, and oxygen compounds. In this case a cobalt/molybdenum or nickel/molybdenum unsulfided (!) catalyst is employed at 700°F to 750°F and 1000 to 1500 psig hydrogen at a space velocity of 1 to 2.

(B) The product from A is then fed to an isomerization unit containing, in this example, a fluoride platinum-on-alumina catalyst, which

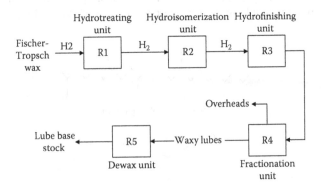

FIGURE 12.5 Possible schematic outline for a GTL plant.
Source: G. P. Hamner, H. A. Boucherand, W. A. Wachter, "Process for the Hydroisomerization of Fischer-Tropsch Wax to Produce Lubricating Oil (OP-3403)," U.S. Patent 4,943,672.

causes some isomerization of the n-paraffins, but without a marked effect on pour point. This is really a pretreat step to reduce the load on the dewax catalyst or (less likely) a solvent dewax unit if that were to be used.

(C)   After isomerization, the product is further hydrotreated to improve final product light stability. In this step, presumably any unsaturates and trace polyaromatics are reduced in level or entirely eliminated.

(D)   The final step is dewaxing, and undoubtedly that will be performed by hydroisomerization using one of ExxonMobil's MSDW catalysts. In this case, dewaxing will be followed by a stabilization hydrofinishing step, which would likely obviate the need for step C.

Examples of GTL base stocks quoted in ExxonMobil patents all have VIs greater than 120, are generally greater than 130, and are as high as 160.

The Shell middle distillate process[7] is similar (Figure 12.6), but the technology and the terminology is different. The FT unit is called the heavy paraffin synthesis unit (multitube reactor) and is followed by the heavy paraffin conversion process, which cracks and isomerizes paraffins into diesel and gasoline ranges. The latter operates at relatively mild conditions of 30 to 50 bars pressure and temperatures of 300°C to 350°C. This unit is reported to hydrogenate any olefins, remove trace oxygen-containing compounds, hydroisomerize n-paraffins, and hydrocrack n-paraffins to isoparaffins of shorter chain length.[7] The proprietary dual-function catalyst shows considerable selectivity in terms of cracking those of higher molecular weight.

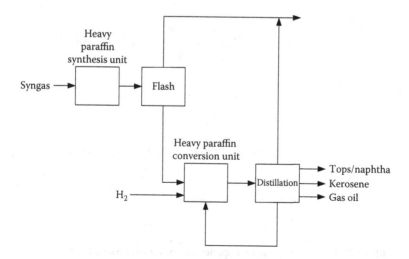

FIGURE 12.6 Schematic of the Shell middle distillate process.
Source: J. Eilers, S. A. Posthuma, and S. T. Hie, "The Shell Middle Distillate Synthesis Process (SMDS)," *Catalysis Letters* 7:253–270 (1990). With permission.

12.5 ^{13}C NUCLEAR MAGNETIC RESONANCE APPLICATIONS TO FISCHER-TROPSCH BASE STOCKS

ExxonMobil has gained valuable insights into the structures of GTL base stocks and this has been made easier because of the uniformity of their origin (i.e., from normal paraffins). Part of their analytical work has been based on ^{13}C nuclear magnetic resonance (NMR), which has the ability to identify a number of types of paraffinic carbons according to their positions and environments along the chain. Their results are outlined in some of their patents as part of their technology development, and I have chosen two to illustrate the information that can be gleaned.

One of these is the "free carbon index" (FCI) concept, which is a measure of the number of carbons in an isoparaffin located at least four (ε) carbons from a terminal carbon and more than three carbons away from a side chain (see Figure 12.7 for examples) and is defined as

(% ε methylene carbons in the ^{13}C NMR spectra)/{(average carbon number from ASTM D2502)/100}.

Generally, high values correspond to high "paraffinic character." Figure 12.7 illustrates this with several examples[19] for a C_{26} alkane with three attached methyl groups and FCIs of 2, 4, and 8, respectively. For the first of these examples, where there are 8ε carbons (structure A in Figure 12.7), the percent ε methylene carbons is (8/26)*100 = 800/26, the average carbon number is 26, and the final

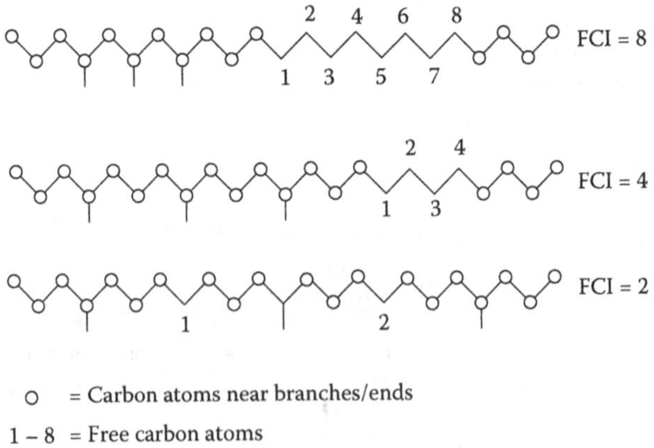

o   = Carbon atoms near branches/ends

1 – 8 = Free carbon atoms

FIGURE 12.7 Illustration of the FCI applied to several isomerized C_{26} paraffins.
Source: W. J. Murphy, I. A. Cody, and B. G. Silbernagel, "Lube Basestock with Excellent Low Temperature Properties and a Method for Making," U.S. Patent 6,676,827.

calculation sequence is $800*(26/100) = 8$ in this case. Low FCI values correspond to alkanes with branches that are separated so that there are few ε carbons (structure C in Figure 12.7), while high values of FCI result for molecules with branches close together such that the rest of the molecule contains significant numbers of carbons (e.g., structure A) and therefore is "n-paraffin like." The measurements of carbon type are made by ^{13}C NMR spectroscopy, since this can distinguish and measure ε-type carbons from the others, and of course this method produces an average value. It would be interesting to know, presumably by separation and further ^{13}C NMR work, just what the distribution of values actually is within GTL base stocks.

Not surprisingly, lower values of FCI for a given alkane correspond to reduced pour points, and ExxonMobil data indicate that this relationship depends on the dewaxing method and catalyst type that follows the hydroisomerization step and on the dewaxing catalyst acidity as well—with the more acidic silica-alumina giving low pour points and low FCI values (Figure 12.8) TON catalysts contain zeolites). Since the FCI value is related to the number of side chains, relationships exist between side chain number and the FCI value and depend again on the mode of dewaxing (Figure 12.9).

A second example of the use of ^{13}C NMR is in its application to characterize the GTL fractions that were judged to be most competitive with PAOs in low temperature properties and base oil oxidative performance. In the particular instance on one of their patents,[20] a C_{20} to C_{40} GTL base stock containing up

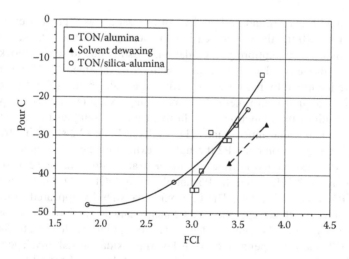

FIGURE 12.8 GTL lube base stocks: FCI versus pour point from ExxonMobil studies on GTL lubes.
Source: W. J. Murphy, I. A. Cody, and B. G. Silbernagel, "Lube Basestock with Excellent Low Temperature Properties and a Method for Making," U.S. Patent 6,676,827.

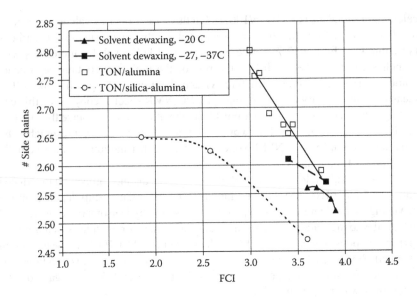

FIGURE 12.9 FCI versus the number of side chains from ExxonMobil studies on GTL lubes.
Source: W. J. Murphy, I. A. Cody, and B. G. Silbernagel, "Lube Basestock with Excellent Low Temperature Properties and a Method for Making," U.S. Patent 6,676,827.

to four alkyl branches and with an FCI of at least 3 was separated by thermal diffusion into 10 fractions of increasing "branchiness." Some of these fractions showed reduced susceptibility to oxidation, possibly due to steric "blocking of reactive hydrogens." These were the fractions of intermediate "branchiness" and were identified by high pressure differential calorimetry (HPDSC) results. (HPDSC is a rapid micro method for measuring oxidation stability, expressed by the length of the induction time. In this case, the samples contained equal quantities of an amine antioxidant.) The results in Table 12.2 show that fractions from thermal diffusion ports P2 through P6 exhibit the best inhibited oxidation stability (longest induction times). Further examination (Table 12.3) of these samples by ^{13}C NMR found the P5 sample has an FCI of 3.0 (and presumably that of P3 is 3.0 or greater). The best oxidative stability appeared to be associated with less than four alkyl attachments and an FCI of 3.0 or greater. Directionally, from the limited data here, this points to fewer attachments, higher FCI, and fewer pendant carbons favoring oxidation stability. The authors of the patent speculate also that with the correct degree of branchiness, steric blocking hinders attack at certain tertiary hydrogens, which otherwise might be expected to react quickly.

TABLE 12.2
HPDSC Induction Times Measured on Samples from Thermal Diffusion Containing Amine Antioxidant

Port Number	Induction Time, Minutes	
	at 170°C	at 180°C
P1	21.0	14.4
P2	32.4	14.8
P3	25.2	15.1
P4	37.4	20.7
P5	34.6	19.0
P6	32.8	16.0
P7	25.1	13.3
P8	16.4	10.1
P9	15.6	10.0
P10	15.3	10.5

Source: H. S. Aldrich and R. J. Wittenbrink, "Lubricant Base Oil Having Improved Oxidative Stability," U.S. Patent 6,008,164.

TABLE 12.3
Compositional Analyses of Selected Fractions from Table 12.2

Port Number	P3	P5	P7	P9
Total number of attachments	3.46	3.14	4.19	3.59
Attachments for C-4 to C-22	1.48	1.54	1.86	1.62
Methyl attachments	2.21	2.36	2.8	2.35
Attachments longer than methyl	1.1	0.93	1.39	1.64
FCI	—	3.0	2.96	2.35
Number of terminal carbons	—	0.4	0.74	0.9
Number of pendent carbons	—	3.19	4.58	4.9
Average length of attachments	1.11	1.0	1.1	1.4

Source: H. S. Aldrich and R. J. Wittenbrink, "Lubricant Base Oil Having Improved Oxidative Stability," U.S. Patent 6,008,164.

REFERENCES

1. T. Sullivan, "Forecast: New Price Leaders in Town," *Lube Report* 5(27) (2005). www.imakenews.com/lng/e_article000423820.cfm?x=b9vsH9b.b1CKBDH,w
2. Lubes N Greases, *2005 Guide to Global Base Oil Refining* (Falls Church, VA: LNG Publishing).
3. J. Rockwell, "History of GTL Technology," Paper LW-01-131, presented at the Lubricants and Waxes meeting, National Petroleum Refiners Association, Houston, Texas, November 8–9, 2001.
4. H. E. Henderson, "Gas-to-Liquids," *Canadian Chemical News* September: 17–19 (2003).
5. R. Freerks, "Early Efforts to Upgrade Fischer-Tropsch Reaction Products into Fuels, Lubricants and Useful Materials," Paper 86d, presented at the American Institute of Chemical Engineers Spring National Meeting, New Orleans, April 2, 2003.
6. T. Sullivan, "GTL Base Oils by 2008?," *Lube Report* 3(42) (2003). www.lubereport.com/e_article000194571.cfm
7. J. Eilers, S. A. Posthuma, and S. T. Hie, "The Shell Middle Distillate Synthesis Process (SMDS)," *Catalysis Letters* 7:253–270 (1990).
8. T. Sullivan, "Sasol, Chevron, Qatar Shake on GTL Base Oils," *Lube Report* 5(10) (2005). www.imakenews.com/lng/e_article000369791.cfm?x=b9vsH9b,b1CKBDH3,w
9. T. Sullivan, "ExMo Rules the Base Oil Roost," *Lube Report* 5(25) (2005).
10. Chevron press release, April 8, 2005.
11. P. Schubert, S. LeViness, K. Arcuri, and B. Russell, "Historical Development of Cobalt-Slurry Fischer-Tropsch Reactor Systems," available at www.fischer-tropsch.org.
12. "ExxonMobil's Advanced Gas-to-Liquids Technology," *Hydrocarbon Asia* July/August:56–63 (2003).
13. T. H. Fleisch, R. A. Sills, and M. D. Briscoe, "2002-Emergence of the Gas-to-Liquids Industry: A Review of Global GTL Developments," *Journal of Natural Gas Chemistry* 11:1–14 (2002).
14. P. V. Snyder, Jr., "GTL Lubricants: The Next Step," Paper LW-99-125, presented at the Lubes and Waxes meeting, National Petroleum Refiners Association, Houston, Texas, 1999.
15. D. L. Yakobson, "Fischer-Tropsch Technology: New Project Opportunities," presented at the Gas-to-Liquids Processing meeting, San Antonio, Texas, May 17–19, 1999.
16. R. Oukaci, "Overview of the Current Status of F-T Technology," Consortium for Fossil Fuel Science, C1 Chemistry Review meeting, Rocky Gap, Maryland, August 4–7, 2002.
17. G. P. van der Laan and A. A. C. M. Beenackers, "Kinetics and Selectivity of the Fischer Tropsch Synthesis: A Literature Review," *Catalysis Reviews—Science and Engineering* 41(3–4):255–318 (1999).
18. G. P. Hamner, H. A. Boucherand, W. A. Wachter, "Process for the Hydroisomerization of Fischer-Tropsch Wax to Produce Lubricating Oil (OP-3403)," U.S. Patent 4,943,672.
19. W. J. Murphy, I. A. Cody, and B. G. Silbernagel, "Lube Basestock with Excellent Low Temperature Properties and a Method for Making," U.S. Patent 6,676,827.
20. H. S. Aldrich and R. J. Wittenbrink, "Lubricant Base Oil Having Improved Oxidative Stability," U.S. Patent 6,008,164.

Index

Printed in the United States
by Baker & Taylor Publisher Services